T0320760

Charge and Field Effects
in Biosystems—2

Charge and Field Effects in Biosystems—2

Edited by
M. J. Allen,
S. F. Cleary, and
F. M. Hawkridge

Virginia Commonwealth University
Richmond, Virginia

Plenum Press • New York and London

Library of Congress Cataloging in Publication Data

International Symposium on Charge and Field Effects in Biosystems—2 (1989: Richmond, Va.)
  Charge and field effects in biosystems—2 / edited by M. J. Allen, S. F. Cleary, and F. M. Hawkridge.
      p.   cm.
  "Proceedings of the 1989 International Symposium on Charge and Field Effects in Biosystems—2, held June 4–9, 1989, in Richmond, Virginia"—T.p. verso.
  Includes bibliographical references.
  ISBN 978-0-306-43401-3
  1. Bioelectrochemistry—Congresses. 2. Electromagnetic fields—Physiological effect—Congresses. I. Allen, M. J. (Milton Joel), date. II. Cleary, Stephen F., date. III. Hawkridge, F. M. IV. Title.
QP517.B53I57   1989   89-26551
574.19′127—dc20   CIP

Proceedings of the 1989 International Symposium on
Charge and Field Effects in Biosystems—2,
held June 4–9, 1989, in Richmond, Virginia

© 1989 Plenum Press, New York
A Division of Plenum Publishing Corporation
233 Spring Street, New York, N.Y. 10013

INTRODUCTION

The success of the first Symposium on Charge and Field Effects in Biosystems, held at the University of Nottingham, England in 1983, has over the years resulted in numerous queries being received by Milton J. Allen regarding the next symposium. It was finally decided, having settled-in after a trek across the Atlantic, that it might now be an appropriate time to finally renew bonds formed at the previous conference.

Again the main objective of the Conference was to bring together those scientists directly or peripherally involved in studies appropriate to the Symposium's structure. The categories assigned to the various areas of interest were:

A. Experimental approaches to the study of charge and energy transfer in biomacromolecular and intact cellular systems.

B. Ion and electron transport properties of biological and artificial membranes.

C. Effects of electrochemical processes and electromagnetic fields on biological systems.

D. Photo-induced bioelectrochemical processes.

E. Applications of bioelectrochemical technology.

The total number of presentations given at this conference were reduced in number in order to allow ample time for each lecture and ensuing discussions.

Overall the 1989 Symposium was as successful and informative as its predecessor held in 1983. The next Symposium is tentatively being scheduled for 1991.

Milton J. Allen
Stephen F. Cleary
Fred M. Hawkridge

ACKNOWLEDGEMENTS

The editors wish to extend their thanks to Gwen Geffert, Diane Holmes and Laura McCullough who assisted in preparing this Symposium volume for publication and the various reviewers who contributed their time in examining the contents of the papers presented.

The Organizing Committee gratefully acknowledges the support and facilities given by Virginia Commonwealth University and also the financial contributions made by:

Virginia Commonwealth University, Department of Chemistry
Medical College of Virginia
Office of Naval Research

CONTENTS

CHARGE AND FIELD EFFECTS IN BIOSYSTEMS 2

PROLOGUE

It may very well be that as a result of Luigi Galvani's experimental studies, commenced in 1786, the areas related to the investigation of charge and field effects on biosystems have evolved. A significant gap in time occurred before the pioneer investigations of Paul Ehrlich were presented in 1885. He demonstrated that different animal organs had the ability to reduce dyes and that this was related to the organs' oxygen requirements. His findings stimulated numerous investigations on the redox potentials of many different biological and biochemical systems using indicator dyes. During this early period, great strides were made by electrophysiologists which led to significant observations and an understanding of ion transport phenomena in plant and animal systems.

Szent Gyargi in 1941 injected a new parameter into the existing areas of study by suggesting that biological systems could be investigated using the concepts of solid-state physics. He indicated that energy in living systems might be transferred by conduction bands. Application of these ideas to studies of isolated proteins and more recently biological systems have been very fruitful.

A formal conference on Electrochemistry in Biology and Medicine, organised by Theodore Shedlovski under the sponsorship of the Electrochemical Society, was first held in New York in 1953. A second such meeting organised by Milton J. Allen and Theodore Shedlovski, was held a few years later. Since then numerous groups have developed conferences on Bioelectrochemistry.

The first Symposium on Charge and Field Effects in Biosystems, which was convened in 1983, was created primarily to loosen the bonds of many previous conferences by expanding the topics to include not only the electrochemistry of biochemical but also metabolically viable biological systems. In addition topics were introduced to include the effects of various types of radiation on living entities, electrophysiology, ion and electron transport phenomena, the 'solid-state' behaviour of biological and artificial membranes, and the applications of electrochemical technology to biochemical and biological problems.

The recently held second Symposium on Charge and Field Effects in Biosystems again brought together scientists from various parts of the world selected for their expertise in one or more of the particular areas encompassed by the conference's topics. The diversity of subject matter stimulated an exchange of ideas and a further education of those scientists whose individual research might be expanded by this exposure.

3

EXPERIMENTAL APPROACHES TO THE STUDY OF CHARGE

AND ENERGY TRANSFER IN BIOMACROMOLECULAR AND

INTACT CELLULAR SYSTEMS

THE CHARGE TRANSLOCATING REACTIONS OF

THE *bc* COMPLEXES AND CYTOCHROME OXIDASE

P.R. Rich, A.J. Moody and R. Mitchell

Glynn Research Institute
Bodmin
Cornwall PL30 4AU

INTRODUCTION

The *bc* complex and cytochrome oxidase are two of the major respiratory electron transfer complexes of mammalian mitochondria. Their catalytic cycles result in the net translocation of a proton across the lipid bilayer, an effect which is in addition to any scalar protonic changes of the chemical reactions. The net translocation causes a charge imbalance across a region of low dielectric strength and so results in a transmembrane electric field, $\Delta\psi$. It should be noted, however, that the field-producing component reactions of the "proton pumping" chemistry need not involve electrogenic proton movements. Transmembrane electron transfer or movements of other charged species can be electrogenic with associated protonic changes being non-electrogenic. In all cases, measurable electrogenicity will only occur if a counter ion is unable to co-migrate, and if the medium through which the charge moves has a sufficiently low dielectric strength (figure 1A). Electrogenic movement of a charged species can occur only along such a "well"; the well interconverts chemical potential field and electric field [1,2]. An effect of electric field which is mediated by such a well should be inducible by a change in

A. Ion Well   B. Ion Channel

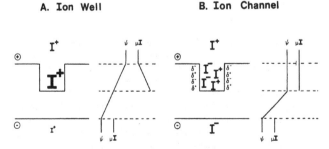

Fig. 1. Schematic representation of the electric (ψ) and chemical (μ) potential fields of (A) an ion-specific well and (B) a non-specific ion channel in a membrane of low dielectric strength.

chemical potential in the appropriate phase in the absence of an electric field and this may be used as a test of the presence of a well. If the dielectric strength of the channel is high, or if counterions can move, then the more trivial case of a non-specific ion channel will occur (figure 1B). Such channels, however, could allow the major electrogenic components to be localised in a very small region of space which does not have to span the 4 nm width of the membrane. This report discusses the chemistry and topology of the electrogenic components which comprise a net charge movement across the lipid bilayer in the reaction cycles of the *bc* complexes and cytochrome oxidase.

ELECTROGENIC REACTIONS OF THE *bc* COMPLEXES

The *bc* complexes from mitochondria, bacteria and chloroplasts are functionally similar. All contain an iron sulphur centre, a *c*-type cytochrome and a two-haem *b* cytochrome. The electron transfer cycle of the enzyme is now well understood. This cycle, together with associated (de)protonation reactions, causes net translocation of a proton across the membrane. An essential feature of this "Q-cycle" [3] is a mechanism of charge movement between two transmembrane quinone binding sites. One of these sites, the *o* site, oxidises quinol with the release of protons into the positive aqueous phase. The first electron lost from the quinol is donated to the high potential iron sulphur centre and *c*-type haem acceptors. The electron remaining on the semiquinone product is transferred, *via* the *b* haems, to the second quinone reactive site, the *i* site, where quinone is reduced in two steps back to quinol with the uptake of protons from the negative aqueous phase.

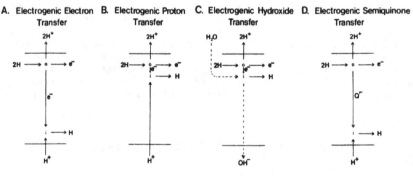

Fig. 2. Some possible charge translocating reactions of the protonmotive reaction cycle of the *bc* complex. At the *o* centre, quinol is oxidised with net proton release. The first electron is donated to high potential acceptors. The second electron is transferred to the *i* centre where it reduces a quinone with net uptake of protons. Within the framework of a Q-cycle mechanism, the electrogenic component(s) could be (A) electron transfer, (B) proton transfer, (C) hydroxide transfer, (D) charged semiquinone movement, or a combination of these.

Figure 2 illustrates the cycle schematically, and emphasises some of the possible electrogenic reactions which might comprise the protonmotive cycle. The possibility that a charged semiquinone can move between the sites [4] (figure 2D) now seems unlikely since studies with chloroplasts on the extent of electric field generation have shown that the low potential conditions which were thought to promote such a reaction indicate that only an uncharged species is moving between the sites [5]. Electrogenic hydroxide movements are also unlikely since the quinones normally associate only with protons and mobile groups which might aid hydroxide conduction through the protein are not apparent. Instead it is becoming likely that the electrogenic steps involve electron transfer and/or protonation changes.

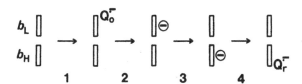

Fig. 3. Schematic representation of the steps which comprise a net charge separation across the membrane in the bc complexes. The semiquinone, Q_o, is produced on oxidation of quinol by the high potential acceptors at the o site. Electron transfer proceeds to the low potential haem, b_L, then to the high potential haem, b_H, and finally to a quinone located in the i site.

Most data on the bc complexes are consistent with the notion that roughly one half of the charge separation arises during electron transfer between the two haems of cytochrome b (step 3 in figure 3). Data include the effects of membrane energisation on the redox behaviour of the b haems [6,7] and the magnitude of the field produced when haem b reoxidation is prevented by inhibitors [8,9]. In the former, the relative midpoint potentials of the two haems are shifted by an imposed electric field by roughly half of the value of the imposed field. In the latter, prevention of haem b reoxidation by inhibitors (step 3 in figure 3) decreases the field generated to about 50% of the uninhibited value (consistent with steps 2 *plus* 3 producing half of the field). A transmembrane pH gradient cannot substitute, hence ruling out an indirect effect of $\Delta\psi$ *via* internal proton wells. Predictions of secondary structure from primary sequence data suggest that the planes of the b haems are roughly perpendicular to the membrane surface with their centres approximately 2 nm apart [10]. EPR data support a similar topology [11].

The remainder of the electrogenicity is likely to arise during the reoxidation of cytochrome b_H by quinone in the i site [6] (step 3 in figure 3). Recently, Robertson and Dutton have confirmed this in bacteria by demonstrating that the reverse reaction is reverse-electrogenic [9], a result consistent with previous demonstrations that reduction of b_L by quinol via the o site (step 2 of figure 3) is not electrogenic [12]. On the basis of the lack of effect of protonation of the i site on the electrogenic span between b_H and Q_i, it was further concluded that it was $b_H \longrightarrow Q_i$ electron transfer, and not a change of protonation state, which was electrogenic [9]. Hope and Rich [13] have also concluded that proton uptake into the i site of the analogous chloroplast bf complex is not itself electrogenic because the extent of electronation-linked protonation change of the site does not affect the extent of electric field which is produced.

The above conclusion is not fully satisfactory from an energetic point of view. In mitochondria for example, at pH 7 the midpoint potential of haem b_H (+90 mV versus SHE) is close to the midpoint potentials of the bound quinone species in the i site (E_{m7} for Q_i/Q_iH_2 = +105 mV vs. SHE [14]). A further complication was noted by Konstantinov et al [15] who

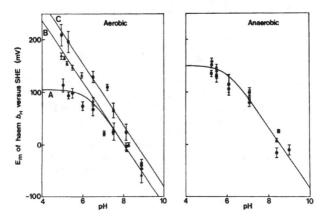

Fig. 4.   Redox Titrations of Cytochrome b_H in Beef Heart Submitochondrial Particles With the Succinate/Fumarate
Couple. Beef heart submitochondrial particles were
resuspended to 4 mg/ml in 50 mM buffer (MES, pH 5-7;
potassium phosphate , pH 7.1 to 8.2; TRICINE, > pH 8.2),
2 mM EDTA, 1 μg/ml valinomycin and 1 μg/ml gramicidin. In
the aerobic titrations 1 mM KCN was also present and in
the anaerobic titrations an atmosphere of nitrogen was
maintained above the sample. Titrations were normally
performed in the oxidative direction after reduction with
1 - 2.5 mM succinate, although reductive titrations gave
comparable data. Cytochrome b_H was monitored at 562 -
0.5·(544 + 580) nm. 50% reduction of b_H was taken to be
the point at which 25% of the dithionite-reducible signal
was developed. Midpoint potentials were calculated from
the succinate/fumarate ratio at this point, and using the
pK values given by Clark [17] with Em_7 set to + 24mV.
Simulated lines are for an n - 1 component with a proton-
atable group of: in the aerobic titrations (A) pK_{ox} 6.25,
pK_{red} > 9; (B) and (C) pK_{ox} < 5, pK_{red} > 9; in the
anaerobic titrations pK_{ox} 6, pK_{red} > 9. ●, control; ▲,
plus 10 μM myxothiazol; ■, plus 10 μM antimycin.

pointed out that the data which showed electrogenic haem b_H oxidation could be caused by a deprotonation of the quinol-oxidising o site and that haem b_H reoxidation by quinone in the i site might not itself be electrogenic. Some support for this idea came from experiments which showed that haem b_H was readily accessible to hydrophilic redox mediators from the negative aqueous phase, suggesting that the haem group is close to this N-phase and from effects of P-phase pH changes on the redox potential of haem b_H [15]. Their interpretation also requires that cytochrome b_H (with an apparent pK_{ox} of 6.25 in unliganded complex, see figure 4A) should be pH-dependent over a wider pH range in the presence of antimycin A and presented data which indicated that antimycin A does indeed lower the pK_{ox} of haem b_H [16].

We reinvestigated the E_m/pH relations of b_H in beef heart sub-mitochondrial particles by aerobic redox titration with the succinate/-fumarate redox couple (figure 4, left). Such titrations have generally been performed aerobically, but in the presence of cyanide in order to inhibit electron flux through cytochrome oxidase. These data confirm that antimycin does apparently lower pK_{ox} of haem b_H from around 6.25 (curve A) to below 5 (curve C), in agreement with the limited data in [6]. However, the data also show that myxothiazol apparently also lowers the pK_{ox} of haem b_H so that it remains pH-dependent between pH 5-9 (curve B). This latter effect is particularly surprising since myxothiazol binds at the o site which is thought to be close only to haem b_L. We repeated the titrations anaerobically and in the absence of cyanide, monitoring cytochrome c redox state concurrently as an indicator of anaerobiosis, data points only being taken at full reduction of the haem c. In this case the results were quite different (figure 4, right). All three titrations were similar, with a possible pK_{ox} on haem b_H only at 6 or below. No large effect of antimycin or myxothiazol is seen. It is concluded that the aerobic titrations are affected by small electron leaks through the system, especially below pH 7 where the apparent midpoint potential can be artefactually raised (antimycin data) or lowered (control data). This conclusion was directly confirmed by demonstrating that an anaerobic addition of either antimycin or myxothiazol to anaerobic submitochondrial particles which had been redox clamped with a 1:1 ratio of succinate:-fumarate at pH 5.6 had very little effect on the redox state of haem b (data not shown).

More conventional anaerobic redox titrations in the presence of redox mediators were also performed. A representative titration at pH 6.4 is shown in figure 5. The data again confirm that antimycin has little or no effect on the redox properties of either b haem. The fit to a combination of two n = 1 components is not good, but these deviations are likely to be real. Their possible origin is under investigation.

Gopher and Gutman showed that acidification of the inner space of inside-out submitochondrial particles did not cause a change in redox poise of b_H when clamped via succinate/fumarate in the presence of antimycin [6], and a similar result was reported by Konstantinov et al [15] with bc complex which had been reincorporated into liposomes. We have found that an external pH pulse given to a similar system poised with quinol/quinone also did not perturb the redox poise of b_H. In contrast to the discussion in [6], all of these experiments might suggest that electronation and protonation of b_H occur from the same membrane side (the Q_o side) in the presence of antimycin, as suggested in [15]. However, the interpretation depends critically on whether thermodynamic equilibrium has been reached. We suggest tentatively that the apparently anomalous effects of antimycin on the redox and protonation properties of haem b_H might arise from a combination of pH-dependent factors which prevent this equilibrium. For example, the equlibration of QH_2/Q with haem b in the presence of antimycin and reduced cytochrome c is known to be extremely slow, and this may explain the results of the pH-pulse experiments above. In the aerobic succinate/fumarate titrations (figure 4), it is likely that the relative activities of succinate dehydrogenase and cyanide-inhibited cytochrome oxidase are altered with decreasing pH so that the extent of kinetic disequilibrium is also pH-dependent.

In conclusion, although not all data are fully consistent, it appears most likely that the electrogenic charge separation in the bc complexes is achieved in two steps. One of these is electron transfer between the two haem groups of cytochrome b. The second step is probably an electrogenic electron transfer from haem b_H to a quinone located in centre i, although doubts do remain both as to whether sufficient energy is available in this step and why the haem b_H seems so readily accessible to hydrophilic redox mediators. There is no strong evidence to date for the participation of large electrogenic proton movements; instead these probably occur electroneutrally in contact with the appropriate aqueous phase. It is the combination of the electrogenic electron transfers with the electroneutral proton exchanges which comprises the net protonmotive activity.

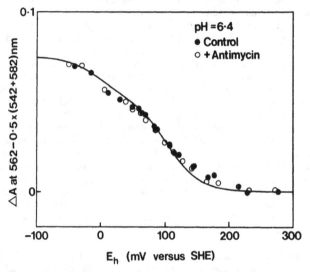

Fig. 5. Anaerobic redox titrations of haem b of beef heart submitochondrial particles. Beef heart submitochondrial particles were resuspended to around 8 mg protein/ml in 50 mM MES and 2 mM EDTA at pH 6.4 which also contained 1 μM valinomycin and 1 μM gramicidin and 400 units/ml catalase. The redox mediators (5-100 μM) were: phenazine methosulphate, phenazine ethosulphate, 1,4-naphtho-quinone-2-sulphonate, diaminodurene, 2,6-dimethyl-benzo-quinone, hexammineruthenium chloride, 2-methyl-naphtho-quinone, trimethyl- and tetramethyl-benzoquinone, 5-hydroxy-naphthoquinone and anthraquinone mono- and di-sulphonate. The reaction mixture was kept under argon, potential was changed with additions of potassium ferri-cyanide or NADH and potential was monitored with a clean glassy carbon surface. Five minutes were allowed for equilibration between each reading. Antimycin when present was 5 μM. The three required wavelengths were monitored cyclically. The theoretical curve is for a combination of two non-interacting n = 1 components of E_m +103 mV (70% of total) and -2 mV (30% of total).

Cytochrome oxidases from a wide variety of organisms also appear to exhibit a remarkable degree of functional and structural similarity. All contain two haems a (a and a_3) and two copper atoms (Cu_A and Cu_B). It is likely that Cu_A is contained in polypeptide subunit II, that the other three metal centres are all in subunit I [18], and that all of the major elements of the catalytic cycle occur within these two polypeptides. Recent work indicates that the same basic mechanistic unit may also occur in the bacterial cytochrome o, which contains a subunit with sequence homology to subunit I.

Understanding of the protonmotive mechanism in oxidase is less well advanced than that of the bc complex. This has arisen for several reasons. Firstly, the net reaction of reduction of molecular oxygen by electrons derived from cytochrome c is itself an electrogenic process, making it more difficult to establish those electrogenic components specifically associated with the additional protonmotive chemistry. Thirdly, it has proved much more difficult to establish definitively the electron transfer route through the enzyme. This latter problem is exacerbated both by the complexity of the intermediates in the oxygen reduction cycle and by the problems in obtaining purified enzyme preparations which have a well-characterised, homogeneous conformation. These difficulties have made possible a wide range of speculations as to the site and mechanism of the protonmotive reactions. Cu_A [19], haem a [20], and the binuclear centre [21,22] have all been invoked as possible participants.

Figure 6 depicts a widely held working model of the protonmotive reactions of the oxidase. Electron transfer from cytochrome c to the Cu_A/haem a acceptors is accompanied by electrogenic uptake of a proton from the negative aqueous phase. Reoxidation of these acceptors by the binuclear centre (Cu_B/haem a_3) results in release of this proton to the positive phase, again via a proton well. Reduction of molecular oxygen by the binuclear centre also involves proton uptake from the negative phase, and these protons have also been suggested to enter the site via a proton well.

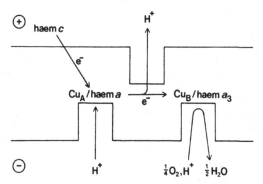

Fig 6. Widely held scheme of electrogenic reactions in cytochrome oxidase. Protonation changes associated with
electronations are depicted as occurring along proton
wells, rather than proton channels (see fig. 1).

Probable electron transfer sequence

It is generally agreed that the assignment of Cu_A/haem a and Cu_B/haem a_3 to the cytochrome c and molecular oxygen 'poles' of the enzyme, respectively, is correct [23]. However, the exact sequence of components on the electron transfer route is still uncertain, the uncertainty beginning with the identity of the primary electron acceptor(s) from cytochrome c. Until recently it was widely believed, principally on the basis of stopped-flow kinetic experiments, that haem a is the primary acceptor in oxidase. However, advances in knowledge of the structure of the enzyme have cast doubt on this assignment. It is thought that Cu_A is bound to subunit II at a site identified by Steffens & Buse [24], while the recent structure modelling work of Holm et al. [18] (see also earlier work [25] and references therein) places haem a in subunit I. Subunit II is known to constitute most of the cytochrome c binding site. While such work is inevitably speculative the possibilities are strongly constrained by the known physical and chemical properties of the redox centres and the conservation of potential ligands to the catalytic metal ions within these centres (see Refs in [18]). In addition, the assignment of Cu_A as the primary acceptor is not inconsistent with at least some of the kinetic data which have been used as evidence in favour of haem a [23,26] if it is remembered that (a) electronic equilibration between Cu_A and haem a may be extremely rapid, and (b) as a result of mutual interactions between the redox centres (see below) the equilibrium distribution of electrons between Cu_A and haem a is strongly dependent on the redox state of the binuclear centre (i.e. haem a has a potential around 120 mV greater than that of Cu_A when Cu_B/haem a_3 is oxidised but haem a and Cu_A are close to equipotential when Cu_B/haem a_3 is reduced) [27].

The site of electron entry into the binuclear centre (Cu_B/haem a_3) is also uncertain. From their modelling work, Holm et al. [18] tentatively placed Cu_B, which is known to be close to haem a_3 (about 3 Å away [28]), on the distal side of this centre to haem a, which implies that electrons enter the binuclear centre via haem a_3. This seems inconsistent with the observation that Cu_B is in rapid equilibrium with Cu_A/haem a in oxidase where haem a_3 is 'clamped' in the oxidised state by cyanide ligation [29-31].

Evidence for electrogenic proton uptake by haem a

In the type of model shown in figure 6, one would expect that at least a part of the proton translocating cycle of reactions could be observed in the presence of inhibitors of the a_3/Cu_B centre. Hinkle and Mitchell [32] showed that in carbon monoxide-inhibited mitochondria, the midpoint potential of haem a, referred to the extramitochondrial redox poise, was a linear function of $\Delta\psi$ to an extent corresponding to about half of the applied potential. This was taken to indicate that haem a is situated approximately halfway through the membrane dielectric layer. If the small pH dependence of haem a is mediated by a proton well from the matrix side in accordance with the model, it would be expected that part of the $\Delta\psi$ effect would result from an induced change in the protonic potential at the bottom of the postulated well. However, the E_m showed no sensitivity at all to changes in intramitochondrial pH at constant external pH [33]. Also we find that in CO- or CN-inhibited mitochondria, the apparent E_m shift following ATP-induced $\Delta\psi$ generation is largely abolished by valinomycin in a potassium-containing medium [34] (also R.M. unpublished observations). Gregory and Ferguson-Miller [35] recently showed that in oxidase vesicles during steady-state turnover the relative reduction levels of cytochrome c and haem a were influenced by $\Delta\psi$ but not by ΔpH (i.e. internal pH). See also [36].

We have recently presented evidence based on the effects of extramitochondrial pH steps on redox poised haem a [37] that the pH sensitivity relates, in intact mitochondria, to the external protonic potential. In cyanide-inhibited mitochondria, at alkaline pH above E_h=280mV there is an additional response to internal pH, and it is this phenomenon which led Artzatbanov and co-workers to draw the conclusion that haem a is in protonic equilibrium with the matrix phase by means of a proton well [38]. However, in CN-inhibited oxidase, haem a exhibits a strong redox interaction with another component (probably Cu_B). The response of the haem to internal pH is easily distinguished from the external pH response in having a much lower rate constant. Even in uncoupled mitochondria, or in isolated enzyme, the response to a step change of pH is distinctly biphasic under appropriate conditions (figure 7).

Since the internal pH effect is seen only in a region of pH and E_h where the interaction energy between haem a and Cu_B becomes pH-sensitive, and is not seen in CO-inhibited mitochondria, where Cu_B appears to be held in the reduced state, we favour the view that this protonation affects haem a in a manner dependent on the redox state of Cu_B or its conformational state. In any event, the kinetics of this protonation response, which appear to be unchanged in the anaerobic, unliganded enzyme (R.M. unpublished work), seem to preclude the involvement of this proton/electron interaction in translocation of protons.

Evidence for electrogenic proton uptake into the binuclear centre

In the working model of cytochrome oxidase shown in figure 6 the binuclear centre is shown to be in protonic contact with the mitochondrial matrix. Konstantinov et al. [39], using oxidase reconstituted into liposomes, have shown that the rate of binding of cyanide to haem a_3 is dependent on the intraliposomal pH. This is certainly consistent with the working model but, as Wikström [40] has pointed out, a remote effect on cyanide-binding kinetics by an acid/base group equilibrating with the matrix cannot be excluded. Nevertheless, Wikström's work [41,42] on protonmotive-force-assisted partial reversal of oxidase has produced direct evidence for protonic contact between the matrix and the site of the oxygen reactions. He has shown that reversed electron transfer from the binuclear centre and water to cytochrome c is strongly and

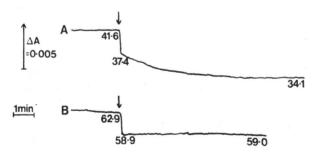

Fig. 7. Time course of 605nm absorbance in aa_3-CN on alkalin-
isation from pH 7.5 --> 8.5 (arrows). KOH was added to a
rapidly-stirred solution of about 1.6μM oxidase in a KCl
medium containing 5mM ferro-ferricyanide and 2μM
cytochrome c at E_h=300mV (A) or 260mV (B). Figures close
to the traces give percentage reduction of haem a.

15

specifically dependent ($2H^+$ output/e^-) on the matrix pH [40]. More recently, Vygodina and Konstantinov [43] have shown that the spectrum of the complex of oxidase with hydrogen peroxide, which may be related to Wikström's putative 'P' (peroxy) intermediate, is also specifically dependent on the matrix pH.

Konstantinov *et al*. [39] claim to have demonstrated that proton uptake by the binuclear centre is electrogenic by showing that the rate of cyanide binding was affected in a similar way by a reverse membrane potential or intraliposomal acidification. Again, this is certainly qualitatively consistent with electrogenic proton uptake, but a definitive conclusion cannot be reached until the quantitative relationships between the effects of $\Delta\psi$ and pH_i, in isolation and in combination, have been measured.

Redox interactions between Cu_A and haem *a*, and Cu_B and haem *a*

It is generally accepted that there is a mutual electronic interaction between the two haems of cytochrome oxidase such that reduction of one lowers the midpoint potential of the other. This is the 'neoclassical' hypothesis [27,44]. However, since its original conception this model has become modified so that additional anticooperative interactions between the copper centres and the haems are included. Goodman [29] noted that an interaction between Cu_B and haem *a* was necessary in order to model her EPR data for cyanide-liganded oxidase. Spectroelectrochemical titrations by Chan and co-workers [45-47] have revealed an interaction between Cu_A and haem *a*. They have also noted that an interaction between Cu_B and haem *a* is necessary to explain the discrepancy in the strength of interaction observed at haem *a* as compared to that observed at haem a_3 ([47], but see [48]). Recently, Wikström [49] concluded that anticooperativity between haem *a* and Cu_B was responsible for the abnormal titration curve of haem *a* in cyanide-liganded oxidase ([38], see also discussion in [47]).

We have found in our own investigations that at neutral pH the titration curve of haem *a* in cyanide-liganded oxidase can be accounted for in terms of a major interaction, probably with Cu_B, and a minor interaction, probably with Cu_A, the strengths being 50-60 mV and 20-30 mV, respectively (figure 8). We have also found that the major interactant is essentially isopotential with haem *a* over the pH range 6-9. If the assignment of Cu_B as this interactant is correct then the potential of Cu_B must, like that of haem *a*, be slightly pH-dependent [50,51]. This is contrary to the long-held view that the midpoint potential of Cu_B is pH-independent [52].

Dielectric location of components

Recently, we showed that imposition of a protonmotive force across the inner membrane of cyanide-inhibited mitochondria had no effect on the redox poise of Cu_A relative to cytochrome *c*, and shifted the redox titration curve of haem *a* by only 50-60mV at the region of maximum effect [53]. We were also able to quantitate the number of electrons which were lost from Cu_B as it became more oxidised on energisation, but were unable to convert this into a change in midpoint potential relative to cytochrome *c* since we did not know the redox state of Cu_B in deenergised mitochondria. However, we can use the data in the analyses of the haem *a* redox titration curves above to deduce that Cu_B and haem *a* have roughly the same redox poise in the de-energised state. Since energisation causes slightly more electrons to be lost from Cu_B than from haem *a* [53], the shift in its E_m relative to cytochrome *c* is only slightly greater than that of haem *a*. The result is important for two reasons. Firstly, the observation that Cu_B

is redox active in the presence of cyanide (which fixes haem a_3 in the oxidised state) confirms that Cu_B is on the cytochrome c side of haem a_3. Secondly, it shows that there is no major electrogenic component of the proton pump cycle between the steps of cytochrome c oxidation and Cu_B reduction. The observed effects of energisation, which arise mostly in response to the $\Delta\psi$ component of protonmotive force, probably indicate the positions of haem a and Cu_B through the membrane dielectric.

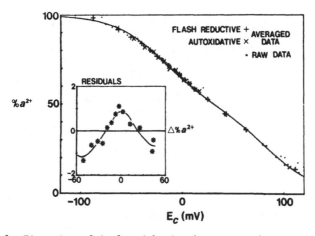

Fig. 8. Titration of isolated bovine heart cytochrome oxidase using flash-induced photoreduction (FIRE, [50]). A representative titration curve for haem a (about 3 μM) at pH 7.0 and 20°C in 0.1 M K-P_i containing 1 mM K-EDTA, 1 mM KCN, 5 μM cyt c, 2 μM PES, and 75 μM riboflavin. E_c is defined as the potential relative to cyt c, the redox indicator, i.e. $58 \cdot \log_{10}(\text{cyt } c^{3+}/\text{cyt } c^{2+})$. The continuous line is a best fit for a single interactant. The fit parameters were: haem a, $E_{mO} = 67$ mV, $E_{mR} = -1$ mV; interactant, $E_{mO} = 68$ mV (all relative to cyt c). E_{mO} and E_{mR} are the midpoint potentials for each component with the other oxidised and reduced, respectively. The difference between these two midpoints is the strength of interaction, in this case 68 mV anticooperative. Note that the residuals obtained on subtraction of the best fit from the data are non-random indicating the presence of a second interactant with E_{mO} some 20-40 mV lower than cyt c.

Since we believe that the order of electron transfer is likely to be c --> Cu_A --> a --> Cu_B, the data argue against an important electrogenic component of the protonmotive chemistry occurring on turnover of Cu_A or of haem a. It should be noted that such data do not preclude during these steps a non-electrogenic protonation of a group which is subsequently involved in the protonmotive chemistry, nor do they preclude an effective contribution to the protonmotive chemistry through the redox interactions which are known to be operative. It would appear, however, that the major energetic step in the pump cycle must be associated with the highly exergonic reactions of oxygen reduction in the binuclear centre.

ACKNOWLEDGEMENTS

Different aspects of this work are supported by the Glynn Research Foundation Ltd, by the SERC (grant GR/F/17605) and by an award from the Wellcome Trust. Experimental work on the bc complex was mostly performed by S.A. Madgwick and A.E. Jeal. R.A. Harper carried out preparative work and produced the figures. Useful discussions, especially on proton wells, with Dr. P. Mitchell have also aided this work.

REFERENCES

1. Mitchell, P. (1968) Chemiosmotic coupling and energy transduction, Glynn Research Ltd, Bodmin.
2. Mitchell, P. (1977) in Symposia of the Society of General Microbiology XXVII. Microbial Energetics (eds.), pp. 383-423.
3. Mitchell, P. (1976) J. Theor. Biol. 62, 327-367.
4. Rich, P.R. and Wikström, M.K.F. (1986) FEBS Lett. 194, 176-182.
5. Moss, D.A. and Rich, P.R. (1987) Biochim. Biophys. Acta 894, 189-197.
6. Gopher, A. and Gutman, M. (1980) J. Bioenerg. Biomemb. 12, 349-367.
7. West, I.C., Mitchell, P. and Rich, P.R. (1988) Biochim. Biophys. Acta 933, 35-41.
8. Jones, R.W. and Whitmarsh, J. (1985) Photobiochem. Photobiophys. 9, 119-127.
9. Robertson, D.E. and Dutton, P.L. (1988) Biochim. Biophys. Acta 935, 273-291.
10. Saraste, M. (1984) FEBS Lett. 166, 367-372.
11. Ohnishi, T., Schägger, H., Meinhardt, S.W., LoBrutto, R., Link, T.A. and von Jagow, G. (1989) J. Biol. Chem. 264, 735-744.
12. Glaser, E.G. and Crofts, A.R. (1984) Biochim. Biophys. Acta 766, 322-333.
13. Hope, A.B. and Rich, P.R. (1989) Biochim. Biophys. Acta, in press.
14. de Vries, S. (1983) Ph.D. Thesis, University of Amsterdam.
15. Konstantinov, A.A. and Popova, E. (1987) in Cytochromes: Molecular Biology and Bioenergetics (Papa, S., Chance, B. and Ernster, L., eds.), pp. 751-765, Plenum Press.
16. Kamensky, Y.A., Artzatbanov, V.Y., Shevchenko, D.V. and Konstantinov, A.A. (1979) Dokl. Acad. Nauk. (USSR) 249, 994-997.
17. Clark, W.M. (1960) Oxidation-reduction potentials of organic systems, Bailliere, Tyndall & Cox, Ltd., London.
18. Holm, L., Saraste, M. and Wikström, M. (1987) EMBO J. 6, 2819-2823.
19. Blair, D.F., Gelles, J. and Chan, S.I. (1986) Biophys. J. 50, 713-733.
20. Wikström, M. and Casey, R.P. (1985) J. Inorg. Biochem. 23, 327-334.
21. Mitchell, P. (1987) FEBS Lett. 222, 235-245.
22. Chance, B. and Powers, L. (1985) Curr. Topics Bioenerg. 14, 1-19.
23. Brunori, M., Antonini, G., Malatesta, F., Sarti, P. and Wilson, M.T. (1988) in Advances in Inorganic Biochemistry Volume 7 (Eichorn, G. and Marzill, L.G., eds.), pp. 93-153, Elsevier, Amsterdam.
24. Steffens, G.J. and Buse, G. (1979) Hoppe-Seyler's Z. Physiol. Chem. 360, 613-619.
25. Wikström, M., Saraste, M. and Penttilä, T. (1985) in The Enzymes of Biological Membranes, Vol. 4 (Martonosi, A.N., ed.pp. 111-148, Plenum Publishing Corporation, .
26. Brunori, M., Antonini, E. and Wilson, M.T. (1981) in Metal Ions in Biological Systems Vol. 13 (Sigel, H., ed.pp. 187-228, Marcel Dekker, New York.
27. Wikström, M.K.F., Harmon, H.J., Ingledew, W.J. and Chance, B. (1976) FEBS Lett. 65, 259-277.
28. Scott, R.A. and Schwartz, J.R. (1986) Biochemistry 25, 5546-5555.
29. Goodman, G. (1984) J. Biol. Chem. 259, 15094-15099.

30. Dervartanian, D.V., Lee, I.Y., Slater, E.C. and van Gelder, B.F. (1974) Biochim. Biophys. Acta 347, 321-327.
31. Johnson, M.K., Eglinton, D.G., Gooding, P.E., Greenwood, C. and Thomson, A.J. (1981) Biochem. J. 193, 699-708.
32. Hinkle, P. and Mitchell, P. (1970) Bioenerg. 1, 45-60.
33. Mitchell, P. and Moyle, J. (1979) in Cytochrome Oxidase (King, T.E., Orii, Y., Chance, B. and Okunuki, K., eds.), pp. 361-372, Elsevier/North-Holland Biomedical Press, Amsterdam.New York.Oxford.
34. Rich, P.R. (1988) Ann. N. Y. Acad. Sci., 550, 254-259.
35. Gregory, L. and Ferguson-Miller, S. (1989) Biochemistry 28, 2655-2662.
36. Moroney, P.M., Scholes, T.A. and Hinkle, P.C. (1984) Biochemistry 23, 4991-4997.
37. Mitchell, R. and Mitchell, P. (1989) Biochem. Soc. Trans., in press.
38. Artzatbanov, V.Yu., Konstantinov, A.A. and Skulachev, V.P. (1978) FEBS Lett. 87, 180-185.
39. Konstantinov, A., Vygodina, T. and Andreev, I.M. (1986) FEBS Lett. 202, 229-234.
40. Wikström, M. (1988) FEBS Lett. 231, 247-252.
41. Wikström, M. (1987) Chemica Scripta 27B, 53-58.
42. Wikström, M. (1988) Chemica Scripta , in press.
43. Vygodina, T. and Konstantinov, A. (1989) Biochim. Biophys. Acta 973, 390-398.
44. Nicholls, P. and Petersen, L.C. (1974) Biochim. Biophys. Acta 357, 462-467.
45. Ellis, W.R., Wang, H., Blair, D.F., Gray, H.B. and Chan, S.I. (1986) Biochemistry 25, 161-167.
46. Wang, H., Blair, D.F., Ellis, W.R., Gray, H.B. and Chan, S.I. (1986) Biochemistry 25, 167-171.
47. Blair, D.F., Ellis, W.R., Wang, H., Gray, H.B. and Chan, S.I. (1986) J. Biol. Chem. 261, 11524-11537.
48. Carithers, R.P. and Palmer, G. (1981) J. Biol. Chem. 256, 7967-7976.
49. Wikström, M. (1988) Ann. N. Y. Acad. Sci., 550, 199-206.
50. Moody, A.J., Mitchell, R. and Rich, P.R. (1989) Biochem. Soc. Trans., in press.
51. Moody, A.J. and Rich, P.R. (1989) Biochem. Soc. Trans., in press.
52. Lindsay, J.G., Owen, C.S. and Wilson, D.F. (1975) Arch. Biochem. Biophys. 169, 492-505.
53. Rich, P.R., West, I.C. and Mitchell, P. (1988) FEBS Lett. 233, 25-30.

ANTIBODIES AFFECT THE STRUCTURE AND FUNCTION

OF IONIC CHANNELS IN A LIPID MEMBRANE

Oleg V. Kolomytkin

Institute of Biological Physics
Academy of Sciences of the USSR
Pushchino, Moscow Region, U.S.S.R.

INTRODUCTION

Antibodies are a promising tool for studying and isolating ion channels of cellular membranes. We can suppose also that monoclonal antibodies are an effective instrument to regulate the dependence of functioning of ionic channel gates on the electric field in the membrane as well as to change the ion-conducting properties of single channels in the open state. However such studies are difficult because the mechanism of antibody - channel interaction as well as the structure of cell membrane channels have not yet been understood enough. The investigation of this mechanism can be facilitated by the study of the action of antibodies on an ion channel with the known structure and properties. Two different ionic pores were chosen to be such channels: (1) the ionic pore formed by the polyene antibiotic amphotericin B and cholesterol in a bimolecular lipid membrane, (2) the ionic pore formed by the polypeptide β-latrotoxin.

The structure and properties of these channels have been described in detail[1-6]. According to currently available conceptions, adding the antibiotic at one (cis) side of the lipid bilayer induces formation of 'half-pores' in the cis-monolayer of the membrane. The 'halfpore' can pierce the membrane for a short time to form an ion channel, its orientation with respect to the membrane surface being not changed. It is believed that this channel is formed by eight molecules of amphotericin B and eight molecules of cholesterol, both oriented normally to the membrane surface. Sterol molecules are situated between the antibiotic molecules. The cylindrical 'halfpore' has a diameter of about 0.8 nm lined with the hydrophilic groups of the amphotericin B lactone ring. If amphotericin B is added at both sides of the lipid bilayer, the amphotericin B and cholesterol molecules form ionic channels which consist of two 'halfpores' occurring in the opposite monolayers of the bilayer and joined together by hydrogen bonds. The properties of β-latrotoxin channels are well known as well[5,6].

The tasks of the present study were: (1) to obtain monoclonal antibodies against amphotericin B and polyclonal antibodies against β-latrotoxin, (2) to define their class, 3) to estimate their effect on the channels formed by amphotericin B and β-latrotoxin in a lipid membrane, (4) to define the localization of the antigen determinants on the channels, (5) to determine the specificity of the antibodies, (6) to estimate the stoichiometric channel-antibody ratio and the structure of the channel-antibody complex, (7) to reveal whether one and the same antibody can increase or block the ion conductance of the amphotericin B channel depending on cholesterol distribution in the membrane. (8) to determine how the antibodies regulate the ionic channel gate controlled by the electric field in the membrane.

MATERIALS AND METHODS

Obtaining of antibodies. Hybridomas and monoclonal antibodies against amphotericin B were obtained according to the conventional technique[7]. The clones were tested with regard to the ability of antibodies isolated from cellular supernatants to change membrane electroconductance in the presence of amphotericin B in the solution at one side of the lipid membrane . In this way three monoclones were selected whose antibodies increased the integral conductance of the membrane. However, the concentrations of these antibodies were taken different to produce an equal effect. In the present study one of these monoclones (HAB4) was used. An antibody-linked immuno-sorbent assay for class-specific antibody activity showed that the class of the antibodies used was IgM.[7]

Antibodies (IgG class) against β-latrotoxin were isolated from the serum of the immunized rabbits by means of the immunosorbent technique[7,8].

Lipid membrane formation. The Teflon cell had two compartments separated by a lavsan film with an aperture for the membrane. The diameter of the aperture was 100 um. Water solution was introduced into the cell. Membranes were assembled by Montal-Mueler's technique from two lipid monolayers[9].

Symmetric and asymmetric bilayers were used. The lipid composition of the symmetric membranes is given in Figure legends. Asymmetric membranes were assembled from two different monolayers. The cis-monolayer was prepared from lipids containing 5% of cholesterol, the trans-monolayer from cholesterol-free phospholipids.

Effect of immunoglobulins on the lipid membrane. Control experiments showed that antibodies when added to one or both compartments up to a concentration of 1 g/l did not increase the conductance of nonmodified bilayers of the composition used.

Channel formation. The channels were formed in a lipid membrane by adding amphotericin B or β-latrotoxin to the water solution surrounding the lipid bilayer.

RESULTS AND DISCUSSION

Effect of specific antibodies on ionic channels formed by adding amphotericin B at two membrane sides. The monoclonal antibodies as well as the polyclonal ones do not affect the electroconductance of ion channels formed by adding amphotericin B on two membrane sides.

The action of specific monoclonal antibodies on the channels formed by adding amphotericin B on one (cis) membrane side. It was naturally to expect that antibodies when added at the cis-side of the membrane would not affect the conductance of 'halfpores', since the orientation of 'half-pores' in the membrane did not practically change and antibodies had no effect on the conductance of ion channels formed by two joined 'halfpores'. This assumption was supported experimentally.

Let us consider the effect of immunoglobulins on a single channel when antibodies were added to the trans-compartment of the cell. Fig.1,a shows the record of jump-like changes in membrane conductance at a small amphotericin B concentration (10 nM) at the cis-side of the lipid bilayer. The kinetics of changeover and the amplitude of electroconductance of such a channel did not change throughout the observation period (15min). After introduction of specific monoclonal antibodies at the trans-side of the membrane with a single channel, a change in the current switching kinetics was observed (Fig.1, record 'b'). Approximately 0.5 h after addition of the monoclonal antibodies to the trans-compartment, the transients ceased and current jumps as shown in Fig.1,b appeared. The kinetics and amplitude of these jumps did not change throughout the observation (15 min). It is seen that the amplitude of such jumps differed little from that of an amphotericin channel without antibodies. At the same time, the probability for this modified channel to be open is much greater than that for the amphotericin channel without antibodies.

It is believed that an ion channel formed by a 'halfpore' is in the open state at time moments when the 'halfpore' pierces the bilayer. If the 'halfpore' is situated entirely in the cis-monolayer of the membrane the channel is closed[1-4]. Based on the above experimental evidence it can be assumed that the interaction of antibodies with the 'halfpore' at the trans-side of the membrane increases the probability for the 'halfpore' to pierce the bilayer. This must be responsible for the conductance increase.

Fig.2 shows the effect of antibodies added to the trans-compartment on the integral current-voltage characteristic of the membrane in the

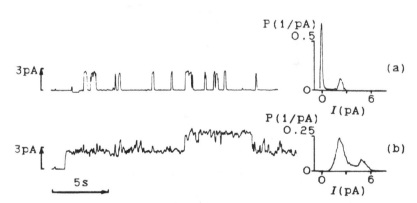

Fig.1. Effect of specific monoclonal antibodies on a single amphotericin B channel in a symmetric membrane. (a) records of trans-membrane current jumps in the presence of 10 nM amphotericin B in the cis-compartment and in the absence of antibodies in the cell; (b) record of transmembrane current jumps 30 min after adding monoclonal antibodies (1 mg/l) to the trans-compartment. The right parts show the probability density functions of current amplitudes during single channel events. The bilayer was formed of bovine brain lipids containing 20% (w/w) cholesterol. The membrane voltage was 0.325 V. The positive voltage sign was in the cis-compartment. Aqueous solution of 2 M KCl/5 mM Tris-HCl was in the cell.

presence of great amounts of amphotericin B in the cis-compartment. Addition of the antibodies at the trans-side of the membrane caused an irreversible increase in membrane conductance by several orders of magnitude; after removal of antibodies and amphotericin B from the cell, the conductance remained practically unchanged throughout the observation time (10 min).

It follows from Fig.2 that the functioning of the amphotericin B channel gate strongly depends on the electric field in the membrane. The channels are open only at high positive membrane voltages. Addition of the antibodies abolishes the strong dependence of functioning of amphotericin B channel gates on the electric field in the membrane.

Measurements of the integral membrane conductance at amphotericin B concentrations below 2 μM and concentrations of monoclonal antibodies [mAb] above 0.7 mg/1 showed that for the initial time period after addition of specific monoclonal antibodies the conductance increment is well approximated by a function:

$$G-g \sim (g)^{\alpha} \cdot [mAb]^{\beta} \cdot t^{\gamma}$$

where g is the membrane conductance before antibody addition, t is the time after antibody addition, G is the membrane conductance at time moments t. $\alpha = 1 \pm 0.2$, $\beta = 3 \pm 0.6$, $\gamma = 3 \pm 0.6$.

Let us consider the mean channel-antibody stoichiometric ratio in the complex responsible for membrane conductance. An amphotericin channel has a rotational eight-fold symmetry[3]. Hence it may be assumed that its valency in the reaction with the antibodies is eight. Fig.3 shows schematically on the same scale a channel with three antibody molecules of IgM class bound to it at the trans-side of the membrane. The channel - antibody size ratio

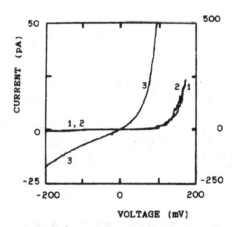

Fig.2. Effect of antibodies on the current-voltage characteristics of a symmetric lipid bilayer formed of brain phospholipids (95% w/w) and cholesterol (5% w/w); (1) in the presense of 1 μM amphotericin B at the cis-side of the membrane; (2) 20 min after adding nonspecific immunoglobulins (100 mg/1) to the trans-compartment of the cell; (3) 20 min after adding specific polyclonal antibodies (25 mg/1) to the trans-compartment. Aqueous solution of 0.1 M KCl/5mM Tris-HCl was in the cell. The membrane voltage increased at a rate of 5 mV/s. The current scale for curves 1 and 2 is shown on the left, for curve 3 on the right.

was taken from an article of De Kruyff, Demel[3] and review of Feinstein, Beale[10]. The antibodies in the Figure are presented so that they do not hide the channel entrance, since the amplitude of channel conductance and its selectivity for potassium and chloride ions change only slightly on binding of antibodies to the channel. It follows from Fig.3 that for the steric reasons no more than three antibody molecules can bind to one channel.

Theoretical analysis showed that under the assumption made the integral conductance increment of the membrane with incorporated channels in the initial time period after antibody addition increases as follows:

$$G-g \sim [Ch]^m \cdot [mAb]^n \cdot t^{m+n-1}$$

where [Ch] is amphotericin B channel concentration in the bilayer. According to the experimental evidence, the conductance rises linearly with increasing g and as a function of the third degree of [mAb] and t with increasing antibody concentration and time. The membrane conductance before antibody addition g is proportional to [Ch]. Hence, it may be assumed that for [Ch] and [mAb] used the complex determining membrane conductance consists of one channel with three antibody molecules bound. It may be assumed that the three antibody molecules shown in Fig.3 can form, due to mutual attraction, a common structure constituting an ion 'halfpore' tightly bound with the 'halfpore' formed of amphotericin B and cholesterol in the membrane.

Characterization of antibodies. A question arises: do antibodies really induce an increase in membrane conductance or the observed increase is due to some minor substances present in the preparation used? To clarify the question, class-specific antibodies were used. The preparation of the mono-clonal antibodies was incubated with class-specific antibodies for 2 hours. Then the incubated mixture was added at the trans-side of the membrane in the presence of 1 μM amphotericin B in the cis-compartment. If the mono-clonal antibodies were incubated with antibodies specific to IgM-class then the membrane conductance was low throughout the observation time (1 h). If the monoclonal antibodies were incubated with antibodies specific to other classes of immunoglobulins, the membrane conductance increased after addition of the incubated mixture to the membrane.

Therefore we can conclude that the increase in membrane conductance

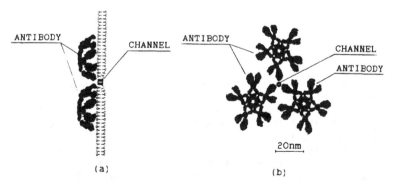

ANTIBODY   CHANNEL   ANTIBODY   CHANNEL

ANTIBODY

20nm

(a)   (b)

Fig.3. Schematic representation of an amphotericin B channel and three antibody molecules bound. On the left, only two anti-body molecules bound to the channel are shown for clarity.

in Figs.1,2 is due to monoclonal antibodies of IgM-class rather than due
to other minor substances in the preparation.

Specificity of the monoclonal antibodies obtained. The control experiment showed that addition of nonspecific immunoglobulins isolated from
a nonimmunised rabbit to the trans-compartment did not change the current-
voltage characteristics of the membrane throughout the observation time
(20 min) (Fig.2).

Another control experiment showed that addition of monoclonal antibodies of IgM class against the histocompatibility antigen to the trans-
compartment at a concentration of 10 mg/l did not increase the membrane
conductance. The evidence obtained points to the high specificity of antibodies isolated from animals immunised by amphotericin B.

The expirements with channels formed of amphotericin B and cholesterol
analogs are also in support of the high specificity of antibody action. It
follows from the expirements with amphotericin B and cholesterol analogs
that both the groups belonging to the trans-end of the amphotericin B
molecule and those situated at the trans-end of the cholesterol molecule
are involved in the formation of the antigenic determinant accessible for
the antibodies.

Effect of monoclonal antibodies on a channel in an asymmetric bilayer.
When cholesterol was present only in the cis-monolayer of the membrane the
action of antibodies on the ion permeability of amphotericin channels was
opposite to that in the case of the symmetric bilayer. Specific monoclonal
antibodies added to the trans-compartment reduced the integral ion conductance of the membrane in the presence of a high amphotericin B concentration
(1 μM) in the cis-compartment (Fig.4).

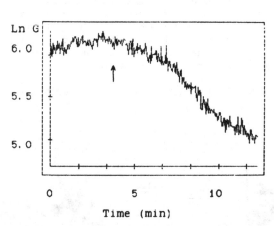

Time (min)

Fig.4. Decrease in the conductance of an asymmetric lipid membrane
       with time in the presence of 1 μM amphotericin B at the cis-
       compartment after adding 1 mg/l of specific monoclonal anti-
       bodies to the trans-compartment of the cell. The arrow shows
       the time moment of adding antibodies. The bilayer was assembled
       of two different monolayers. The cis-monolayer was formed of
       95% (w/w) brain phospholipids and 5% (w/w) cholesterol, the
       trans-monolayer of asolectin without cholesterol. The membrane
       voltage was 20 mV, the positive sign was in the cis-compartment.
       The ordinate is the logarithm of the conductance measured in pS.
       0.1 M KCl/ 5 mM Tris-HCl was in the cell.

The conductance block by antibodies seems to occur as follows. An antibody binds to the open channel at the trans-side of the membrane. Cholesterol migrates from the channel to the trans-monolayer of the membrane, because of which the channel is destroyed. A channel fragment containing the antibiotic and possessing no transmembrane ion conductance is left bound to the antibody. We showed theoretically that if one takes the binding of antibodies with the channel to be the limiting stage and the quantity of amphotericin B in the membrane to be constant, the membrane conductance after antibody addition must decrease as follows: $G \sim EXP(-K \cdot [mAb] \cdot t)$ where K is a constant. The exponential decrease in conductance observed in the experiment (Fig.4) is evidence that the above blocking mechanism is true.

An alternative mechanism for blocking the conductance by antibodies may work on the principle of a plug closing the entry to the channel. However such a mechanism is unlikely, since in this case the activation of conductance by antibodies in case of a symmetric lipid bilayer is hardly to explain.

To summarise, the experiments with monoclonal antibodies showed that a single amphotericin channel can bind 1 to 3 antibody molecules on the trans- side of the membrane. The same antibody when bound to the channel can increase and decrease the channel conductance in dependence of the symmetrical or asymmetrical distribution of cholesterol in the membrane. In both cases the antibodies abolish the strong dependence of functioning of amphotericin B channel gates on electric field in the membrane.

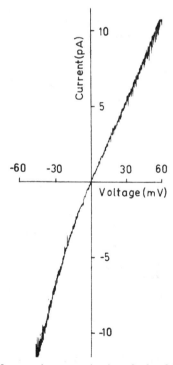

Fig.5. The current-voltage characteristic of the lipid bilayer in the presence of 4±1 mg/l β-latrotoxin at one side of the membrane. The membrane voltage was lowered at a rate of 1.5 mV/s.

The most essential channels of the natural membranes are proteins. Therefore the channel formed by the polipepide β-latrotoxin is a more adequate model of the native membranes in comparison with the amphotericin B channel. Let us consider the influence of specific antibodies on the β-latrotoxin channel gate.

Effect of the antibodies against β-latrotoxin on ionic channels formed by β-latrotoxin. Fig.5 shows the current-voltage characteristic of a bilayer in the presence of β-latrotoxin at one (cis) side of the membrane. The current jumps seen in the Figure were due to the random closing and opening of the channel during membrane voltage scunning. For the lipid composition used, the membrane current-voltage characteristics averaged over a great number of channels differed little from the linear one for the voltage interval studied (-70 to 70 mV).

Fig.6 shows the current-voltage characteristic of the bilayer in the presence of both β-latrotoxin at one (cis) side and specific antibodies at the opposite (trans) side of the membrane. The introduction of specific antibodies produced a decrease in the mean absolute value of transmembrane current at voltages below -20 mV. No marked effect of antibodies on the ion current through latrotoxin channels was observed at positive voltages. Negative voltages resulted in characteristic current fluctuations at frequencies of about 3 Hz which were much higher than those in the absence of antibodies (0.02 Hz).

The control experiment showed that nonspecific immunoglobulins had no effect on latrotoxin channels.

Based on the experimental evidence we can conclude that for β-latro-toxin-formed ionic channels, antibodies were obtained which, being bound to the channel, create a strong dependence of channel conductance on the electric field in the membrane, which was not observed in the absence of antibodies.

The main conclusion of the present investigation. To summarise the results of the experiments with amphotericin B and β-latrotoxin channels we can say that antibodies are an effective instrument to regulate the

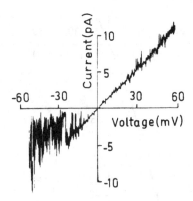

Fig.6. The current-voltage characteristic of the lipid bilayer in the presence of both 4±1 mg/l β-latrotoxin at the cis-side and 1.5 g/l specific antibodies at the trans-side of the membrane. The membrane voltage was lowered at a rate of 1.5 mV/s.

dependence of functioning of ionic channel gates on the electric field in the membrane. The obtained evidence is of interest in elucidating the general features of interaction of antibodies with the ionic channels of cellular and model membranes.

ACKNOWLEDGEMENT

The author wish to thank Drs. Yu.A. Manzygin and N.V. Swyatukhina for preparing hybridomas, Drs. S.K. Kasymov and S.I. Salikhov for isolation of β-latrotoxin and antibodies against it, Profs. L.N. Ermishkin and I.G. Akoev for their helpful discussions in the course of this investigation.

REFERENCES

1. T. E. Andreoti, Kidney Int. 4:337 (1973).
2. A. Finkelstein and R. Holz, in: "Membranes", C. Eisenman, ed., vol.2, Dekker inc., New York (1973).
3. B. De Kruyff and R. A. Demel, Polyene antibiotic-sterol interactions in membranes of acholeplasma laidlawii cells and lecitin liposomes. III. Molecular structure of the polyene antibiotic-cholesterol complexes, Biochim. Biophys. Acta 339:57 (1974).
4. L. N. Ermishkin, Kh. M. Kasumov and V. M. Potzeluev, Single ionic channels induced in lipid bilayers by polyene antibiotics amphotericin B and nystatine, Nature 262:698 (1976).
5. A. Finkelstein, L. L. Rubin and M. C. Tzeng, Black widow spider venom: effect of purified toxin on lipid membranes, Science 193:1009 (1976).
6. O. V. Krasilnikov, V. I. Ternovsky and B. A. Tashmukhamedov, Investigation of the channel formation properties of black widow venom, Biofizica 27:72 (1982).
7. Zelig Eshhar, "Hybridoma Technology in the Biosciences and Medicine", T. A. Springer, ed., Plenum Press, New York and London (1985).
8. A. E. Gurvich and E. V. Lekhtchind, Porous cellulose beads as matrix for an immunosorbent, Bull. Exp. Biol. Med. 12:752 (1981).
9. M. Montal and P. Mueller, Formation of bimolecular membranes from lipid monolayers and a study of their electrical properties, Proc. Nat. Acad. Sci. USA 69:3561 (1972).
10. A. Feinstein and D. Beale, in: "Structure and Function of antibodies", L.E. Glynn and M.W. Steward, eds., John Wiley and Sons, Chichester, New York, Brisbane, Toronto (1981).

RESONANCE RAMAN SPECTROSCOPIC CHARACTERIZATION OF THE OXIDATION OF THE

HORSERADISH PEROXIDASE ACTIVE SITE

James Terner, Andrew J. Sitter and John R. Shifflett

Dept. of Chemistry
Virginia Commonwealth University
Richmond, Virginia 23284-2006

INTRODUCTION

Enzymes containing heme prosthetic groups (heme enzymes) are an extensive class of biological catalysts. These enzymes possess differing reactivities and specificities, even though they contain a very similar active
site heme group, which is in many cases an unmodified iron-bound
protoporphyrin IX. Some heme enzymes of current interest include cytochrome
P-450, cytochrome oxidase, tryptophan pyrollase, prostaglandin synthase,
secondary amine mono-oxygenase, ligninase and the various peroxidases and
catalases [1-3]. The ability of the heme enzymes to mediate reactions
involving oxygen is of importance to industrial applications and health
related areas [4].

One of our aims has been to contribute to an understanding of the
structure and function of the highly oxidized heme states which occur during
the catalytic mechanisms of the heme enzymes. The activated enzyme intermediates are amenable to study by resonance Raman spectroscopy because of
strong resonance enhancement of the heme [5]. Using rapid acquisition techniques we have been able to obtain highly detailed vibrational spectra of
short-lived reaction intermediates and excited states [6]. With the aid of
isotopic substitutions and normal coordinate analysis, vibrational bands can
be assigned to specific structures. Correlation of structural features to
mechanistic aspects is greatly facilitated by crystal structures which are
available for a number of heme enzymes [7], and recent investigations of
enzyme active sites which involve site-directed mutagenesis [8-10].

The catalytic sequences of the heme enzymes involve a number of intermediate enzymatic states, of which the most easily observable spectroscopi-
cally are those of peroxidase and catalase [11]. A well characterized
peroxidase is isolated from horseradish root. This enzyme is typical of
most plant peroxidases and is obtainable in large quantities from biochemical suppliers at a moderate price. The physiological intermediates of
horseradish peroxidase are known as compounds I and II [11,12]. Compound I
is formed by a two-electron oxidation of the resting enzyme by peroxide and
contains an Fe(IV) porphyrin π-radical cation heme [13]. A one-electron
reduction of compound I results in the compound II intermediate which
contains an Fe(IV) (ferryl) heme [11,12].

The ferryl/ferric redox potential for isoenzyme C was determined to be 1.06 V at pH 6 [14]. However, at pH 11.8 the redox potential was found to be lowered to 0.52 V, such that a mild oxidant, ferricyanide, was found to be able to oxidize the ferric enzyme directly to compound II [14-16], even though the the formation of horseradish peroxidase compound II (the ferryl state) is normally preceded by the formation of compound I [17]. We have been able to rationalize these observations in terms of oxidation of a ligated ferric hydroxyl group facilitated through base catalysis by a distal histidine [18].

THE STRUCTURE OF FERRYL HEME

Resting horseradish peroxidase is a brown colored protein, containing an Fe(III) heme. Upon reaction with peroxide it forms the green colored intermediate known as compound I, which is two oxidation equivalents above the resting enzyme. A one electron reduction of compound I, via an oxidizable substrate, results in a red colored intermediate known as compound II. Compound II is an Fe(IV) heme, one oxidation equivalent above the resting enzyme [11]. While compound I of horseradish peroxidase is formally an Fe(V) heme, it contains an Fe(IV) with another electron removed from the highest occupied molecular orbital of the porphyrin π-system, resulting in a porphyrin π-radical cation [13]. While aspects of the heme structure of compound II are being unraveled, structural aspects of compound I have been more difficult to determine [19].

Horseradish peroxidase compound II is one oxidation equivalent above the resting enzyme. For many years, while it was known that the heme iron was in the iron(IV) state, the identity of the distal ligand was undetermined. Proposed heme structures were Fe(IV)=O, Fe(IV)-OH, Fe(IV)-OOH, or an Fe(II)-oxene, among others. Of these a prevalent feeling was that Fe(IV)=O was likely since oxo-ligation is known to stabilize transition metal ions in higher oxidation states [20]. However the Fe(IV)=O formalism was not universally accepted since it was not believed that an Fe(IV) group could be stable in aqueous solution or within a protein [21]. Stabilization of a d^4 metal ion by oxo ligation is uncommon. Additionally, published EXAFS data indicated a single bond [12].

The first observations of resonance Raman Fe(IV)=O vibrations in heme enzyme intermediates, were reported by our laboratory [22,23]. The Fe(IV)=O stretching vibrations were identified by isotopic substitution studies on isoenzymes of horseradish peroxidase compound II [22] and oxidized (ferryl) myoglobin [23]. Our initial reports were rapidly confirmed by subsequent reports from other laboratories [24,25]. The direct observations of the Fe(IV)=O vibrations are allowing the elucidation of long standing questions on the postulated structures of activated enzymatic intermediates in the heme enzymes [26].

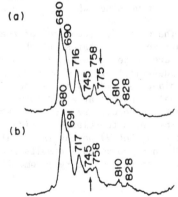

Figure 1. Resonance Raman spectra of horseradish peroxidase compound II isoenzyme C, using 4067 Å excitation, formed with (a) unlabelled H_2O_2 and (b) $H_2{}^{18}O_2$, showing the shift of the Fe(IV)=O stretch from 775 (a) to 745 cm^{-1} (b) upon ^{18}O-substitution [22].

Using isotopic labelling, we demonstrated that the iron atom of the heme was bound to an oxygen atom which was derived from hydrogen peroxide which had reacted with the heme. In horseradish peroxidase compound II, the Fe(IV)=O frequency was found at 779 cm^{-1}, undergoing an ^{18}O induced shift to 743 cm^{-1} [22], as shown in Figure 1. These frequencies were too high for a single bonded oxygen. The Fe(III)-OH resonance Raman frequency in alkaline myoglobin is 490 cm^{-1} [27], while the Fe-O$_2$ resonance Raman stretching mode in hemoglobin was at 567 cm^{-1} [28]. Additionally we demonstrated that the peroxidase model, ferryl myoglobin, also contained an Fe(IV)=O group, though interestingly, at a significantly higher frequency, 797 cm^{-1} [23]. Our oxyferryl assignment was subsequently confirmed by Hashimoto et al. [24] with ^{54}Fe-labelling studies. However, since the O-O peroxide vibration is close to the Fe(IV)=O frequencies, it was necessary to perform double labelling experiments. For example, the iron bound peroxide stretch in resonance Raman spectra of hemerythrin occurs at 844 cm^{-1} [29] and at 742 cm^{-1} for hemocyanin [30]. Using isotopically mixed H$_2$O$_2$, we observed only two Fe=O bands, with an intensity ratio equal to the ratio of isotopes in the total H$_2$O$_2$ sample, confirming the Fe(IV)=O assignment [31].

THE IMPORTANCE OF HYDROGEN BONDING TO Fe(IV)=O

Horseradish peroxidase contains protoporphyrin IX at the active site [11], in common with many other heme proteins such as hemoglobin, myoglobin, catalase, the b-type cytochromes, and cytochrome P-450; and a histidine imidazole at the fifth (axial) coordination position in common with hemoglobin, myoglobin and others. Though these heme proteins contain the same prosthetic group at the active site, their functions are quite different. It has been felt that the unique functionalities are determined in large part by the differing protein structures which surround the heme. Much effort has been focused on developing a description of the structures of the amino acid pockets which surround the heme groups [32,33]. Resonance Raman spectroscopy has been a very effective direct probe of the structures of the heme groups in heme proteins [5] and in horseradish peroxidase in particular [22,24,25,31,34-42]. It had been hoped that resonantly enhanced heme vibrations would be sensitive to structures surrounding the heme and could thus serve as probes of the effects of the surrounding structures on the heme group itself. The Fe(IV)=O vibration in oxidized horseradish peroxidase provided such an opportunity.

We found that the Fe(IV)=O vibrational band of horseradish peroxidase compound II could assume either of two frequencies depending on pH [43]. At neutral pH, the Fe(IV)=O stretching frequency is found at 776 cm^{-1}, confirmed by an ^{18}O-induced shift to 743 cm^{-1} as we had reported earlier [22]. At alkaline pH, the Fe(IV)=O stretching frequency assumes a higher value, 788 cm^{-1}, which was confirmed by an ^{18}O-induced frequency shift to 756 cm^{-1}. The pK of this transition is 8.5 for isoenzymes B and C in agreement with the previously reported pK of the heme-linked ionization of compound II of isoenzyme C [44,45]. A pK of 6.9 was found for isozyme A-1.

When the measurements were performed in D$_2$O, additional effects were observed [43]. At pD values above the pK, no significant frequency variations (that we could detect with the available resolution) were observable for either set of isoenzymes. At pD values below the pK the Fe(IV)=O vibrations were shifted. A 4 cm^{-1} shift to higher energy was seen for isoenzymes B and C. The up-shift is possibly due to kinematic coupling to a deuterated N-H motion, but is more probably a weakened hydrogen bond due to deuterium substitution. At pH values below the pK the Fe(IV)=O oxygen atom appeared to be hydrogen bonded to an exchangeable proton of an amino acid residue, since it was sensitive to substitution of the amino acid hydrogen by a deuteron. At pH values above the pK, the amino acid was unprotonated, and the Fe(IV)=O stretching frequency was apparently insensitive to deuterium substitution. The Fe(IV)=O resonance Raman bands showed remarkable intensity enhancements in D_2O relative to H_2O.

The existence of ionizable groups in the proximity of the heme group of horseradish peroxidase has been known for many years. The distal ionizing groups have been suggested to be histidines in both isoenzyme C and the A isoenzymes [32]. The kinetics of the reactions of horseradish peroxidase compound II are pH dependent and are characterized by pK values of 6.9 and 8.5 for isoenzymes A-1 and C [46,47]. It is not unreasonable that these pK values are due to histidine. The pK of histidine is well known to be dependent on differences in neighboring charged groups. Examples are the histidines in active sites of enzymes such as alkaline phosphatase and ribonuclease [48,49].

The ionizations of the distal histidines are therefore likely to be responsible for the observed frequency shifts of the Fe(IV)=O groups of both isoenzymes. The low energy value of the Fe(IV)=O frequency, at 779 cm^{-1} for isoenzymes A-1 and A-2, and 775 cm^{-1} for isoenzymes B and C, is likely to be due to hydrogen bonding of the Fe(IV)=O group to a protonated distal group such as histidine. Above the pK of the ionizing group, the Fe(IV)=O frequency is high, at 789 cm^{-1}, presumably due to the deionization of the distal group and lack of hydrogen bonding. Hydrogen bonding from the protonated amino acid residue appears to be coincident with higher oxidative activity of the Fe(IV) heme at pH values below the pK of the amino acid residue, suggesting a scheme for the reversion of the Fe(IV)=O heme of compound II to the Fe(III) heme of the resting enzyme as shown below:

Myoglobin in the Fe(III) heme state (metmyoglobin) can be made to react with hydrogen peroxide to form a compound known as ferryl myoglobin which contains an Fe(IV) heme. This compound is similar in structure to horseradish peroxidase compound II, however its peroxidative activity is much lower. Yamazaki et al. [32] has suggested that the interaction of a distal base with the sixth ligand is weak in myoglobin, but very strong in peroxidases. Our data corroborates this view. We observed no shifting of the Fe(IV)=O stretching frequency of ferryl myoglobin as pH is varied from pH 6 to pH 12, contrary to the observations for horseradish peroxidase compound II. Additionally, the ferryl myoglobin Fe(IV)=O resonance Raman frequency is high (797 cm^{-1}), showing no detectable sensitivity to deuterium substitution, suggesting a lack of hydrogen bonding to the oxo-group by a distal amino acid group. Thus it appears that hydrogen bonding to the oxygen of the oxyferryl groups in horseradish peroxidase compound II plays an important role in peroxidase activity.

In addition to peroxides, horseradish peroxidase can also be activated by oxidants such as HOCl, HOBr, $NaClO_2$, $KBrO_3$, and KIO_4 [50,51]. Because of functional parallels to the enzyme chloroperoxidase, there has been considerable interest in the structure of the chlorite activated form of horseradish peroxidase, known as compound X [52-54]. A chlorite derived chlorine atom is known to be retained by compound X, and has been proposed to be located at the heme active site [52]. Both horseradish peroxidase and chloroperoxidase will chlorinate a substrate such as monochlorodimedone, upon reaction with chlorite. Chloroperoxidase, however, will perform a chlorination reaction in the presence of hydrogen peroxide and chloride ion, whereas horseradish peroxidase will not [55]. Compound X has been proposed to be the electrophilic halogenating intermediate in the catalysis of the chlorite reaction [53,54]. Several proposals have been made regarding the structure of the heme active site in compound X [52,55]. The heme active site of compound X has been proposed to contain an Fe(IV)-OCl heme based on the incorporation of ^{36}Cl into the compound X structure, with a stoichiometry of one chlorine atom per enzyme molecule [52].

When maintained at pH 10.7, compound X (formed from horseradish peroxidase isoenzymes B and C) is especially stable and can even be subjected to gel filtration [52]. A search for a resonance Raman Fe(IV)-OCl vibration was undertaken but was not successful. Instead a polarized band at 787 cm^{-1} was observed for compound X at pH 10.7 [22] which was remarkably similar to the Fe(IV)=O frequencies of compound II formed above pH 9 [43]. When compound X was formed at pH 10.7 with $NaCl^{18}O_2$, the 787 cm^{-1} band shifted to 756 cm^{-1} [31] demonstrating that the band is due to an Fe(IV)=O group and that the oxygen atom of the Fe(IV)=O group is derived from chlorite. These frequencies are the same as are observed for the alkaline form of compound II formed with H_2O_2 and $H_2^{18}O_2$ [43].

The chlorinating oxidant therefore appears not to be an Fe(IV)-OCl group of compound X at alkaline pH. Though the chlorinating group is associated with the enzyme, it appears to be at a location other than the heme. It has been shown that halogenation reactions catalyzed by chloroperoxidase are the same as reactions that occur with molecular halogen or hypohalous acid, with respect to reaction products or stereoselectivity [56]. However, it has been demonstrated the chloroperoxidase halogenating agent is electrophilic, enzyme bound, and does not involve the formation of a molecular halogen intermediate [57,58].

The distal heme ligand of compound X is thus an Fe(IV)=O group, rather than an Fe(IV)-OCl, at least above pH 8. The optimal activity of compound X, however, occurs between pH 4 and 7. When compound X formed at alkaline pH, was adjusted to acidic pH in the absence of a halogen acceptor, a significant amount of a green compound I type intermediate was detected by electronic absorption spectroscopy [52,59]. Though a compound II type (typical for compound X) absorption has been detected to be transiently present at acidic pH [59], the compound I type intermediate may be the actual species that contains the activated heme Fe(IV)-OCl group. Compound X has been shown to be formed before compound I in the chlorite reaction. It has been proposed that an oxidizing substance, remaining from the oxidation of the resting enzyme with chlorite, might be present to account for the further oxidation of compound X to compound I at low pH [59]. Identification of this species has not been made, except that it is believed not to be chlorine dioxide [52,59] or hypochlorite [53,55,60], though chlorine monoxide has been suggested as a possibility [59]. Since the oxygen atom of the compound X is chlorite derived, it is possible that the remaining chlorite fragment, e.g. chlorine monoxide, remains in the heme pocket, in proximity of the heme group, but not yet bound to the heme group itself.

While compound X has been shown to behave as a true halogenating intermediate, it appears that the chlorination activity is latent rather than active, above pH 8. Actual chlorination activity stored in compound X above pH 8, becomes active upon adjustment to the required pH range, resulting in a structural change in the intermediate. The active halogenating intermediate may therefore contain a modified form of the compound X heme, or may be a halogenated form of compound I, similar to chloroperoxidase which appears to perform the halogenation reaction through a halogenated intermediate that is derived from compound I [57].

HYDROXIDE LIGATION

The pH dependent changes in heme coordination and spin state, of horseradish peroxidase and other ferric heme proteins, have been known for many years [61]. A readily apparent transition occurring above pH 7 is known as the "alkaline transition". The alkaline transition occurs under relatively alkaline conditions for many heme proteins. Cytochrome c has a number of transitions, one of which has a pK of 12.8 [62]. For horseradish peroxidase isoenzyme C the transition point is pH 10.9 [63]. However, some heme proteins and peroxidase isoenzymes have transitions much closer to neutrality, indicating a likelihood of physiological function. Horseradish peroxidase isoenzyme A-1 has a pK of 9.3 [64]. Met-hemoglobin [65] and met-myoglobin [66] have pK values near 9.0. Turnip peroxidase isoenzyme P_7 has a pK of 8.4 [67].

The peroxidase alkaline transitions are characterized by a transition from predominantly five-coordinate high-spin hemes at neutral pH to six-coordinate low-spin hemes at alkaline pH. Many of the heme protein alkaline forms such as those of hemoglobin and myoglobin contain a hydroxide ligand at the distal (sixth) position of the heme iron. However, spectroscopic and kinetic studies on alkaline horseradish peroxidase have argued against iron-hydroxide ligation [68-72]. In particular, resonance Raman studies failed to identify an iron-hydroxide mode in alkaline horseradish peroxidase [72] that had been previously identified for hemoglobin and myoglobin [73,74]. The above cited studies [68-72] led to conclusions that alkaline horseradish peroxidase does not contain a hydroxide-ligated iron heme. We believe those conclusions to be in error. We were able to identify and confirm the existence of Fe(III)-OH vibrations in horseradish peroxidase isoenzymes through extensive isotopic substitution studies [18].

Figure 2. Resonance Raman spectra of horseradish peroxidase isoenzyme C, at alkaline pH, locating the Fe(III)-OH group at 503 cm^{-1} (a), and verifying by ^{18}O-substitution at 484 cm^{-1} (b) [18].

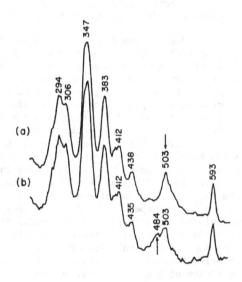

Resonance Raman spectra of the alkaline forms of isoenzymes of horse-radish and turnip peroxidases show characteristic porphyrin vibrational frequencies for low-spin six-coordinate heme complexes [18,42]. This is in contrast to the alkaline forms of aquomet-myoglobin and aquomet-hemoglobin whose resonance Raman spectra are characteristic of mixed spin species [75] in agreement with magnetic measurements [76]. The hydroxyl-iron vibrations of the alkaline forms of methemoglobin and metmyoglobin are believed to be enhanced with red laser excitation due to a metal to ligand charge transfer as determined from the Raman excitation profile [73]. Reported attempts to observe similarly enhanced resonance Raman Fe(III)-OH modes for alkaline horseradish peroxidase were not successful [72]. Nonetheless, we were able to observe the Fe(III)-OH vibrations using Soret excitation (4131 Å). In the resonance Raman spectra of the alkaline form of horseradish peroxidase isoenzyme C a band assignable to the Fe(III)-OH stretch at 503 cm^{-1} (Fig. 2a) shifts to 484 cm^{-1} (Fig. 2b) upon ^{18}O-substitution. Equivalent results were obtained for isoenzyme A-1 of horseradish peroxidase and turnip peroxidase isoenzyme P$_7$ [18]. The observed ^{18}O-induced Fe(III)-OH band shifts are a few wavenumbers below those expected from simple harmonic oscillator calculation indicating minimal coupling to other out-of-plane motions. The data is comparable (though somewhat higher in frequency) to that previously reported for the alkaline forms of aquomet-myoglobin and aquomet-hemoglobin which are known to be authentic ferric-hydroxyl complexes.

To further substantiate these results we compared horseradish peroxidase reconstituted with ^{54}Fe and ^{57}Fe hemes, and through deuterium substitution. The most significant frequency shifts were observed for an out-of-plane mode near 380 cm^{-1} and the Fe(III)-OH mode. The Fe(III)-OH mode of isoenzyme C exhibited a downshift from 504 cm^{-1} for the ^{54}Fe heme to 502 cm^{-1} for the ^{57}Fe heme. The out-of-plane mode showed shifts from 380 cm^{-1} for ^{54}Fe to 378 cm^{-1} for ^{57}Fe for both isoenzymes A-1 and C of horse-radish peroxidase. Sensitivity of the 380 cm^{-1} out of plane mode to ^{18}O-substitution at alkaline pH was also evident. The 2 cm^{-1} ^{54}Fe to ^{57}Fe downshifts for the Fe(III)-OH mode are within the values expected from a simple harmonic oscillator calculation and are comparable to the magnitude of the shift which was previously reported for the iron substituted heme Fe(III)-OH of met-myoglobin [74]. When the samples were suspended in D$_2$O additional significant effects were observed. Suspension of isoenzyme C in D$_2$O resulted in an upshift of the Fe(III)-OH band from 503 to 509 cm^{-1}. The out-of-plane mode at 381 cm^{-1} split into two bands at 379 and 385 cm^{-1} providing additional evidence that an exchangeable hydrogen is involved in out-of-plane motion involving the heme iron. At neutral pH the 380 cm^{-1} band retains its iron sensitivity but loses its D$_2$O sensitivity.

The hydroxide complexes of met-myoglobin and met-hemoglobin exist in thermal spin state equilibria due to the small energy differences between the high- and low-spin species [77]. At room temperature, 70% of hydroxymet-myoglobin [76,77] and 45% of hydroxymet-hemoglobin [76] is high-spin. It has been reported that alkaline horseradish peroxidase is 7% high-spin [76]. We estimate any high-spin component to be less than 5%. The Fe(III)-OH stretching frequencies of the peroxidase hydroxy-complexes are significantly higher than those observed for the hemoglobin and myoglobin complexes. A tabulation of known heme protein Fe(III)-OH frequen-cies, given in Table I, shows that the spin-state distributions are related to the value of the iron-ligand stretching frequency. The fraction of high spin component decreases as the Fe(III)-OH frequency increases. Apparently the energy differences separating the spin-states in the hydroxyl complexes are small enough that small energy differences in the iron to sixth ligand bond strength can correlate with changes in the spin-state distribution. In the case of turnip and horseradish peroxidases, the strength of the

Table I

Listing of Fe(III)-OH resonance Raman stretching frequencies in cm^{-1} with spin-state distributions of alkaline forms of met-myoglobin (Mb), met-hemoglobin (Hb), horseradish peroxidase isoenzyme C (HRP C) and A-1 (HRP A-1), and turnip peroxidase P_7 (TP$_7$).

	$\nu_{Fe(III)-OH}$	% high-spin	reference
Mb	490	70%	73,74
Hb	497	45%	27
HRP C	503	< 5%	18
HRP A-1	516	< 5%	18
TP$_7$	514	< 5%	18

metal-ligand interaction may be large enough to shift the spin-state equilibrium to low-spin.

As mentioned previously, the physiological function of horseradish peroxidase involves two intermediates known as compounds I and II, which are two and one oxidation equivalents respectively, above the resting enzyme [1,11]. The formation of compound II (ferryl state) is usually be preceded by the formation of compound I, however one-electron oxidation of the resting enzyme to the ferryl state is known [14,17]. Electrochemical studies showed a significant decrease in the ferryl/ferric redox potential of the enzyme at alkaline pH that was not observed for the compound I/compound II redox potential [14-16]. The ferryl/ferric redox potential for isoenzyme C was measured to be 1.06 V at pH 6. However, at pH 11.8 the redox potential was found to be lowered to 0.52 V, such that a mild oxidant, ferricyanide, was able to oxidize the ferric enzyme directly to the compound II type ferryl state at alkaline pH [14]. Resonance Raman spectra of the oxidation product showed a band at 788 cm^{-1} assignable as an Fe(IV)=O stretch on the basis of ^{18}O-substitution, which resulted in a shift of the 788 cm^{-1} band to 756 cm^{-1}. The 788 cm^{-1} value is the frequency of horseradish peroxidase compound II found at alkaline pH [43], and confirms that a compound II type heme can be formed by a one electron oxidation under mild conditions, though this ferryl heme is different from the enzymatically active form of compound II found at neutral pH [22,43] which exhibits an Fe(IV)=O frequency of 779 cm^{-1} for isoenzyme C.

The lowering of the ferryl/ferric redox potential at alkaline pH is likely due to a number of contributing factors. Conversion from a high-spin to a low-spin heme is no longer required since the iron is already low-spin. An oxygen atom is already coordinated to the heme as an Fe(III)-OH. Ligation of iron(III) tetraphenylporphyrins by the strongly basic oxyanions, methoxide and hydroxide, has been shown to favor one-electron oxidation of the iron rather than the porphyrin [78,79]. The increase of the ligand field strength of the ligated hydroxide is likely facilitated by hydrogen bonding to a distal histidine, for which there is considerable evidence [32,43,80]. The deuteration sensitivity of our data is consistent with an iron hydroxyl group that donates a hydrogen bond to the distal histidine. Our data show deuterium induced upshifts in the Fe(III)-OH resonance Raman frequency of 6 cm^{-1} for horseradish peroxidase isoenzyme C and 4 cm^{-1} for the A-1 isoenzyme. Deuteration shifts of M-OH stretching vibrations which are due mainly to the increase in mass typically exhibit downshifts on the order of 20 cm^{-1} [81], however a hydroxyl group involved in a hydrogen bonding interaction would experience a modified potential field. Small (2 to 4 cm^{-1}) deuteration up or downshifts have been reported for M-OH stretching vibrations involved in hydrogen bonding [82].

Thus, the peroxidase alkaline forms exist in a pH region where the distal histidine imidazole is able to accept a hydrogen bond from a proton of a iron-ligated hydroxyl group. This interaction would give the hydroxyl-ligand partial oxo character resulting in the high Fe(III)-OH frequencies that are observed for the alkaline peroxidases. These frequencies are 10 to 25 cm^{-1} higher than the corresponding Fe(III)-OH frequencies of alkaline met-hemoglobin and met-myoglobin [18]. By contrast, the Fe(IV)=O frequencies of ferryl horseradish peroxidase [22,43] are 10 to 20 cm^{-1} lower than the Fe(IV)=O frequency of ferryl myoglobin [23]. The heme Fe(III)-OH is a hydrogen bond donor and the Fe(IV)=O group is a hydrogen bond acceptor. These frequency differences are consistent with a more favorable hydrogen bonding interaction with a distal histidine for horseradish peroxidase relative to myoglobin. More effective hydrogen bonding would serve to increase the effective ligand field strength of the hydroxyl group resulting in the observed low-spin peroxidase hemes. The increased π-donation into the heme iron would be expected to have significant effects which contribute to the lowering of the ferryl/ferric redox potential. The stabilization of higher versus lower oxidation states of the heme iron by increasing electron density into the iron atom has been discussed extensively as it pertains to increased electron donation resulting from hydrogen bonding of the proximal imidazole [83-89]. Electron donation would be expected to destabilize the iron d-electron energy levels, facilitating removal of an electron upon oxidation. The facile oxidation of the heme Fe(III)-OH to Fe(IV)=O can be diagrammed with the distal histidine acting as a base catalyst forming a hypothetical imidazolium intermediate (or equivalent structure), reverting to a non-hydrogen bonded oxo-ferryl heme as shown below:

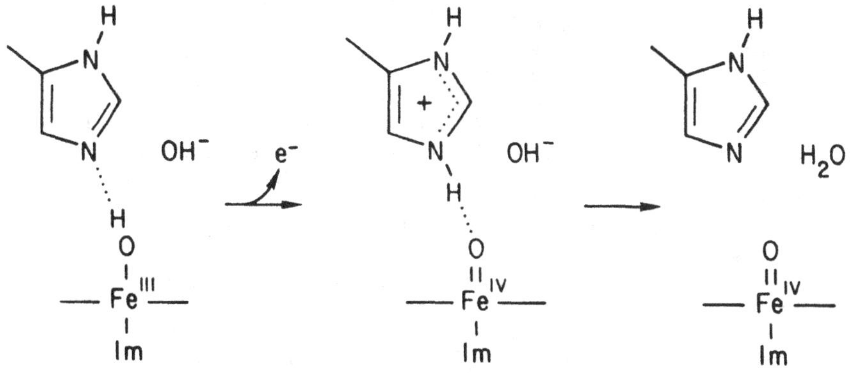

While most peroxidases studied to date exhibit similar ferryl state Fe(IV)=O frequencies, we have observed that cytochrome c peroxidase and lactoperoxidase, exhibit low Fe(IV)=O frequencies, 753 and 745 cm^{-1} respectively, indicative of variations in the heme pocket from that typified by horseradish peroxidase [90]. Data for the oxidized intermediate, compound ES of cytochrome c peroxidase, is shown in Figure 3. It is hoped that such studies will contribute to an understanding of how the reactivities of the various heme enzymes are tuned by the structures surrounding the heme.

Figure 3. Resonance Raman spectra of compound ES of cytochrome c peroxidase, showing the shift of the Fe(IV)=O frequency from 753 cm⁻¹ (a) to 725 cm⁻¹ (b) upon ¹⁸O-substitution. From reference 90.

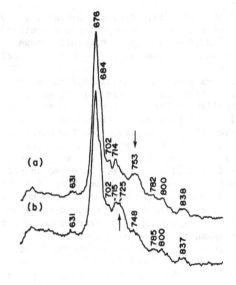

Acknowledgements   This work was supported by NIH Grant GM34443, NIH Small Instrument Grant GM39172, NSF Instrumentation Grant CHE-851347, Virginia Commonwealth University - American Cancer Society Grant IN-105K, the Jeffress Memorial Trust, Petroleum Research Fund (administered by the American Chemical Society), and Research Corporation. A loan of stable isotope material was provided by the NIH (RR02231) - USDOE/OHER Stable Isotope Program at Los Alamos. J. Terner is an Alfred P. Sloan Fellow.

REFERENCES

1. H.B. Dunford (1982) Adv. Inorg. Biochem. 4, 41-68
2. J.T. Groves (1979) Adv. Inorg. Biochem. 1, 119-145
3. J.H. Dawson (1988) Science 240, 433-439
4. T.J. McMurry and J.T. Groves (1986) in Ortiz de Montellano, P.R. (ed.) "Cytochrome P-450, Structure Mechanism and Biochemistry" pp. 1-28
5. T.G. Spiro (1974) Accts. Chem. Res 7, 339-345
6. J. Terner and M.A. El-Sayed (1985) Accts. Chem. Res. 18, 331-338
7. F.C. Bernstein, T.F. Koetzle, G.J.B. Williams, E.F. Meyer, M.D Brice, J.R. Rodgers, D. Kennard, T. Shimanouchi, and M. Tasumi (1977) J. Mol. Biol. 112, 535-542
8. L.A. Fishel, J.E. Villefranca, J.M. Mauro, J. and Kraut (1987) Biochemistry 26, 351-360
9. D.B Goodin, A.G. Mauk, and M. Smith, (1987) J. Biol. Chem. 262, 7719-7724
10. J.M. Mauro, L.A. Fishel, J.T. Hazzard, T.E. Meyer, G. Tollin, M.A. Cusanovich, and J. Kraut (1988) Biochemistry 27, 6243-6256
11. W.D. Hewson and L.P. Hager, (1979) in "The Porphyrins" D. Dolphin (ed.), Vol. VII, Academic Press, NY, pp. 295-332
12. B. Chance, L. Powers, Y. Ching, T. Poulos, I. Yamazaki and K.G. Paul (1984) Arch. Biochem. Biophys. 235, 596-611
13. D. Dolphin and R.H Felton (1974) Acc. Chem. Res. 7, 26-32
14. Y. Hayashi and I. Yamazaki (1979) J. Biol. Chem. 254, 9101-9106
15. Y. Hayashi and I. Yamazaki (1979) in "Biochemical and Clinical Aspects of Oxygen", Caughey, W.S. ed., Academic Press, N.Y. pp. 157 - 165
16. I. Yamazaki, M. Tamura and R. Nakajima (1981) Mol. Cell. Biochem. 40, 143-153
17. P. George (1953) Science 117, 220-221

18. A.J. Sitter, J.R. Shifflett and J. Terner (1988) J. Biol. Chem. 263, 13032-13038
19. W.A. Oertling and G.T. Babcock (1988) Biochemistry 27, 3331-3338
20. C.A. Reed (1982) in "Biological Chemistry of Iron" H.B. Dunford et al. (eds.), D. Reidel Publishing Co. pp. 25-42
21. J. Peisach, W.E. Blumberg, B.A. Wittenberg and J. Wittenberg (1968) J. Biol. Chem. 243, 1871-1880
22. J. Terner, A.J. Sitter, and C.M. Reczek (1985) Biochim. Biophys. Acta 828, 73-80
23. A.J. Sitter, C.M. Reczek, and J. Terner (1985) Biochim. Biophys. Acta 828, 229-235
24. S. Hashimoto, Y. Tatsuno, and T. Kitagawa (1984) Proc. Jap. Acad. 60B, 345-348
25. R. Makino, T. Uno, Y. Nishimura, T. Iizuka, M. Tsuboi, and Y. Ishimura, (1986) J. Biol. Chem. 261, 8376-8382
26. G. Simonneaux, W.F. Scholz, C.A. Reed and G. Lang, (1982) Biochim. Biophys. Acta 716, 1-7
27. S.A. Asher and T.M. Schuster (1979) Biochemistry 18, 5377-5387
28. H. Brunner (1974) Naturwiss. 61, 129-131
29. J.B.R. Dunn, D.F. Shriver and I.M. Klotz (1973) Proc. Natl. Acad. Sci. USA, 70, 2582-2584 (1973).
30. J.S. Loehr, T.B. Freedman, and T.M. Loehr (1974) Biochem. Biophys Res. Commun. 56, 510-515
31. A.J. Sitter, C.M. Reczek and J. Terner (1986) J. Biol. Chem. 261, 8638-8642
32. I. Yamazaki, T. Araiso, Y. Hayashi, H. Yamada, and R. Makino (1978) Adv. Biophys. 11, 249-281
33. P. Nicholls and G.R. Schonbaum in "The Enzymes" P. Boyer et al. (ed.), 8 p. 147 Academic NY (1963)
34. G. Rakshit and T.G. Spiro, (1974) Biochemistry 13, 5317-5323
35. G. Rakhit, T.G. Spiro, and M. Uyeda (1976) Biochem. Biophys. Res. Commun. 71, 803-808
36. J. Teroaka and T. Kitagawa (1981) J. Biol. Chem. 256, 3969-3977
37. R.H. Felton, A.Y. Romans, N.-T. Yu, and G.R. Schonbaum, (1976) Biochim. Biophys. Acta 434, 82-89
38. R.D. Remba, P.M. Champion, D.B. Fitchen, R. Chiang, and L.P. Hager (1979) Biochemistry 18, 2280-2290
39. J. Teroaka, T. Ogura, and T. Kitagawa (1982) J. Amer. Chem. Soc. 104, 7354-7356
40. T. Kitagawa, S. Hashimoto, J. Teroaka, S. Nakamura, H. Yajima, and T. Hosoya (1983) Biochemistry 22, 2788-2792
41. A. Desbois, G. Mazza, F. Stetzkowski, and M. Lutz (1984) Biochim.Biophys. Acta 785, 161-176
42. J. Terner and D.E. Reed (1984) Biochim. Biophys. Acta 789, 80-86
43. A.J. Sitter, C.M. Reczek, and J. Terner (1985) J. Biol. Chem. 260, 7515-7522
44. J.E. Critchlow and H.B. Dunford (1972) J. Biol. Chem. 247, 3703-3713
45. H.B. Dunford and M.L. Cotton (1975) J. Biol. Chem. 250, 2920-2932
46. B.B. Hasinoff and H.B. Dunford (1970) Biochemistry 9, 4930-4939
47. Y. Hayashi and I. Yamazaki (1978) Arch. Biochem. Biophys. 190, 446-453
48. J.D. Otvos and D.T. Browne (1980) Biochemistry 19, 4011-4021
49. O. Jardetsky and G.K.C. Roberts (1981) "NMR in Molecular Biology" Academic Press, NY
50. B. Chance (1952) Arch. Biochem. Biophys. 41 425-431
51. P. George (1953) J. Biol. Chem. 201, 413-426
52. S. Shahangian and L.P. Hager (1982) J. Biol. Chem. 257, 11529-11533
53. L.P. Hager, P.F. Hollenberg, T. Rand-Meier, R. Chiang, and D. Doubek (1975) Ann. N.Y. Acad. Sci. 244, 80-93
54. R. Chiang, T. Rand-Meier, R. Makino, and Hager, L.P. (1976) J. Biol. Chem. 251, 6340-6346

55. P.F. Hollenberg, T. Rand-Meier and L.P. Hager (1974) J. Biol. Chem. 249, 5816-5825
56. H. Yamada, N. Itoh and Y. Izumi (1985) J. Biol. Chem. 260, 11962-11969
57. F.S. Brown and L.P. Hager (1967) J. Amer. Chem. Soc. 89, 719-720
58. R.D. Libby, J.A. Thomas, L.W. Kaiser and L.P. Hager (1982) J. Biol. Chem. 257, 5030-5037
59. W.D. Hewson and L.P. Hager (1979) J. Biol. Chem. 254, 3182-3186
60. W.D. Hewson and L.P. Hager (1979) J. Biol. Chem. 254, 3175-3181
61. D. Keilin and T. Mann (1937) Proc. Roy. Soc. London B122, 119-133
62. H. Theorell and Å. Åkesson (1941) J. Amer. Chem. Soc. 63, 1804-1820
63. H.A. Harbury (1957) J. Biol. Chem. 225, 1009-1024
64. S. Marklund, P.-I. Ohlsson, A. Opara and K.-G. Paul (1974) Biochim. Biophys. Acta 350, 304-313
65. P. George and G.I.H. Hanania (1953) Biochem. J. 55, 236-243
66. P. George and G. Hanania (1952) Biochem. J. 52, 517-523
67. D. Job, J. Ricard, and H.B. Dunford (1977) Arch. Biochem. Biophys. 179, 95-99
68. N. Epstein and A. Schejter (1972) FEBS Letters 25, 46-48
69. T. Iizuka, S. Ogawa, T. Inubushi, T. Yonezawa and I. Morishima, I. (1976) FEBS Letters 64, 156-158
70. I. Morishima, S. Ogawa, T. Inubishi, T. Yonezawa and T. Iizuka, T. (1977) Biochemistry 16, 5109-5115
71. J.S. de Ropp, G.N. La Mar, K.M. Smith and K.C. Langry (1984) J. Amer. Chem. Soc. 106 4438-4444
72. J. Teroaka and T. Kitagawa (1981) J. Biol. Chem. 256, 3969-3977
73. S.A. Asher, L.E. Vickery, T.M. Schuster and K. Sauer (1977) Biochemistry 16, 5849-5856
74. A. Desbois, M. Lutz and R. Banerjee (1979) Biochemistry 18, 1510-1518
75. T.C. Strekas and T.G. Spiro (1974) Biochim. Biophys. Acta 351, 237-245
76. P. George, J. Beetlestone and J.S. Griffith (1964) Rev. Mod. Phys. 36, 441-458
77. J. Beetlestone and P. George (1964) Biochemistry 3, 707-714
78. W.A. Lee, T.A. Calderwood and T.C. Bruice (1985) Proc. Natl. Acad. Sci. USA 82, 4301-4305
79. M.A. Phillippi and H.M. Goff (1982) J. Amer. Chem. Soc. 104, 6026-6034
80. G.R. Schonbaum, R.A. Houtchens W.S. and Caughey (1979) in "Biochemical and Clinical Aspects of Oxygen" Caughey, W.S. (ed.) Academic Press, New York, pp. 195-211
81. T.M. Loehr and R.A. Plane, R.A. (1968) Inorg. Chem. 7, 1708-1714
82. B.M. Sjöberg, J. Sanders-Loehr and T.M. Loehr (1987) Biochemistry 26, 4242-4247
83. T.L. Poulos and B.C. Finzel (1984) Peptide and Protein Reviews 4, 115-172
84. T.G. Traylor and R. Popovitz-Biro (1988) J. Amer. Chem. Soc. 110, 239-243
85. M.M. Doeff, D.A. Sweigart and P. O'Brien (1983) Inorg. Chem. 22, 851-852
86. P. O'Brien and D.A. Sweigart (1985) Inorg. Chem. 24, 1405-1409
87. R. Quinn, J. Mercer-Smith, J.N. Burstyn and J.S. Valentine (1984) J. Amer. Chem. Soc. 106, 4136-4144
88. T.G. Traylor, W.A. Lee, and D.V. Stynes (1984) J. Amer. Chem. Soc. 106, 755-764
89. F.A. Walker, M.-W. Lo, and M.T. Ree (1976) J. Amer. Chem. Soc. 98, 5552-5560
90. C.M. Reczek, A.J. Sitter and J. Terner (1989) J. Mol. Struct. (in press)

SURFACE-ENHANCED RESONANCE RAMAN SCATTERING FROM CYTOCHROMES C AND P-450 ON BARE AND PHOSPHOLIPID-COATED SILVER SUBSTRATES

Bernard N. Rospendowski, Vicki L. Schlegel,
Randall E. Holt, and Therese M. Cotton

Department of Chemistry
University of Nebraska
Lincoln, NE 68588-0304

ABSTRACT

Surface-enhanced resonance Raman scattering (SERRS) studies of cytochrome c (cyt c) have demonstrated that the native integrity of this heme-containing protein is preserved in the presence of citrate-reduced sols. Cyt c SERRS spectra exhibit similar spin and oxidation markers as the resonance Raman scattering (RRS) signals displayed by the native protein solution. Previous SERRS investigations of cyt c showed that the protein denatures at the substrate surface. This hemoprotein denaturation was attributed to an electrostatic interaction between the porphyrin and the SERRS-active substrate. The results presented in this study suggest that the citrate ions prevent the biomolecule from directly interacting with the silver-sol surface. The native state of cyt c is also retained with the addition of phosphatidylcholine and/or phosphate buffer. Although the SERRS bandshifts correlate with the RRS spectra, the relative intensities of specific cyt c frequencies change in the presence of adsorbed citrate, phosphate buffer, and phosphatidylcholine. These compounds may influence the orientation of the protein at the substrate surface, which would account for the spectral intensity differences. In order to study the orientation effects further, additional studies of cyt c adsorbed onto silver electrodes and thin silver films were conducted. A comparison of the SERRS relative band intensities and frequency bandshifts shows that the spectra of the protein adsorbed onto phospholipid-coated substrates closely correspond to the RRS cyt c solution spectra. In contrast to the behavior exhibited by cyt c, cytochrome P-450 (cyt P-450) adsorbed onto a phosphatidylcholine silver-sol substrate leads to a low to high spin-state conversion. The shifts in the spin-state markers are ascribed to a strong interaction of the cyt P-450 with the phospholipid coating.

INTRODUCTION

Many biological systems are found to be intimately associated with membranes *in vivo*.[1,2] This is true of proteins involved in the electron-transport chain[3] and those responsible for the light-harvesting properties of plants and algae.[4]

Isolation of these proteins requires their removal from the membranous environment for *in vitro* studies. To investigate the effects of the membrane on the protein, reconstitution of the protein in artificial membranes, natural membranes or vesicles is often performed. Spectroscopic techniques can then be used to probe the effect of the membrane on the protein.

Resonance Raman scattering (RRS) has become a viable and powerful method for studying biomolecules. This technique is particularly useful for the analysis of proteins containing a chromophore because the electronic and vibrational structure of the prosthetic group is selectively interrogated by the RRS effect.[5] If the protein's chromophore is involved in the biological function, RRS spectroscopy can be used to study the structure and the function of the biomolecule. The Raman scattering signals from the surrounding apoprotein are only observed when UV resonance Raman spectroscopy is employed. However, changes in the apoprotein may be transmitted to the chromophore via interactions with nearby amino acid residues. These interactions may alter the spectroscopic properties of the chromophore. For example, cyt c RRS spectra exhibit vibrational frequency shifts and differences in relative band intensities when the electronic structure of the chromophore is perturbed. In addition, the RRS signals displayed by the biomolecule reflect environmental changes resulting from a steric reorganization of the protein envelope. The preferred orientation of the protein at the membrane interface may also be probed by RRS. In the later case, changes in relative intensities rather than frequency shifts are expected.

A non-random orientation of the protein may induce an orientational effect on the chromophore RRS. The tensorial nature of the Raman scattering process is responsible for this effect.[5] The polarizability tensor involves nine terms of the type α_{ij}, where i and j are combinations of Z, X and Y, and i refers to the direction of the dipole induced by the incident electric field vector along j. The direction of the principal rotation axis is defined by Z. Perpendicular to the Z axis are the mutually orthogonal axes, X and Y. A particular orientation of the chromophore with respect to a polarized incident beam results in the selective enhancement of specific Raman modes that scatter along the direction of the largest α_{ij}. For example, consider a molecule oriented such that the principal (Z) axis is aligned along the direction of incident Z-polarized light. If the largest contribution to the scattering tensor is α_{ZZ}, then the normal modes that transform as α_{ZZ} are selectively enhanced. Scattering by totally symmetric modes is involved in most of these cases. The orientation of a molecule can be deduced by assuming that the totally symmetric modes are enhanced by using Z-polarized light and that α_{ZZ} provides the largest contribution to the scattering tensor. Consequently, the principal axis of the molecule would be aligned with the incident electric dipole vector.

The effect of orientation on RRS from the heme chromophores in single crystals of myoglobin has been demonstrated recently[6]. The results presented in this study suggest that RRS has considerable potential for obtaining orientational information relating to the chromophore within a membrane. Moreover, if x-ray crystallography data is available regarding the disposition of the chromophore within the biomolecule, it may be feasible to ascertain the overall orientation of the protein in the membrane system. As discussed previously, in order to acquire such orientational information from RRS, the membrane must be in a known alignment relative to the incident beam. This might be accomplished by the intrinsic orientational effects present in Langmuir-Blodgett monolayers, or alternatively, by self-assembled films of lipids.[7,8,9] The protein of interest can be analyzed by depositing the sample onto these model membrane surfaces. In the case of intrinsic membrane proteins, it should be possible to incorporate the protein directly into the lipid monolayer. The ultimate goal would be to examine intact membranes for comparison purposes. Nonetheless, the use of RRS for the study of intact protein membranes remains limited. A primary disadvantage associated with this technique is that intense fluorescence often accompanies RRS from proteins embedded in membranes. In addition, relatively high concentrations (10^{-5} M) of the biological sample is required to obtain good quality RRS spectra.

Recently, surface-enhanced resonance Raman scattering (SERRS) spectroscopy has been successfully applied to a number of biological molecules containing chromophores[10,11,12,13] as well as intact membrane proteins and preparations.[14,15] The application of SERRS to the analysis of protein systems is advantageous for several reasons: first, protein concentrations (10^{-9} M) comparable to, or even lower than, those found in membranes can be probed by SERRS. Second, fluorescence is often quenched when biomolecules are adsorbed on metal surfaces.[11] Third, the SERRS-active metal substrates can be used to mimic a membrane, either directly or by deposition of a monolayer of lipid onto the surface. In the present account, SERRS studies of two different proteins, cytochromes c and P-450, are presented. The effect of adsorbed lipids on silver substrates will also be discussed. As a prelude to the present study, previous SERRS studies of these proteins will be reviewed briefly.

SERRS from Heme-Containing Proteins. Cytochromes c and P-450 are heme-containing proteins that undergo a change in redox state as part of their biological function. Cyt c transfers electrons in the respiratory chain between the enzyme complexes cytochrome reductase and cytochrome oxidase.[16] In contrast, the cyt P-450 protein functions as a metabolizing enzyme that catalyzes substrate hydroxylation.[17] Essentially, cyt P-450 is involved in a complex catalytic cycle that results in the insertion of an oxygen atom into the substrate and the concomitant formation of H_2O. Both the microsomal hydroxylase system and the mitochondrial cyt c system are deeply embedded in the membrane matrix. It is therefore desirable to determine the effect of the membrane on the redox properties of the protein or, in the case of cyt P-450, on substrate binding. Changes in the spin- and coordination-state about the central-iron atom of these heme-containing proteins are expected.[18,19] Environmental factors that presumably induce these changes include structural alterations of the apoprotein contiguous with the membrane, high electric field effects at the protein/membrane interface, and the presence of charged lipid molecules constituting the membrane surface. Furthermore, structural alterations may result in protein unfolding and exposure of the heme to the electric-field gradient present at the membrane surface. Spin- and coordination-state changes at the heme molecule are reflected in RRS or SERRS spectra as shifts in the frequencies.[13,15,20,23] Low- and medium-frequency bands have been correlated with more subtle interactions of the heme group with its surroundings.[21]

Cotton et al. were the first to apply SERRS to the study of cyt c and myoglobin.[10] In a latter study, cyt c_3 was also examined by this group. Hildebrandt and Stockburger analyzed the behavior of cyt c adsorbed onto SERRS-active silver sols and silver electrodes.[13] From their results[13b], it was concluded that cyt c exists in two states at the substrate surface. State I cyt c was characterized as an essentially native six-coordinate low-spin species (6cls). The State II cyt c configuration resulted in a reversible equilibrium between a five-coordinate high-spin (5chs) and six-coordinate low-spin species (6cls).

The above authors stressed that the potential of the silver electrode is critical in determining the observed states of cyt $\underline{c}$ by SERRS. State I results only when cyt $\underline{c}$ is adsorbed onto an electrochemically roughened silver electrode from solution at < -0.2 V vs standard calomel electrode (SCE). The heme group is either oxidized or reduced depending on the potential applied to the electrode following adsorption of the protein. State II is generated when cyt $\underline{c}$ is adsorbed from solution at potentials > -0.2 V vs. SCE. The changes in the heme spin- state with adsorption potential was attributed to perturbations in the electric field across the double layer as a function of potential. In the State II orientation, the authors suggested that the heme crevice interacts with the silver surface resulting in a thermally dependent Fe-S$_{methionine}$ bond cleavage equilibrium. It is this reversible ligation change that is responsible for the observed 5chs and 6cls equilibrium.

Recent SERRS investigations of cyt c and cyt P-450 adsorbed onto silver electrodes at open circuit showed that the native states of the proteins remain intact when the experiments are conducted under liquid N_2 temperatures, i.e. 77 K.[22] The low temperature SERRS spectrum of cyt $\underline{c}$ exhibits a typical 6cls state as described previously. However, at room temperature, the biomolecule exists in a characteristic 5chs state. In order to address the possible cause of the spin-state change with temperature, further SERRS studies of cyt $\underline{c}$ as a function of laser irradiation time were performed. The use of a conventional diode-array detector allowed for the acquisition of a spectrum by summing 25 1-second scans. The electrode/cyt $\underline{c}$ system was continuously irradiated over a 25 minute period and the spin-state changes were subsequently monitored at specific time intervals. After only 25 seconds, the spectrum changes from a 6cls state to a mixed spin state. The denatured 5chs state is evident at prolonged laser irradiation period of 25 minutes. Similar experiments conducted with cyt P-450 exhibited the same spin-state conversion with laser irradiation time. By submerging the electrode in liquid N_2, the native state of the cyt P-450 is preserved as verified by RR studies of this protein. These results suggest that the changes in the spin states of cyt $\underline{c}$ and cyt P-450 at the electrode surface is caused by a photodegradation effect. Under low-temperature conditions, photodamage to the proteins is avoided and it is, therefore, possible to obtain information regarding the native structure of the heme-containing molecules.

As in the initial potential-dependent SERRS from silver electrodes, the control of Ag sol conditions and pH is critical for SERRS studies of cytochromes. Kelly et al.[11] and Wolf et al.[23] studied two structurally related cyt P-450 proteins, i.e. PB$_{3a}$ and PB$_{3b}$ which are equivalent to the forms b and e described by Ryan et al.[24] The authors verified that the subtle differences between the proteins and their biological activities are preserved on a sol surface. Following careful control of the sol preparation procedure and solution pH, both the spin-state equilibria and substrate-induced low to high spin state-state change were observed by SERRS. This work demonstrated that sol-induced spin-state conversion could be minimized by using specifically prepared, pH controlled, sols. Furthermore, the formation of the biologically inactive cyt P-420 was avoided. Evidence of biological activity and the preservation of spin-state differences between the two isozymes, cytochromes P-450 PB$_{3a}$ and PB$_{3b}$, excluded the possible formation of cyt P-420. SERRS from the high spin-state cyt P-450 MC1a and low spin-state cyt P-450 LM$_2$ has been exhibited.[25] No significant perturbation of the spin states of these proteins was observed following adsorption on the sol surface. Low to high spin-state conversion was clearly displayed in the SERRS spectrum of the substrate bound cyt P-450 LM$_2$ biomolecule.

Hester et al. showed that the low-spin nature of hemoglobin and the biologically significant R to T hemoglobin transition upon oxygen uptake are preserved on citrate-reduced silver sols.[12] However, Smulevich and Spiro[26] reported denatured hemoglobin SERRS spectra resulting from extraction of the heme group from the protein envelope. Borohydride-reduced sols were used as the SERRS-active substrates in the latter studies.[26] These results suggest that adsorbed citrate ions prevent the prosthetic group of the heme-containing proteins from closely interacting with the silver surface. Direct contact of the protein with the positive silver ions can cause denaturation of the biomolecule. Citrate ions adsorbed onto sol surfaces appear to stabilize the colloids as well.[25]

As noted previously, another form of hemoprotein denaturation involves a low to high spin-state transition. This spin-state conversion can be minimized or prevented by controlling the pH with phosphate buffer. Addition of phosphate buffer results in the adsorption of PO_4^{3-} ions on the surface of the colloid. These ions act as an auxiliary spacer to the citrate ions already present on the Ag surface. Concurrent with this spacer effect, the presence of phosphate buffer may influence the orientation of the protein at the sol surface. In the absence of buffer or pH control, the protein appears to unfold upon binding to the surface. This leads to a relatively more exposed heme site, and hence a low to high spin-state conversion. Phosphate and citrate ions may induce the biomolecule to orient in a favorable configuration at the surface thereby preventing denaturation of the protein matrix. A similar explanation was proposed by Hildebrandt et al. to account for the existence of State I and State II cyt $\underline{c}$ configurations at potential-controlled electrode surfaces.[13b] The orientation of cyt $\underline{c}$ can be probed by SERRS. The spectra would exhibit relative band intensity differences rather than frequency shifts compared with the solution RRS. The changes in relative band intensity can be interpreted in terms of the selection rules for surface-enhanced Raman scattering as described by Creighton[27] and summarized below.

Surface Selection Rules and SERRS from Chromophoric Proteins. Information regarding the orientation of an adsorbate showing surface enhancement may be obtained from surface selection rules. These rules arise from the preferential amplification of the incident and scattered electric fields along particular directions. An effective polarizability, α_{eff}, has been derived by Creighton for a molecule adsorbed on an isolated small metal sphere (where the diameter, d, of the sphere is less than the wavelength of the exciting radiation), and is described by:

$$\alpha_{eff} = \frac{9}{(\epsilon_s + 2)(\epsilon_i + 2)} \begin{bmatrix} \alpha_{11} & \alpha_{12} & \epsilon_i \alpha_{13} \\ \alpha_{21} & \alpha_{22} & \epsilon_i \alpha_{23} \\ \epsilon_s \alpha_{31} & \epsilon_s \alpha_{32} & \epsilon_s \epsilon_i \alpha_{33} \end{bmatrix}$$

where ϵ_s is the dielectric constant of the metal at the scattered frequency, and ϵ_i is the dielectric constant of the metal at the incident frequency. The α_{11}, α_{12}.... terms etc. are the Raman polarizability tensor elements of the molecule directed along the orthogonal axes. The axes are defined by the plane of the surface (1,2) and surface normal (3). Thus, the greatest preferential enhancement will occur for modes that transform as α_{33} for $|\epsilon| > 1$. Modes which scatter via α_{13} or α_{23} will be enhanced over those that scatter via α_{12} by a factor $|\epsilon|$. The modes that scatter via α_{33} will be enhanced by a factor $|\epsilon|^2$. A generalized view of this effect is: modes which transform as tensor elements involving perpendicular (3) scattering are selectively enhanced in preference of those modes that do not. To take into account the effect of aggregates of metal spheres, refinement of equation (I) leads to a red shift for the dipole resonance condition (ϵ_s or ϵ_i equal to 2). Therefore, the prominence of vibrations that scatter via tensor elements having a perpendicular field involvement is expected to increase with longer wavelength excitations.

Deviation of the surface spectrum from a normal or resonance Raman spectrum is anticipated when a molecule is adsorbed onto SERRS-active substrates. For unperturbed biological pigments, it is desirable for the electromagnetic (EM) component of the SERRS mechanism to prevail.[10] Alternatively, an interaction of the chromophore with the metal surface is required for an appearance of a charge-transfer contribution to the enhancement. This would necessitate direct contact of the chromophore with the metal surface and probable concomitant denaturation of the protein. Such denaturation effects may be observed in SERRS of biological species possessing deeply buried chromophores such as glucose oxidase.[28] In the absence of denaturation only very weak signals would be expected.

Interpreting the SERRS spectrum of an adsorbed protein is difficult because both resonance Raman and surface-enhancement contributions to the spectrum must be considered. In addition, the effect of chromophore orientation with respect to the surface can have a significant effect on relative band intensities. For proteins adsorbed onto SERRS-active metal colloids, a random orientation of the chromophore with respect to the incident beam direction can be assumed. However, a fixed orientation of the protein with respect to the metal surface may be adopted. Therefore, surface selection rules resulting in the preferential enhancement of particular chromophore bands are operative. This would result in a SERRS spectrum which differs from the solution RRS spectrum of the protein. Determination of the protein's orientation from the selective enhancement of particular bands is also feasible. These bands may be assigned to modes of particular symmetry. For example, consider a molecule possessing D_{4h} symmetry, scattering from modes with A_{1g} symmetry may be selectively enhanced compared with scattering from B_{1g} or B_{2g} modes. Also, scattering from modes that involve a particular motion with respect to the surface may be selectively enhanced such as the out-of-plane bending modes in a porphyrin adsorbed flat on the surface. Thus, information regarding the orientation of the prosthetic group with respect to the surface can be obtained by SERRS.

Furthermore, the overall orientation of the protein at the surface may be deduced by assuming that the disposition of the chromophore in the surface-bound protein is the same as that in the quaternary structure. A layer of lipid predeposited onto the colloidal particles may influence the protein to orient differently with respect to the surface, as opposed to a non-lipid layered surface. Alternatively, a random alignment of the proteins may exist at the lipid surface.

Because roughened metal films and electrodes are fixed in space, these surfaces may promote further order in the orientation of the protein relative to the incident radiation. However, the surface roughness required to generate the SERRS effect may militate against an ordered protein condition. Monolayer coverage of proteins adsorbed on smooth films or electrodes may exhibit better orientational order. However, the surface selection rules corresponding to the smooth substrates would differ for the unenhanced scattering.[29]

RESULTS AND DISCUSSION

In the following discussion, the influence of adsorbed citrate, phosphate buffer and phosphatidylcholine on the SERRS spectrum of cytochromes $\underline{c}$ and P-450 are examined. The silver substrates used in this study included citrate sols, electrodes, and thin silver films on glass slides, prepared by vacuum deposition.

1) SERRS from Cytochrome $\underline{c}$. Figure 1A and 2A illustrate the SERRS high- and low-frequency regions of 1×10^{-7} M cyt $\underline{c}$ adsorbed onto silver sols. Citrate-reduced silver colloids containing 10 mM phosphate buffer, pH 7.5, were used as the SERRS-active substrates. Before addition of the protein and buffer, the absorption maximum of the sol was 408 nm. The reported spectra were recorded by using 413.1 nm excitation. The SERRS spectra exhibited by cyt $\underline{c}$ in the presence of phosphatidylcholine are shown in Figure 1B and Figure 2B. The cyt c/phospholipid system was prepared by adding 3×10^{-6} M phosphatidylcholine to the silver sol prior to the adsorption of the protein. The RRS spectrum of a 5×10^{-5} M cyt $\underline{c}$ solution in 10 mM phosphate buffer, pH 7.5, was obtained by using the 413.1 nm excitation. Cyt $\underline{c}$ RRS signals in the high-wavenumber and low-wavenumber regions are shown in Figure 3A and Figure 3B, respectively.

The SERRS spectra displayed by cyt $\underline{c}$ adsorbed on the phospholipid-free sol is discussed first. A comparison of the high-wavenumber regions of the adsorbed protein with the solution RRS signals shows that the bands corresponding to v_{10} (B_{1g}) and v_{30} (B_{2g}) at 1636 cm^{-1} and 1176 cm^{-1}, respectively, are less intense in the sol spectrum. In the low-wavenumber region, the bands at 568 cm^{-1} and 399 cm^{-1} are not as strong in the SERRS spectrum compared to these frequencies displayed in the solution spectrum. The set of bands between 300 cm^{-1} and 500 cm^{-1} are highly sensitive to interactions between the heme and nearby protein matrix.[13,21c] Perturbation of the apoprotein is reflected by changes in the relative intensity and/or frequency of these bands. The sensitivity of the low-frequency bands to protein structure is attributed to a decrease in the heme symmetry resulting from interactions with the protein envelope. The low frequency modes are thought to be comprised principally of in-plane pyrrole bending and pyrrole carbon-substituent in-plane wagging. Therefore, in the presence of an isotropic electric field, Raman scattering by these modes will reveal symmetry changes taking place because of changes in the heme environment. However, these modes may be sensitive to an anisotropic electric field that may be present at the surface of the sol particles. Under D_{4h} symmetry, the 568 cm^{-1} band was tentatively assigned to pyrrole in-plane bending (E_u) mode. The 399 cm^{-1} band and the 378 cm^{-1} were assigned to a pyrrole folding mode and to an out-of-plane d-C_bS, respectively. Although these assignments are not certain at the present, the decrease of the in-plane mode intensities with respect to out-of-plane mode intensities indicates a relatively flat orientation of the heme plane to the metal surface. As discussed in the Introduction, scattering from modes involving motion perpendicular to the plane of the substrate should undergo preferential enhancement compared to those that involve motion parallel to the surface.

Further evidence for a nearly parallel orientation of the heme group relative to the surface is provided by the relatively weak enhancement of the 1636 cm^{-1} band. Under D_{4h} symmetry, this mode transforms as $\alpha_{x^2-y^2}$. Scattering via this polarizability element is not expected to be selectively enhanced for a flat orientation of the haem group.

In the absence of phosphate buffer, striking differences are observed between the SERRS spectrum (Figure 4) and solution RRS spectrum (Figure 5c) using 514.5nm excitation. Figure 4 reveals the effect of pH on the SERRS of cyt $\underline{c}$. SERRS spectra at pH 7.0 (Figure 4A) and pH 8.5 (Figure 4B) are displayed. Comparing the SERRS from cyt $\underline{c}$ at neutral pH with the solution RRS, the most

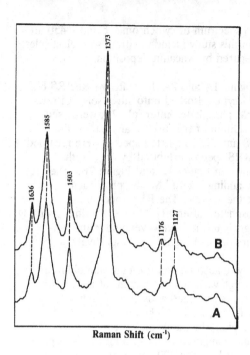

Figure 1.
High wavenumber SERRS spectrum
of 10^{-7} M cytochrome c Fe(III)
in 10 mM phosphate buffer on a
silver sol, in the absence (A)
and presence (B) of phosphatidylcholine.
λ_{ex}=413.1 nm.

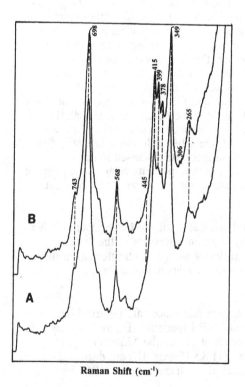

Figure 2.
Low wavenumber SERRS spectrum
of 10^{-7} M cytochrome c Fe(III)
in 10 mM phosphate buffer on a
silver sol, in the absence (A)
and presence (B) of phosphatidylcholine.
λ_{ex}=413.1 nm.

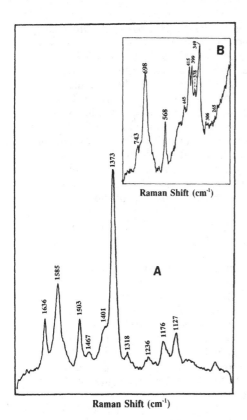

Figure 3.
High (A) and low (B) wavenumber RRS
spectrum of 5 x 10^{-5} M cytochrome c
Fe(III) in 10 mM phosphate buffer.
λ_{ex} =413.1 nm.

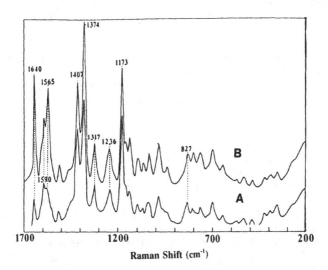

Figure 4.
SERRS spectrum of 10^{-7} M cytochrome c
Fe(III), in the absence of phosphate,
on a silver sol at pH 7.0 (A) and at
pH 8.5 (B).
λ_{ex} =514.5 nm.

significant difference is the relative intensity of the v_{10} (B_{1g}) band at approximately 1640cm^{-1} The relative intensity of the band is considerably reduced in the SERRS spectrum, whereas the relative intensity of bands attributable to A_{1g} modes, eg. v_4 (1374cm^{-1}) and v_2 (1590cm^{-1}), is increased. These relative intensity differences may be rationalized on the basis of a relative flat orientation of the heme plane with respect to the metal surface and the operation of surface selection rules. This is corroborated by the large relative intensity decrease of v_{21} (A_{2g}) band at 1313 cm^{-1} in both SERRS spectra with respect to this frequency displayed in the RRS spectrum.   Raman scattering from v_{21} involves substantial in-plane C_m-H bending. If the heme was oriented flat to the surface, this mode would not be strongly enhanced. Because scattering from the v_{21} mode is weaker than that of the solution RR spectrum, surface selection rules are operative. C_m-H motion is purely in-plane, therefore, no contribution from α_{zz} is possible for this coordinate displacement. The change in the relative intensity of SERRS bands upon increasing the pH from 7.0 (Figure 4A) to 8.5 (Figure 4B) may be due to the protein adopting a different orientation at the higher pH.

SERRS spectra of 1 x 10^{-7} M cyt c acquired by using 413.1 nm excitation and the 514.5 nm excitation depict the spectral changes of cyt c when the protein was adsorbed onto phosphatidylcholine-coated silver sols. The intensity of the v_{10} band at 1636 cm^{-1} is stronger in the presence of phospholipid compared to the relative intensity exhibited by this shift in the absence of phosphatidylcholine as illustrated in Figure 1. In the low-frequency region, the bands at 568 cm^{-1} and 399 cm^{-1} are stronger when cyt c is adsorbed onto a phospholipid-coated sol compared to adsorption onto phospholipid-free sols as shown in Figure 2. Similar effects are exhibited in the spectra obtained with the 514.5 nm excitation. Figure 5A shows the v_{10} band at 1636 cm^{-1} increases in relative intensity in the presence of phospholipid. Significantly, both of the bands assigned to A_{1g} modes at 1373 cm^{-1} (v_4) and 1586 cm^{-1} (v_2) appear weaker in the SERRS spectrum of the phospholipid system as compared to the solution RRS spectrum. An impediment to the interpretation of this effect, is the accidental degeneracy of v_2/A_{1g} and v_{19}/A_{2g} at 1586 cm^{-1}. The loss in relative band intensity may be attributed to decreased scattering from either or both of these modes.

The SERRS spectra of cyt c in the presence of phospholipid corresponds closely to the solution RR spectra. Whether this is a result of a more random orientation of the cyt c protein on the phospholipid coated sol surface, or to the adoption of a more vertical orientation can only be determined by further measurements.

Citrate-reduced silver sols appear to preserve the native structure of the protein while generating intense, surface-enhanced signals. The SERRS of the heme-containing protein cannot belong to the State II adsorbed cyt c described by Hildebrandt and Stockburger[13b] because there is essentially no high-spin protein component in the reported spectra. Moreover, the six-coordinate form of State II is not exhibited. If the spectrum resulted from the purely 6cls State II species of cyt c, then the frequencies of the spin- and oxidation- state marker bands should be different from those seen in the native cyt c solution RRS. As discussed previously, the SERRS signals of cyt c compare with the RRS spectrum obtained in the present study.

The frequencies of the principal oxidation- and spin-state marker bands quoted by Hildebrandt and Stockburger for the solution RRS spectrum of native ferric cyt c do not agree with the values acquired in this work. For example, v_4 appears at 1369 cm^{-1} in the previous study, whereas it is observed at 1373 cm^{-1} in

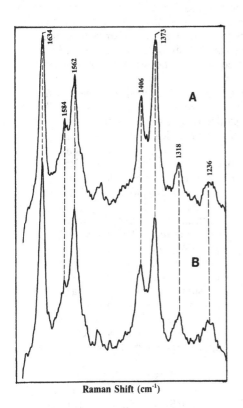

Raman Shift (cm⁻¹)

Figure 5.
High wavenumber SERRS spectrum
of 10^{-7} M cytochrome c Fe(III)
in 10 mM phosphate buffer on a
siver sol, in the presence (A)
and absence (B) of phosphatidylcholine.
λ_{ex}=514.5 nm.

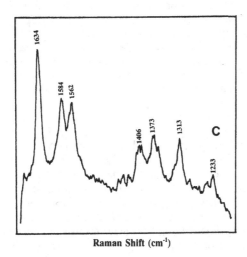

Raman Shift (cm⁻¹)

Figure 5C
RRS spectrum of 5 x 10^{-7} M cytochrome c
Fe(III) in 10 mM phosphate buffer.
λ_{ex}=514.5 nm.

the reported RRS spectrum. Furthermore, discrepancies of -3 or -4 wavenumbers are observed for other bands: v_{10}, 1636 cm^{-1} (1633 cm-1); v_2, 1586 cm^{-1} (1582 cm^{-1}) and v_3, 1503 cm^{-1} (1500 cm^{-1}). The values in brackets are taken from reference 13b. Notwithstanding the fact that the resolution is an order of magnitude less than that of the previous study, the observed peak positions are very reproducible.

In order to immobilize the protein on a surface fixed in space, silver-island films and electrochemically roughened silver electrodes were employed to study the SERRS of cyt c. Figure 6 shows cyt c adsorbed onto a silver electrode at 77 K in the absence (C) and in the presence (B) of phospholipid. The spectra were obtained by using the 406.7 nm excitation. The low temperature conditions prevented the formation of the high-spin species. Interestingly, the SERRS spectrum of cyt c adsorbed onto the phospholipid-coated electrode exhibited similar spectral changes that were observed from phospholipid-coated silver sols. The 1636 cm^{-1}, v_{10}, band is clearly enhanced in the presence of phospholipid compared to the corresponding band in the spectrum of the lipid-free systems. This further suggests that the protein changes orientation in the presence of phospholipid. A comparison of the RRS, sol/SERRS, and electrode/SERRS spectra illustrated in Figures 3A, 1A, and 6B, respectively, shows that the relative intensity of the bands at 1636 and 1507 cm^{-1} decreases when the phospholipid-coated electrode was used as the SERRS-active substrate. This effect may be may attributable to differences in substrate surface roughness and/or to a fixed orientation of the substrate with respect to the laser beam. Therefore, should orientational order prevail at the electrode surface, the observed SERRS may be different from that seen from translationally unhindered sol particles.

Vapor-deposited silver-island films were also used as the SERRS-active substrate. The diameters of the particles constituting the silver-island films depend critically upon the deposition conditions.[30] The SERRS spectrum of cyt c adsorbed onto a silver film that was initially precoated with phosphatidylcholine is shown in Figure 6A. The SERRS signals displayed by the protein/thin film system are weaker compared to the spectra obtained from the other substrates employed in this study. The lipid was adsorbed onto the silver films by dipping the substrate into a hexane/phosphatidylcholine solution. Thus, multilayers of phosphatidylcholine may have deposited on the metal substrate resulting in a decrease in EM enhancement. In future studies, Langmuir-Blodgett films will be deposited on the surface. It will then be possible to control exactly the number of lipid layers and the orientation of the lipid chains at the surface.

The spectrum in Figure 6C differs from those shown in Figures 6A, 6B, and Figure 1B. However, the cyt c/silver film SERRS signals appear to closely resemble the solution RR spectrum in relative band intensities and bandshifts . Although the poor S/N renders such comparisons dubious, a structured lipid multilayer may have the effect of lessening the orientational order of the adsorbed protein induced by the silver surface. Alternatively, as the surface enhancement is diminished with distance from the surface, the operation of surface selection rules may also become less effective. In either case, a spectrum similar to the solution spectrum should result. Polarization data may resolve the question of random vs. ordered alignment of the protein on the lipid coated thin film. Work is in progress to address this problem.

2. SERRS from Cytochrome P-450

Liver microsomal cyt P-450 is an intrinsic membrane protein that exists in multiple forms *in vivo*. In contrast to the behavior displayed by cyt c, this

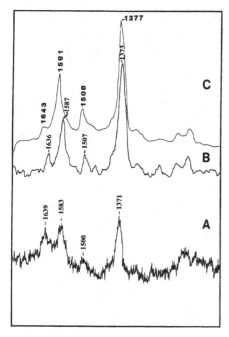

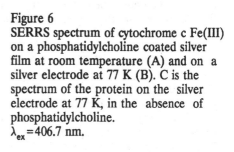

Figure 6
SERRS spectrum of cytochrome c Fe(III) on a phosphatidylcholine coated silver film at room temperature (A) and on a silver electrode at 77 K (B). C is the spectrum of the protein on the silver electrode at 77 K, in the absence of phosphatidylcholine.
λ_{ex}=406.7 nm.

Raman Shift (cm^{-1})

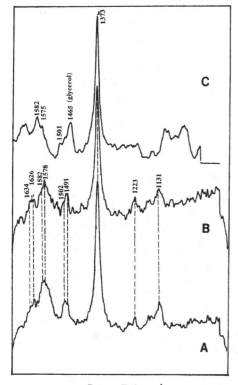

Raman Shift (cm^{-1})

Figure 7.
SERRS spectrum of cytochrome P-450b Fe(III) in 10 mM phosphate buffer on a silver sol, in the absence (A) and presence (B) of phosphatidylcholine. C is the spectrum of the protein on a silver electrode at 77K.
λ_{ex}=413.1 nm.

hemoprotein interacts strongly with phosphatidylcholine. Recently, it was reported that cyt P-450 PB IIB1 extracted from rat liver appears to be anchored to liposomes via a single peptide that spans the membrane.[31] A strong interaction with these lipids could lead to structural modifications of the protein and consequently, affect the heme. Previous optical and ESR studies showed that a low to high spin-state transition is induced upon protein binding to the membrane surface.[32] Whether the spin-state transition is caused by an electrostatic interaction with the charge residing on the phospholipid or by protein conformational changes has yet to be determined. However, a relationship between the negative charge of the membrane and the degree of high spin state species was established. It was shown that the acidic phospholipids favor the high spin-state transition.

SERRS studies of liver microsomal cyt P-450 adsorbed on silver colloids showed that low to high spin-state conversion was minimized or eliminated.[23] This procedure was used to determine the effect of phospholipids on the SERRS spectrum of liver microsomal cyt P-450b. The protein was obtained from phenobarbitol pretreated male rats. Figure 7 shows the SERRS spectra exhibited by cyt P-450 on phospholipid-free (A) and phospholipid-coated (B) silver sols. Figure 7C depicts the solution RRS spectrum for comparison. The spectra were recorded by using the 413.1 nm excitation. Based upon the similarities between the SERRS frequencies to those acquired by using RRS, cyt P-450 adsorbed onto the silver surfaces appears to retain its native structure in the absence of the phospholipid coating. The low spin-state marker at 1501 cm^{-1} (v_3) appears as a resolved peak in both the SERRS spectrum and in the solution RRS spectrum. In the solution spectrum, the band is partially obscured by the strong glycerol peak at 1465 cm^{-1}. Addition of glycerol to the buffer solubilizes the protein. The low spin-state marker at 1582 cm^{-1} (v_2) is more resolved in the RR spectrum than the corresponding band in the SERRS spectrum. This may be attributed to low to high spin-state conversion upon adsorption of the protein on the colloidal surface or to activation of the A_{2g} mode v_{19} in the SERRS spectrum.

Preadsorbing a layer of phosphatidylcholine onto the silver sol leads to a low to high spin-state conversion as shown in Figure 7B. The high spin-state band is exhibited at 1492 cm^{-1} while the low spin 1501 cm^{-1} band appears as a shoulder. The poorly resolved band at approximately 1582 cm^{-1} in Figure 7A appears as a resolved peak at 1578 cm^{-1}. This v_2 frequency has been correlated with five coordinate high spin-state heme.[20,23]

In summary, it is apparent that the presence of phosphatidylcholine causes a low to high spin-state transition when cyt P-450 is adsorbed on silver sols. These results contrast from those obtained for cyt c adsorbed onto a phospholipid-coated silver sol-surface. In the latter case, no low to high spin-state conversion was observed (*vide supra*). Rather, a different orientation of the cyt c protein appears to result from the presence of the lipid. On the other hand, these studies indicate a strong interaction between cyt P-450 and the lipid. It has yet to be determined if reorientation accompanies this interaction. However, the similar relative intensities of the SERRS bands and solution RR shifts suggest a random orientation of the protein on the sol surface. Alternatively, a more vertical orientation of the heme with respect to the surface may exist. If the heme plane is parallel to the Z-direction, defined by the surface normal, the surface selection rules predict equally enhanced scattering from the in-plane modes, irrespective of the mode symmetry. Scattering from out-of-plane modes should be very weak.

It has already been shown that substrate induced low to high spin-state conversion of adsorbed liver microsomal cyt P-450 can be detected by using

SERRS[11b, 18]. In this study, another biological phenomenon was observed on a sol surface: the low to high spin conversion of a cyt P-450 induced by the presence of phospholipid.

CONCLUSIONS

The above studies demonstrate that the native integrity of cyt c can be preserved by adsorbing the protein onto citrate-reduced sols. With the addition of phosphatidylcholine, the SERRS spectra of this biomolecule adsorbed onto silver sols, electrodes, and thin films closely resembles the solution RR spectra in bandshifts and relative band intensities. However, bandshifts from low to high spin-states are apparent in the SERRS spectra of cyt P-450 in the presence of phospholipid-coated silver sols. These results indicate a strong interaction of the lipid with this protein. Thus, the use of SERRS to investigate a mimetic membrane on silver substrates appears to be a viable technique. Information pertaining to protein structure and function and to membranae interaction effects on structure are elicited by SERRS. Future application of SERRS to a wide variety of biological systems seems to be entirely feasible.

ACKNOWLEDGEMENTS

The authors are grateful for the support of the National Institutes of Health (GM 35108).

REFERENCES

1. Martonosi, A.N, in The Enzymes of Biological Membranes, Plenum Publishing Co., Vol. 4, New York, 1985, p. 608.
2. Nabiev, I. R., Efremov, R.G., and Chumanov, G.D. J. Biosci. 8, 363, (1985).
3. Dolphin, D., Ed., in The Porphyrins, Academic Press, Vol. VII, New York, 1987.
4. Amisz, J., Ed., Photosynthesis, Elsevier, Vol. 15, New York, 1987.
5. Felton, R.H., and Yu, N.T., in The Porphyrins, Dolphin, D. Ed., Academic Press, Vol. III, New York, 1978, p. 347.
6. Sage, J.T., Morikis, D., Champion, P.M. J. Chem. Phys. 90, 3015, (1989).
7. Cotton, T.M., Uphaus, R.A., Mobius, D.J. J. Phys. Chem. 90, 6071, (1986).
8. Sonderman, J. Liebigs Ann. Chem. 749, 183, (1971).
9. Kim, J.-H., Cotton, T.M., Uphaus, R.A. J. Phys Chem. 92, 5575, (1988).
10. Cotton, T.M., in Spectroscopy of Surfaces, Clark, J.H., Hester, R.E., Eds., John Wiley & Sons Ltd., New York, 1988, p. 91.
11. a) Kelly, K., Rospendowski, B.N., Smith, W.E., and Wolf, C.R. FEBS Lett., 222, 120, (1987). b) Kelly, K., Rospendowski, B.N., Smith, W.E., Wolf, C.R., in Cytochrome P-450: Biochemistry and Biophysics, Schuster, I., Ed., Taylor and Francis, London, 1989, p.375.
12. a) de Groot, J., and Hester, R.E. J. Phys. Chem. 19, 1693, (1987). b) de Groot, J., Hester, R.E., Kaminaka, S., Kitagawa, T. J. Phys. Chem. 92, 2044, (1988).
13. a) Hildebrandt, P., and Stockburger, M. J. Phys. Chem. 90, 6017, (1986). b) Hildebrandt, P., and Stockburger, M., in Vibrational Spectra and Structure, in press.
14. Seibert, M., Cotton, T.M., Metz, J.G. Biochim Biophys. Acta 934, 235, (1988).
15. Picorel, R., Holt, R.E., Cotton, T.M., Metz, J.G. J. Bio. Chem. 263, 4374, (1988).
16. Dickerson, R.E., Timkovich, R., in The Enzymes, Boyer, P., Ed., Academic Press, Vol. 4, New York, 1975, p. 397.

17. Ortiz de Montellano, P.R., Ed., Cytochrome P-450, Structure , Mechanism and Biochemistry, Plenum Press, New York, 1986.
18. Hildebrandt, P., Rudiger, G., Stier, A., Stockburger, M., and Taniguchi, H. FEBS Lett. 227, 76, (1988).
19. Mack, D.M., Bruno, G.V., Griffin, B.W., Peterson, J.A. J. Biol. Chem. 257, 5372, (1982).
20. Parthasarathi, N., Hansen, C., Yamaguchi, S., Spiro, T.G. J. Am. Chem. Soc. 109, 3865, (1987).
21. a) Choi, S., Spiro, T.G., Langry, K.C., Smith, K.M. J. Am. Chem. Soc. 104, 4337, (1982). b) Choi, S., Spiro, T.G., Langry, K.C., Smith, K.M., Budd, D.L., La Mar, G.N. J. Am. Chem. Soc. 104, 4345, (1982). c) Valance, W.G. and Strekas, T.C. J. Phys. Chem. 86, 1804, (1982).
22. Cotton, T.M., Schlegel, V.L., Holt, R.E., Swanson, B., Ortiz de Montellano, P., in Proceedings of SPIE Conference Biomolecular Sectroscopy, Los Angeles, CA, 1989.
23. Wolf, C.R., Miller, J.S., Seilman, S., Burke, M.D., Rospendowski, B.N., Kelly, K., Smith, W.E. Biochemistry 27, 1597, (1988).
24. Ryan, D.E., Thomas, P.E., Riek, L.M., Levin, W. Xenobiotica 12, 727, (1982).
25. Rospendowski, B.N., unpublished results.
26. Smulevich, G., and Spiro, T.G. J. Phys. Chem. 89, 5168, (1985).
27. Creighton, J.A., in Spectroscopy of Surfaces, Clark, R.H., Hester, R.E., Eds., John Wiley & Sons Ltd., New York, 1988, p. 37.
28. Holt, R.E., Cotton, T.M., J. AmChem. Soc. 109, 1841, (1987).
29. Campion, A., Hallmark, V.M. Chem. Phys. Letts. 110, 561, (1984).
30. Schlegel, V.L., Cotton, T.M., in press.
31. Vergères, G., Winterhalter, K.H., Richter, C. Biochemistry 28, 3650, (1989).
32. Ruckpaul,K., Rein, H., Blanck, J., Ristau, O., Coon, M.J., in Biochemistry, Biophysics and Regulation of Cytochrome P-450, Gustafsson, J.-A. et al. Eds., Elsevier/North-Holland Biomedical Press, Amsterdam, 1980, p539.

INTERHEME INTERACTIONS IN A TETRAHEME PROTEIN, <u>DESULFOVIBRIO</u>
<u>VULGARIS</u> MIYAZAKI CYTOCHROME $\underline{c}_3$, AS MONITORED BY MICROSCOPIC
REDOX POTENTIALS

Hideo Akutsu, Kejun Fan, Yoshimasa Kyogoku[1]
and Katsumi Niki

Department of Physical Chemistry, Faculty of
Engineering, Yokohama National University, Hodogaya
Yokohama 240, Japan, [1]Institute for Protein
Research, Osaka University, Suita, Osaka 565, Japan

INTRODUCTION

Cytochrome $\underline{c}_3$ is an electron transfer protein found in
sulfate-reducing bacteria. This protein carries four heme
groups in a single polypeptide, the molecular weight of which
is about 14,000. Its midpoint potential is as low as -300 mV
vs. NHE (normal hydrogen electrode) (Niki et al., 1979 and
1984). The crystal structure was reported for those from
<u>Desulfovibrio vulgaris</u> Miyazaki and <u>Desulfovibrio desulfuricans</u>
Norway. Their structures were found similar to each other.
Both 5-th and 6-th ligands of all hemes are the imidazol rings
of histidine residues. Because of the high density of heme
population, the nature of the interheme interactions has been
one of the major interests. It was discussed in the studies
of Mossbauer spectroscopy (Ono et al., 1975), x-ray crystallography (Pierrot, et al., 1982; Higuchi et al., 1984),
electrochemistry (Niki et al., 1984), NMR (Santos et al., 1984)
and Raman spectroscopy (Verma et al., 1988). We have developed
new methods to determine the macroscopic and microscopic redox
potentials of a tetraheme protein, taking advantage of the high
resolution of its NMR spectrum. They will be briefly shown in
this paper. And the results will be discussed from the view
point of interheme interactions. A preliminary report of a
part of this paper was published elsewhere (Fan et al., 1988).

MATERIALS AND METHODS

Cytochrome $\underline{c}_3$ was purified from <u>Desulfovibrio vulgaris</u>
Miyazaki according to the reported method (Yagi and Maruyama,
1971). The purity index (A_{552}(reduced)/A_{280}(oxidized)) of the
sample was larger than 3.0. Cytochrome $\underline{c}_3$ was dissolved in a
30 mM phosphate buffer solution (p^2H 7.0) and a trace amount of
hydrogenase was added as a redox catalyst. The exchangeable
protons in the sample were replaced with deuterons. 500 MHz ^1H
NMR spectra were measured with a JEOL GX-500s NMR spectrometer.
Partial reduction of cytochrome $\underline{c}_3$ was performed by controlling

the partial pressure of hydrogen under an argon atmosphere. Absorption spectra with various applied potentials were obtained with an optically transparent thin layer electrode (OTTLE) cell and Shimazu spectrophotometer UV-240. The Working and reference electrodes were 500-mesh gold minigrid and Ag/AgCl elctrode, respectively. All potentials in this paper are referred to the normal hydrogen electrode (NHE). All experiments were conducted at 30°C, using the buffer mentioned above.

RESULTS

^{1}H NMR Spectra of Dv. Miyazaki Cytochrome c_3

^{1}H NMR spectra of cytochrome c_3 are presented in Fig. 1 for a variety of redox stages (to different extents of partial reduction). The top spectrum represents the fully oxidized state. As can be seen in the low field region, many heme methyl signals were observed separately, which were labeled in

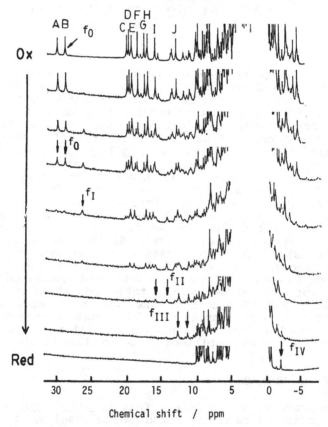

Fig. 1. 500 MHz ^{1}H NMR spectra of Dv. Miyazaki cytochrome c_3 at a variety of redox stages at 30°C (Fan et al., 1988). Top; fully oxidized, bottom; fully reduced, f_i; i-electron reduced state. Some of the heme methyl signals were labeled alphabetically on the top. Partial reduction of cytochrome c_3 was performed by controlling the partial pressure of hydrogen in the presence of hydrogenase under an argon atomosphere.

the spectrum. Since the heme methyl signals are sensitive to the oxidation states, we can use these signals to monitor the redox process. During the entire reduction process, five sets of spectra were observed. They were labeled by f_i (i=0-IV) in the spectra. Althogh the intensities of signals changed dramatically, their positions remained the same throughout the reduction proceess. As indicated by Santos et al.(1984) and Kimura et al. (1985), these five spectra represents the five macroscopic oxidation states. They are defined as the fully-oxidized, one-, two-, three- and four-electron reduced states, which are schematically shown in Fig. 2. From this observation, we can conclude that the intermolecular electron exchange rate of <u>Dv.</u> Miyazaki cytochrome $\underline{c}_3$ is slower and its intramolecular electron exchante rate is faster than the NMR time scale (ca. 10^5 s^{-1}) as in the case of <u>D. gigas</u> (Santos et al., 1984) but different from the case of <u>Desulfovibrio desulfuricans</u> Norway (Guerlesquin et al., 1985).

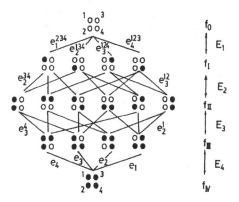

Fig. 2. Correlation among macroscopic oxidation states (f_i), macroscopic redox potentials (E_i), molecular redox species and microscopic redox potentials (e_i). Open and closed circles show the oxidized and reduced hemes, respectively. $e_i{}^{jkl}$ denotes the microscopic redox potential of heme i with hemes j, k and l in the same molecule being oxidized.

Determination of Macroscopic Redox Potentials

The molar fraction of cytochrome $\underline{c}_3$ in the i-electron reduced state, f_i, is a function of the equilibrium potential E and formal redox potentials $E_i{}^\circ$, which can be shown by the Nernstian equation

$$E = E_i{}^\circ + (RT/F)\ln(f_{i-1}/f_i) \quad \ldots\ldots(1).$$

This formal redox potential will be referred to as the macroscopic redox potential, because it has nothing to do with the oxidation state of each heme, but wih that of molecule. Although the macroscopic potentials are not directly involved in the electron transfer process, these values play an important role in the estimation of the formal redox potential of each heme as shown later. They have been determined by electrochemical techniques and EPR, using computational analysis. Since the relative intensities of NMR signals belonging to the different oxdation states give the molar fraction in each oxidation state, the macroscopic redox potentials also can be determined at hith resolution by NMR, provided that the equilibrium potential is known.

The equilibrium potential of each NMR spectrum was determined by mesuring the absorption spectrum of the solution in the NMR tube before and after each NMR measurement. The relationship between the absorption spectrum and the potential was determined by using an optically transparent thin layer electrode cell. The molar fraction of cytochrome c_3 in each oxidation state was plotted as a function of the equilibrium potential in Fig. 3. The solid lines were obtained by nonlinear least-square fitting, using equation (1). The cross point of two curves of near-by oxidation states represents a macroscopic redox potential. The best-fit values of the macroscopic redox potentials were -260, -312, -327 and -369 (+1) mV vs. NHE at 30°C, which were in good agreement with the repoted

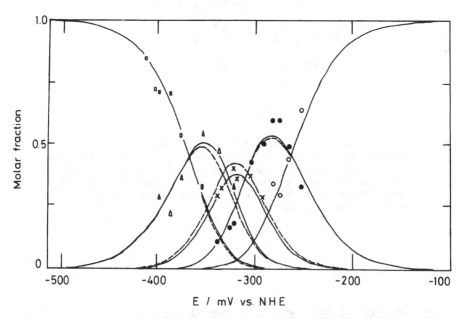

Fig. 3. The molar fraction of each oxidation state as a function of equilibrium potential (E) obtained from NMR spectra. o; fully oxidized, •; one-electron reduced, x; two-electron reduced, Δ; three-electron reduced, □; fully reduced state. Solid lines represent the best-fit curves in the non-linear least-square fitting. Broken lines were obtained from the estimated microscopic redox potentials.

values (Niki et al., 1984). The most important feature of this method is that the macroscopic formal redox potentials can be obtained directly from the observations without any assumptions. Fig. 3 also provides the direct evidence for the conclusion that the redox process of cytochrome c_3 proceeds by a four-consecutive one-electron transfer mechanism (Niki et al., 1984).

Estimation of Microscopic Redox Potentials

Since the electron transfer either within a molecule or between molecules takes place between particular two hemes, the redox potential of each heme plays a crucial role in the electron transfer process. However, the efforts to determine the formal redox potential of each heme (referred to as the microscopic redox potential, hereafter) encounter the great difficulties, since there are 32 microscopic redox potentials in the whole redox process, as shown in Fig. 2. The conventional electrochemical techniques have no resolution to determine so many parameters. On the other hand, NMR has much higher resolution. Many of 16 heme methyl signals were observed separately and even the same methyl group gives different signals in different oxidation states, as can be seen in Fig. 1. Furthermore, the chemial shift of the heme methyl signal in the intermediate oxdation state is a function of microscopic redox potentials. Thus, NMR is a promising method to tackle this problem.

If we can determine the chemical shifts of heme methyl groups belonging to four different hemes in the five oxidation states, we will end up with the nine independent observations, which are still far from enough to determine 32 micrscopic redox potentials. In order to reduce the parameter, we have to introduce some assumptions. Santos et al. (1984) have carried out a pioneering work in this analysis. They introduced interacting potentials I_{ij} defined as

$$I_{ij} = e_i - e_i^{\ j} = e_i^{\ k} - e_i^{\ jk} = e_i^{\ kl} - e_i^{\ jkl} \quad \ldots \ldots (2)$$

where $e_i^{\ jkl}$ denotes the microscopic redox potential of heme i with hemes j, k and l in the same molecule being oxidized. This enable us to reduce the parameters to ten, which are still more than the number of independent observations. In order to meet this, we used two of the macroscopic redox potentials, which were determined independently in this work. The residual two macroscopic parameters were used to check the justification of the results.

The chemical shifts of heme methyl signals in the five oxidation states were determined by saturation transfer method for eight heme methyl groups (A through H in Fig. 1) belonging to four different heme groups. Using these data, the ten parameters were analytically determined. The detail of analysis will be published elsewhere. The estimated interacting potentials and microscopic redox potentials are ginven in Table 1. The molar fraction of cytochrome c_3 in the five oxidation states also can be calculated using microscopic redox potentials and the equilibrium potential. It has been carried out to check the justification of the results. The result is shown in Fig. 3 by broken lines. A very good agreement between the solid and broken lines shows that the

Table 1. The Estimated Microscopic Formal Redox Potentials and Interacting Potentials of Dv. Miyazaki Cytochrome c_3 (mV)

For heme 1.*

e_1^{234}	OOOO + e⁻ = ROOO	-270
e_1^{34}	OROO + e⁻ = RROO	-265
e_1^{24}	OORO + e⁻ = RORO	-291
e_1^{23}	OOOR + e⁻ = ROOR	-305
e_1^4	ORRO + e⁻ = RRRO	-286
e_1^3	OROR + e⁻ = RROR	-300
e_1^2	OORR + e⁻ = RORR	-326
e_1	ORRR + e⁻ = RRRR	-321

For heme 2.*

e_2^{134}	OOOO + e⁻ = OROO	-328
e_2^{34}	ROOO + e⁻ = RROO	-323
e_2^{14}	OORO + e⁻ = ORRO	-285
e_2^{13}	OOOR + e⁻ = OROR	-339
e_2^4	RORO + e⁻ = RRRO	-280
e_2^1	OORR + e⁻ = ORRR	-296
e_2^3	RROO + e⁻ = RROR	-334
e_2	RORR + e⁻ = RRRR	-291

For heme 3.*

e_3^{124}	OOOO + e⁻ = OORO	-340
e_3^{24}	ROOO + e⁻ = RORO	-361
e_3^{14}	OROO + e⁻ = ORRO	-297
e_3^{12}	OOOR + e⁻ = OORR	-347
e_3^4	RROO + e⁻ = RRRO	-318
e_3^2	ROOR + e⁻ = RORR	-368
e_3^1	OROO + e⁻ = ORRR	-304
e_3	RROR + e⁻ = RRRR	-325

For heme 4.*

e_4^{123}	OOOO + e⁻ = OOOR	-303
e_4^{23}	ROOO + e⁻ = ROOR	-338
e_4^{13}	OROO + e⁻ = OROR	-314
e_4^{12}	OORO + e⁻ = OORR	-310
e_4^3	RROO + e⁻ = RROR	-349
e_4^2	RORO + e⁻ = RORR	-344
e_4^1	ORRO + e⁻ = ORRR	-321
e_4	RRRO + e⁻ = RRRR	-356

Interacting Potentials

$I_{12} = 5$, $I_{13} = -21$, $I_{14} = -35$, $I_{23} = 43$, $I_{24} = -11$, $I_{34} = -7$

*O and R denote oxidized and reduced hemes, respectively.

results are reasonable. Therefore, it can be concluded that the interheme interactions clearly affect the redox potentials of each hemes in different oxidation states. Furthermore, it should be indicated that although most interacting potentials are negative or close to zero, one of them shows relatively large positive value.

Assignment of Heme Groups

In order to elucidate the interheme interactions in terms of physicochemical and structural parameters, the assingnments of NMR signals of heme methyl groups to those of the crystal structure are crucial. We are conducting nuclear Overhauser effect (NOE) experiments to establish the assignments. It turned out so far that the hemes with the highest and lowest redox potentials in the fully oxidized state (designated as hemes 1 and 3 in this paper, respectively) are the hemes I and IV in the crystal structure of Dv. Miyazaki cytochrome c_3 (Higuchi et al., 1984). This assignment is in accord with the tentative assignment by Niki et al. (1984) and contradicts that by Gayda et al. (1988). Our assignment on the heme with the highest redox potential is also consistent with that of D. desulfuricans Norway cytochrome c_3 on the basis of chemical modification (Dolla et al., 1987).

DISCUSSION

This is the first time that a whole set of the microscopic formal potentials of a cytochrome c_3 was estimated. Our results contradict the conclusion on the basis of the EPR experiments that the interheme interactions in <u>Dv.</u> Miyazaki cytochrome c_3 are negligible (Gayda et al., 1987). Our results showed that analysis on the basis of any noninteracting model is not appropriate. Santos et al. (1984) reported the relative differences of the microscopic redox potentials of <u>D. gigas</u> cytochrome c_3. However, the relative differences of the macroscopic redox potentials calculated from them are different from the recent results obtained by polaropraphy (Niviere et al., 1988). In cotrast, our results are in good agreement with the electrochemical observations. We developed their method by introducing new formulation and macroscopic redox potentials, which enabled us to solve the problem analytically under certain assumptions.

Interheme interactions in cytochrome c_3 have been discussed on the basis of a variety of observations. Mossbauer experiments provided evidence for heme-heme interactions, which were suggested to include not only dipole interactions but also a superexchange interaction in the fully oxidized state (Ono et al., 1975). A resonance Raman study showed the existence of vibrational exiton coupling among the four hemes (Verma et al., 1988). It was suggested that the repulsive forces play a significant role in the interactions in the fully oxidized state, while dispersion forced become more dominant in the fully reduced state. This was discussed in connection with the changes in the interiron distances induced by the reduction. our results are in accord with these conclusions as to the existence of interheme interactions. Furthermore, our results showed that the interheme interactions are not identical for all heme pairs. The mechanism of the interactions should be investigated at atomic resolution.

Our results showed the existence of one highly positive interacting potential, I_{23}. This means that the presence of an electron at heme 2 (or 3) makes the reduction of heme 3 (or 2) much easier. If an electrostatic interaction is the major factor in the interacting potential, only negative one would be expected. Therefore, a positive interaction should be elucidated in terms of other factors including structural

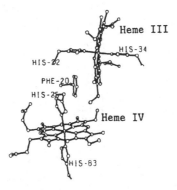

Fig. 4. Conformation of hemes III and IV in the crystal structure (Niki et al., 1984).

parameters. It is known that the orientations of the four hemes in the crystal structures are almost identical for cytochromes $\underline{c}_3$ from $\underline{Dv.}$ Miyazaki (Higuchi et al., 1984) and $\underline{Dd.}$ Norway (Pierrot et al., 1982) in spite of their low sequential homology. Especially, two of the four hemes are located very close to each other at almost right angle and the aromatic ring of the conserved phenylalanine residue (Phe20) is intervening them in a way almost parallel to the planes of heme III and imidazole ring ligated to heme IV as shown in Fig. 4. A specific interaction between these hemes was anticipated by crystallographers. Our NOE experiments showed that heme 3 is heme IV. Although the assignment of heme 2 is not established yet, it is strongly suggested that the positive interacting potential, I_{23}, is connected with the specific conformation of hemes III and IV.

Using the microscopic redox potentials, we can obtain the molar fraction of each redox species in the each oxidation state. They are shown in Fig. 5. This gives an idea that which redox species is dominant at any intermediate redox stage. The redox potential of each heme also changes significantly with the oxidation states of the other hemes. As can be seen in Table 1, the range of the variation is about 60 mV. This suggests that the four hemes can become either electron donors or electron acceptors depending on the environmental potential. These two factors would play roles

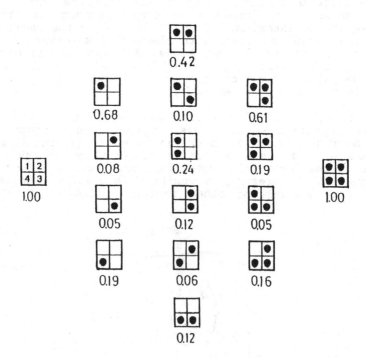

Fig. 5. The molar fraction of each redox species in each macroscopic oxidation state obtained from the estimated microscopic redox potentials. From left to right columns; fully oxidized, one-electron, two-electron, three-electron, four-electron (or fully) reduced states. Closed circles represent electrons.

in regulating the direction of the electron flow in the electron transport system of sulfate reducing bacteria. This regulation mechanism would be also useful in switching the electron flow in a molecular device by changing the applied potential.

ACKNOWLEDGEMENTS

We are grateful to Prof. N. Yasuoka, Himeji Institute of Technology, for the kind gift of hydrogenase. This work was partly supported by Ciba-Geigy Foundation for the Promotion of Science.

REFERENCES

Dolla, A., Cambillau, C., Bianco, P., Haladjian, J. and Bruschi, M., 1987, Structural assignment of the heme potentials of cytochrome c_3, using a specifically modified arginine, Biochem. Biophys. Res. comm., 147: 818.

Fan, K., Akutsu, H., Niki, k., Higuchi, N. and Kyogoku, Y., 1988, ^{1}H-NMR investigation of macroscopic redox states of cytochrome c_3, J. Chem. Soc. Jpn., No.4: 512.

Gayda, J. P., Yagi, T., Benosman, H. and Bertrand, P., 1987, EPR redox study of cytochrome c_3 from Desulfovibrio vulgaris Miyazaki, FEBS Letters, 217: 57.

Guerlesquin, F., Bruschi, M. and Wuthrich, K., 1985, ^{1}H-NMR studies of Desulfovibrio desulfuricans Norway strain cytochrome c_3, Biochim. Biophys. Acta, 830: 296.

Higuchi, Y., Kusunoki, M., Matsuura, Y., Yasuoka, N. and Kakudo, M., 1984, Refined structure of cytochrome c_3 at 1.8 A resolution, J. Mol. Biol., 172: 109.

Kimura, K., Nakajima, S., Niki, K. and Inokuchi, H., 1985, Determination of formal potential of multihemeprotein, cytochrome c_3 by ^{1}H nuclear magnetic resonance, Bull. Chem. Soc. Jpn., 58: 1010.

Niki, K., Yagi, T., Inokuchi, H. and Kimura, K., 1979, Electro-chemical behavior of cytochrome c_3 of Desulfovibrio vulgaris, strain Miyazaki, on the mercury electrode, J. Am. Chem. Soc., 101: 3335.

Niki, K., Kawasaki, Y., Nishimura, N., Higuchi, Y., Yasuoka, N. and Kakudo, M., 1984, Electrochemical and structural studies of tetra-heme proteins from Desulfovibrio -Standard potentials of the redox sites and heme-heme interaction, J. Electroanal. Chem., 168: 275.

Niviere, V., Hatchikian, E. C., Bianco, P. and Haladjan, J., 1988, Kinetic studies of electron transfer between hydrogenase and cytochrome c_3 from Desulfovibrio gigas. Electrochemical properties of cytchrome c_3, Biochim. Biophys. Acta, 935: 34.

Ono, K., Kimura, K., Yagi, T. and Inokuchi, H., 1975, Mossbauer study of cytochrome c_3, J. Chem. Phys., 63: 1640.

Pierrot, M., Haser, R., Frey, M., Payan, F. and Astier, J. P., 1982, Crystal structure and electron transfer properties of cytochrome c_3, J. Biol. Chem., 257: 14341.

Santos, H., Moura, J. J. G., LeGall, J. and Xavier, A. V., 1984, NMR studies of electron transfer mechanisms in a protein with interacting redox centres: Desulfovibrio gigas cytochrome c_3, Eur. J. Biochem. , 141: 283.

Verma, A. L., Kimura, K., Nakamura, A., Yagi, T., Inokuchi, H. and Kitagawa, T., 1988, Resonance Raman Studies of Hydrogenase-catalyzed reduction of cytochrome c_3 by Hydrogen. Evidence for heme-heme interactions, J. Am. Chem. Soc., 110: 6617.

Yagi, T. and Maruyama, K., 1971, Purification and properties of cytochrome c_3 of Desulfovibrio vulgaris, Miyazaki, Biochim. Biophys. Acta, 243: 214.

ELECTRON TRANSFER REACTIONS OF IRREVERSIBLY ADSORBED CYTOCHROMES

ON SOLID ELECTRODES

Maryanne Collinson, James L. Willit, and Edmond F. Bowden*

Department of Chemistry
North Carolina State University
Raleigh, North Carolina 27695-8204

INTRODUCTION

Within the expanding research area of protein electrochemistry[1-3], the study of irreversibly adsorbed proteins and enzymes has begun to receive increasing attention[4-18] . Studies of this type have significant potential for contributing to both fundamental and technological progress. In particular, the following research areas are expected to benefit directly from in-depth electrochemical investigations of adsorbed electroactive proteins:

1) biological electron transfer
2) protein adsorption at solid/liquid interfaces
3) amperometric biosensors
4) bioelectrosynthesis

The attractiveness of this approach for studying biological electron transfer results from the treatment of irreversibly adsorbed proteins as protein/electrode complexes,[15] an electrochemical analogy of the widely studied protein/protein electron transfer complexes.[19-21] Electron transfer in both the electrochemical and the conventional reaction situations is "intramolecular".[22] The absence of diffusional contributions to the reaction rates of such systems results in a simplified kinetic analysis due to the absence of association/dissociation work terms. The electrochemical approach has an added advantage, namely, that the free energy of reaction is controllable in a continuous fashion through the applied potential.[22] Drawbacks of the electrochemical approach, at present, include uncertainties associated with the heterogeneous nature of solid electrode surfaces and with the orientation of the irreversibly adsorbed proteins.

Protein adsorption at solid/liquid interfaces, although a widely studied phenomenon, is not yet well understood due to the chemical complexity of the systems and the analytical difficulty of gaining molecular information about the adsorbed species.[23] Electrochemical studies of irreversibly adsorbed redox proteins may be able to provide some unique contributions to this area due to the additional analytical capability arising from the electroactivity of the protein. Furthermore, the effect of electric fields on protein adsorption/desorption processes can be studied through variation of the electrode potential.

In the area of amperometric biosensors,[24,25] direct electronic coupling between irreversibly adsorbed redox enzymes and solid electrodes may lead to the development of molecular-layer biosensors with fast response times. Promising advances have been reported both in understanding direct electron transfer[1-18] and in immobilizing active enzyme species on solid electrode surfaces.[11,13,14,18] The same principle of immobilized enzyme electrocatalysis may also lead to useful bioelectrosynthetic applications.[11,14]

A substantial part of our research effort[15-17] on irreversibly adsorbed protein electrochemistry has been directed towards the cytochromes.[26,27] The monohemic cytochromes can be thought of as model systems with excellent potential for developing molecular-level descriptions of protein electrochemistry. In this respect, the key features of cytochromes include their small size, relatively simple structure and function, stability, and, for many species, the availability of crystallographic structures. Furthermore, the heme group is an attractive chromophore for spectroelectrochemical investigations.

Cytochrome $\underline{c}$ on conductive tin oxide has been shown by us to be an excellent system for studying irreversibly adsorbed redox proteins.[15,16] The system is quite stable and reproducible, and the irreversibly adsorbed cytochrome $\underline{c}$ appears to be in a native conformation, as discussed below.

The purposes of this chapter are threefold. First, previous results on the cyclic voltammetry of cytochrome $\underline{c}$[15,16] are summarized and updated with some new data on competitive polyion effects and probe protein measurements. Second, initial results on measurements of cytochrome $\underline{c}$ adsorption using long optical pathlength thin-layer spectroelectrochemistry[28,29] are presented. This method is shown to hold substantial promise for determining adsorption isotherms on electrodes and for investigating the influence of the interfacial electric field. Third, a discussion of the orientation of adsorbed cytochrome $\underline{c}$ on tin oxide electrodes is presented.

EXPERIMENTAL

Horse heart cytochrome $\underline{c}$ (Sigma, Type VI) was chromatographically purified[31] and used without freezing or lyophilization. A molar absorptivity of 105,000 $M^{-1}cm^{-1}$ at 408 nm[30] was used to determine the concentration of cytochrome $\underline{c}$. Baker's yeast cytochrome $\underline{c}$ peroxidase (CCP) was extracted and purified using established procedures[32,18] and was stored under liquid nitrogen. Thawed CCP samples had an absorbance purity index ($A_{408\ nm}/A_{280\ nm}$) of 1.29 in pH 7 phosphate.[33] The Compound I form of CCP was prepared by addition of hydrogen peroxide to solutions of ferric CCP. Water was purified with a Milli-Q system with an Organex-Q final stage (Millipore, Bedford, MA). Poly-L-lysine hydrobromide (MW = 4200-6000) and sodium poly-L-glutamate (MW = 43,000) were purchased from Sigma Chemical Company. All other chemicals were reagent grade.

Two types of thin-film tin oxide material were used with satisfactory results. Fluorine-doped tin oxide, deposited by spray pyrolysis on 5 mm thick glass, was donated by PPG Industries, Pittsburgh, PA. Antimony-doped tin oxide, deposited by chemical vapor deposition on 3 mm thick glass, was purchased from Delta Technologies, Ltd, Stillwater, MN (part #CG-100SN-P). Pretreatment of electrodes consisted of sequential 20 minute sonications in detergent solution (Alconox), 95% ethanol, and (twice) Milli-Q water, followed by overnight equilibration in the electrolyte to be used for electrochemistry.

The all-glass cyclic voltammetry cell was similar in design to that of Koller and Hawkridge.[34] The tin oxide working electrode was mounted horizontally on the bottom of the cell and had a projected area of 0.32 cm^2. All potentials in this paper refer to the Ag/AgCl (1.00 M KCl) reference electrode used in these experiments; the potential of this electrode is +0.23 V vs NHE. The auxiliary electrode was a platinum wire coiled around the reference electrode capillary tube. Conventional electrochemical instrumentation was employed. The procedure for adsorbing cytochrome c was similar to previous descriptions.[15,16] In the present study, a freshly equilibrated electrode was exposed to a 30 μM ferricytochrome c, 4.4 mM phosphate, pH 7.0 solution for 30 minutes, followed by copious rinsing with Milli-Q water. The cell was then filled with electrolyte only, and cyclic voltammograms were acquired. All experiments were performed at ambient temperature, 20-24 °C.

Spectroelectrochemistry was performed using a long optical pathlength thin-layer cell (LOPTLC) based on the Zak-Porter-Kuwana design.[28] A simple schematic of the cell geometry is shown in Figure 1. The thin-layer volume is defined by the pathlength (c), the thickness (a), and the height (b). For clarity the dimensions are not to scale; the pathlength is 1.51 cm whereas the thickness, which is defined by strips of Teflon tape (B), is typically 0.016 cm. The optically monitored cell volume is approximately 15 μl. A detailed description of the cell construction will appear elsewhere.[35] Solution inlet/outlet and electrode placement were handled in a similar fashion as previously reported[28] except that the reference and auxiliary electrodes share a common port, namely, the solution exit. Front and rear glass windows were sealed to the cell body using thin (<0.1 mm) gaskets prepared from General Electric RTV silicone rubber adhesive, and retainer plates. Electrical contact was made with an alligator clip attached to a part of the tin oxide electrode that extended above the top of the cell body. Sample injections were made through a Hamilton valve using a glass syringe and a septum. The cell was mounted on a translation stage, designed and fabricated in-house, in the sample compartment of a Varian 2300 UV/Vis/NIR spectrophotometer. The cell was aligned to provide maximum light throughput to the PMT.

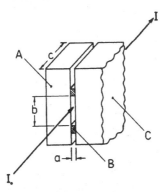

Fig. 1. A simplified view of the LOPTLC geometry. (A) Tin oxide electrode; (B) Teflon tape strips of nominal 0.005 inch thickness; (C) portion of the Lucite plastic cell body. Details of solution inlet/outlet, reference and auxiliary electrode placement, and optical windows are not shown; see text. (I$_o$) Intensity of incident light; (I) Intensity of transmitted light. The thin-layer volume is defined by: (a) the thickness, typically 0.016 cm; (b) the height, typically 0.6 cm; and (c) the pathlength, 1.51 cm. Note: the drawing is not to scale.

RESULTS AND DISCUSSION

Cyclic Voltammetry of Irreversibly Adsorbed Cytochromes

The cyclic voltammetric behavior of irreversibly adsorbed horse heart cytochrome c on tin oxide thin film electrodes has been described previously.[15,16] Figure 2 shows some typical cyclic voltammograms for irreversibly adsorbed cytochrome c on tin oxide. The electrolyte is 24 mM phosphate, pH 7.0, which has an ionic strength of 55 mM. An important feature of the CV's shown is the near-coincidence of the surface formal potential of adsorbed cytochrome c with the solution value, +0.26 V vs NHE.[26] This observation is a strong thermodynamic indication that both redox forms of adsorbed cytochrome c exist in near-native conformations. Indeed, the formal potential of adsorbed cytochrome c actually exhibits a small negative shift from the solution value, as seen in Figure 2, which is characteristic of native cytochrome c when it binds to membranes or other proteins.[36] Additional spectroscopic experiments are planned to further evaluate the degree to which adsorbed cytochrome c can be considered to be native.

Cyclic voltammetry has also been used to semi-quantitatively determine cytochrome c surface coverage by integrating the faradaic peaks. In Figure 2, the voltammetrically determined surface concentration is 5 pmol/cm^2 of projected electrode area. This value is within the range of coverage typically observed, 3-10 pmol/cm^2. Low ionic strengths ($\leq$0.1 M) and pH values of 7-8.5 have been found to result in higher cytochrome coverage, whereas pH 6 and/or high ionic strengths result in low or negligible coverage.[15,16] Increasing the ionic strength to 0.5 M has been found to result in rapid desorption of cytochrome c at pH 7. These results are compatible with electrostatic attraction being the primary mode of binding of cytochrome c to the tin oxide surface. This view is consistent with the isoelectric points of cytochrome c and tin oxide, namely, 10[27] and 5.5,[37] respectively, as previously discussed.[15] Thus, at pH 7-8.5, the tin oxide surface will be anionic whereas cytochrome c will be cationic.

Additional support for the proposed electrostatic binding comes from polyion inhibition studies. The addition of a polycation, polylysine,

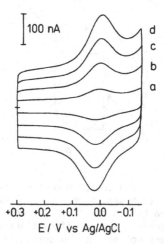

Fig. 2. Cyclic voltammetry of irreversibly adsorbed cytochrome c. Buffer: 24 mM potassium phosphate, pH 7.0, 55 mM ionic strength. Tin oxide was from Delta Technologies, Ltd. Scan rates in mV/s: (a) 10; (b) 30; (c) 50; (d) 70.

to the buffer resulted in immediate and total quenching of any electro-chemical response due to adsorbed cytochrome c. Furthermore, if a newly cleaned and equilibrated tin oxide electrode was first exposed to a polylysine solution and then rinsed, subsequent adsorption of cytochrome c did not occur. Presumably, the polylysine binds more effectively to the tin oxide surface than cytochrome c at pH 7. On the other hand, the presence of a polyanion, polyglutamate, was found to have only minor effects on the adsorption and voltammetry of cytochrome c. A fuller description of these results will be presented elsewhere.[38]

The electron transfer kinetics of irreversibly adsorbed cyto-chrome c have been evaluated from cyclic voltammograms using a method developed by Laviron.[39] In his method, the anodic-cathodic peak separa-tion is measured as a function of potential sweep rate, whereupon the "intramolecular" electron transfer rate constant, k_{et}, can be extracted from a working curve based upon Butler-Volmer kinetics and the Langmuir isotherm. Electron transfer kinetic results for adsorbed cytochrome c under a variety of solution conditions have been reported.[15,16] We wish to draw attention to two broad observations which apply to all the kinetic data acquired to date. First, the observed rate constants are surprisingly small, typically falling in the range of 1 to 10 s^{-1}.[15,16] Second, an anomalous trend is observed in which the apparent value of k_{et} increases with potential scan rate. The small rate constants could be due to a long electron transfer distance resulting from non-optimum orientation of cytochrome c and/or space charge tunneling within the tin oxide. A large activation barrier as a possible source of the kinetic slowness has been ruled out using variable temperature kinetic experi-ments.[40] The anomalous dependence of the rate constant on scan rate may be due, in part, to the presence of a distribution of rate constants arising from the heterogeneous nature of the tin oxide surface. Other possible causes,[15,16] however, remain under active consideration.

The voltammetric behavior of irreversibly adsorbed cytochrome c-552 on EOPG has been briefly investigated.[17] These initial results have revealed some similarities to those discussed for cytochrome c on tin oxide. For cytochrome c-552, the electron transfer rate constant was also found to be small, ca. 1 s^{-1}, and a similar anomalous trend of rate constant with scan rate was observed. Thus, it appears that some gener-al features may apply to the adsorption and electron transfer behavior of small redox proteins on solid electrode surfaces.

Probe Protein Experiments: Electrocatalytic Reduction of Cytochrome c Peroxidase by Adsorbed Cytochrome c

To understand the adsorption and voltammetric behavior of adsorbed redox proteins, a determination of protein orientation will be necessary. This is a major unsolved problem in protein adsorption science.[23] One approach we are pursuing is the use of "probe proteins". The strategy relies on the fact that protein-protein electron transfer complexes gener-ally form through interaction of specific complementarily charged regions of their surfaces.[19-21] Cytochrome c is known to bind through its posi-tively charged heme edge region to negatively charged surface patches on protein partners.[27] For example, it is known that a stable electron transfer complex must form between ferrocytochrome c and the highly oxi-dized Compound I form of cytochrome c peroxidase (CCP) in order for elec-tron transfer between these two proteins to occur.[41] Therefore, some information on the orientation of adsorbed cytochrome c on tin oxide can potentially be gained by studying its binding and electron transfer reac-tivity with a diffusing protein such as CCP.

Some results are shown in Figure 3. Cytochrome c was first adsorbed and cyclic voltammograms were acquired in the presence of only the buffer/electrolyte, 8.5 mM phosphate at pH 7.0. The dashed trace is a typical CV of adsorbed cytochrome c. The Compound I form of CCP was then added to the solution, and additional CV's were acquired, including the solid trace shown in Figure 3. The increased cathodic current and decreased anodic current are characteristic of an electrocatalytic reaction.[42] This catalytic current is attributed to the reduction of diffusing CCP Compound I by adsorbed ferrocytochrome c. Control experiments showed that Compound I was not reducible in this potential range at a bare tin oxide surface in the absence of adsorbed cytochrome c. In order for electron transfer between CCP and adsorbed cytochrome c to occur, it is presumed that a functional protein/protein complex[41] must form. For this complex to form, the adsorbed cytochrome c must be oriented so that its exposed heme edge is at least partially accessible to diffusing CCP. This result apparently argues against adsorbed cytochrome c oriented with its exposed heme edge directly facing the electrode surface. However, it is also possible that some reorientation of adsorbed cytochrome c may be induced by the binding forces between the two proteins. Full details of these experiments, including an analysis of the electron transfer kinetics attendant to the CCP/cytochrome c reaction, will be presented elsewhere.[38]

Spectroelectrochemical Determination of Cytochrome c Adsorption on Tin Oxide

In this section initial spectroelectrochemical results for cytochrome c adsorption on tin oxide using a long optical pathlength thin-layer cell (LOPTLC) are presented. The development of LOPTLC's is due primarily to Kuwana and co-workers,[28,29] who have described quantitation of adsorbed species.[29] LOPTLC spectroelectrochemistry is a high surface/volume method involving longitudinal optical monitoring of a long,

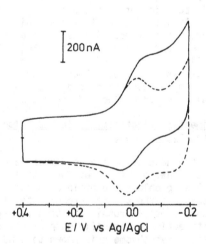

$$\text{200 nA}$$

E / V vs Ag/AgCl

Fig. 3. Cyclic voltammetry demonstrating the electrocatalytic reduction of the Compound I form of cytochrome c peroxidase by adsorbed cytochrome c. Tin oxide electrode was from PPG. Buffer: 8.5 mM potassium phosphate, pH 7.0, 19 mM ionic strength. Scan rate: 50 mV/s. The dashed trace was acquired for irreversibly adsorbed cytochrome c in buffer alone; surface concentration was 8 pmol/cm^2 of projected electrode area. The solid trace was subsequently acquired after addition of 48 μM CCP Compound I (the solution also contained a minor amount of the ferric form of CCP, which cannot transfer electrons with cytochrome c).

thin solution volume trapped between a planar electrode and a parallel wall. The adsorption of cytochrome c on a tin oxide electrode surface can be determined by spectrophotometric monitoring of the decrease in solution concentration that occurs following injection of a known solution into the thin layer.

Figure 4 shows initial results for five sequential injections of a 5.0 µM ferricytochrome c solution into the LOPTLC containing an open-circuited, equilibrated tin oxide electrode. In this experiment, absorbance at 408 nm was monitored. The dashed line at the top of the figure is the predicted absorbance that would result from an unperturbed 5.0 µM cytochrome c solution in the LOPTLC. Trace (a), the absorbance-time response following the first injection of solution, shows a sharp decrease of some 30-40% due to cytochrome c adsorption. The present instrumental set-up did not allow absorbance monitoring during the first 1/2 minute following solution injection; a dashed line has been added to the righthand side of the figure to approximately depict the response in this region. The absorbance-time responses for the next 4 sequential injections of 5.0 µM ferricytochrome c solution at the same electrode are shown as traces (b) thru (e). For clarity, traces (b) thru (e) have been redrawn with sample injection set to t = 0. With each additional injection, further adsorption of cytochrome c is indicated by a reduction in absorbance. The amount of adsorption progressively decreases, however, presumably because saturation coverage is being approached.

At the end of each absorbance-time trace, a Soret region spectrum was acquired before injection of the next volume of solution. These spectra, shown in Figure 5, confirm the interpretation in the above paragraph, namely, that the absorbance is due to solution cytochrome c and that with each additional injection, less cytochrome c adsorbs.

The total amount of cytochrome c adsorbed at the end of trace (e) in Figure 4 was estimated by summing the absorbance decreases for the five injections and assuming that this total absorbance change (ΔA) represented the amount adsorbed. If it is further assumed that cytochrome c adsorbs solely on the tin oxide electrode, then Γ, the corresponding surface concentration, can be calculated using the following expression:[29]

$$\Gamma = V \cdot \Delta A / \varepsilon \ell A$$

where V is the volume of the thin-layer, ε is molar absorptivity, ℓ is the optical pathlength, and A is electrode area. For the data shown in Figures 4 and 5, an estimated value of Γ = 35-50 pmol/cm^2 was determined[43] using a molar absorptivity of 105,000 M^{-1}cm^{-1} for ferricytochrome c.[30] This estimated value for cytochrome c coverage, which represents the total surface concentration, can be compared to the electroactive coverage, typically found to be 3-10 pmol/cm^2.[15,16] The large difference between these two values of coverage may indicate that not all of the adsorbed cytochrome c is electroactive and/or that undesirable adsorption of cytochrome c on surfaces other than the tin oxide electrode is occurring. More recent evidence indicates that cytochrome c does adsorb on lucite surfaces to some degree, which would contribute to the absorbance drops seen in Figures 4 and 5. Work is currently underway to eliminate adsorption contributions due to non-electrode surfaces. Once this is accomplished, an accurate comparison of total and electroactive cytochrome c surface coverage on these electrodes will be feasible.

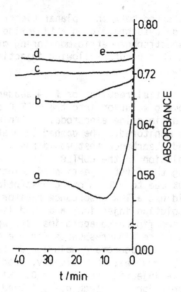

Fig. 4. Long optical pathlength thin-layer spectroelectrochemistry of the adsorption of cytochrome c on tin oxide. Traces (a) through (e) are absorbance-time traces that resulted from five sequential injections of 5.0 μM ferricytochrome c solution into the thin-layer cavity of the LOPTLC shown in Figure 1. Wavelength: 408 nm. Buffer: pH 7.0, 4.4 mM potassium phosphate, 10 mM ionic strength. Tin oxide electrode was from Delta Technologies, Ltd. The horizontal dashed line at the top of the figure is the absorbance calculated for a 5.0 μM ferricytochrome c solution using a molar absorptivity of 105,000 $M^{-1}cm^{-1}$ at 408 nm and a pathlength of 1.5 cm.

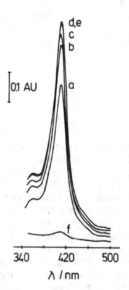

Fig. 5. Soret region absorbance spectra acquired with the LOPTLC in conjunction with the data in Figure 4. Scans (a) through (e) were acquired after absorbance-time traces (a) through (e), respectively, in Figure 4. Scan (f) was subsequently acquired after displacing the ferricytochrome c solution with buffer (4.4 mM phosphate, pH 7.0).

Orientation of Adsorbed Cytochrome c on Tin Oxide

At present the orientation of adsorbed cytochrome $\underline{c}$ on tin oxide is undetermined. The effects of pH, ionic strength, and polyions on voltammetrically determined surface coverages indicate, however, that cytochrome $\underline{c}$ binds to tin oxide electrostatically via some portion of its positively charged front hemisphere.[26,27]

Other experiments suggest further that cytochrome $\underline{c}$ may not be oriented with its exposed heme edge directly facing the electrode surface. Two key pieces of evidence, as discussed above, are the small value of the electron transfer rate constant, which suggests a long electron transfer distance, and the electrocatalytic activity of the diffusing probe protein CCP, which suggests accessibility of the heme edge of adsorbed cytochrome $\underline{c}$. An attractive possibility that could account for these results would involve binding of cytochrome $\underline{c}$ via its left or left-front surface. This surface region of cytochrome $\underline{c}$ has a high density of positive charge. Furthermore, the dipole vector of cytochrome $\underline{c}$ crosses the left-front surface, making a 30° angle with the heme plane.[44] Further experiments are planned to evaluate this proposed orientation, including probe protein measurements in conjunction with long optical pathlength thin-layer spectroelectrochemistry. Consideration will also be given to the presence of a distribution of orientations, which is almost certain to exist on heterogeneous electrode surfaces.

CONCLUSIONS

Irreversibly adsorbed cytochromes are very useful systems for investigating protein adsorption and electronic coupling between proteins and solid electrodes. In particular, the system comprised of irreversibly adsorbed cytochrome $\underline{c}$ on conductive tin oxide holds great promise for achieving progress in these areas. Irreversibly adsorbed cytochrome $\underline{c}$ on tin oxide appears to function in a normal fashion and can be treated as a protein/electrode complex, by analogy to the well-studied protein/protein complexes. The absence of diffusional components from the reaction conveys an added simplicity to the analysis of the associated electron transfer kinetics.

Initial results using long optical pathlength thin-layer spectroelectrochemistry have demonstrated the potential for establishing adsorption isotherms for cytochromes on electrodes as well as for comparing the total surface coverage of cytochrome $\underline{c}$ on tin oxide with the voltametrically determined electroactive surface coverage.

The binding of cytochrome $\underline{c}$ to tin oxide appears to be primarily electrostatic in nature and undoubtedly involves the positively charged front hemisphere of the molecule. Small electron transfer rate constants, probe protein results using cytochrome $\underline{c}$ peroxidase, and consideration of the molecular structure of cytochrome $\underline{c}$ further suggest that the preferred binding site may be on the left to left-front surface.

ACKNOWLEDGEMENTS

We would like to thank Mr. William Patrick for design and fabrication of the translation stage for the LOPTLC and Dr. John Sopko of PPG for providing fluorine-doped tin oxide material. Support of this work through a Biomedical Research Support Grant (NIH No. RR7071) is gratefully acknowledged.

REFERENCES

1. G. Dryhurst, K.M. Kadish, F. Scheller, and R. Renneberg, "Biological Electrochemistry," Volume 1, Academic Press, New York (1982), pp 398-521.

2. E.F. Bowden, F.M. Hawkridge, and H.N. Blount, in "Comprehensive Treatise of Electrochemistry," Volume 10, S. Srinivasan, Yu. A. Chizmadzhev, J. O'M. Bockris, and E. Yeager, eds., Plenum Press, New York (1985), pp. 297-346.

3. F.A. Armstrong, H.A.O. Hill, and N.J. Walton, Quart. Rev. Biophys. 18:261 (1986).

4. R.E. Holt and T.M. Cotton, J. Am. Chem. Soc. 109:1841 (1987).

5. T.M. Cotton, S.G. Schultz, and R.P. Van Duyne, J. Am. Chem. Soc.

6. C. Hinnen, R. Parsons, and K. Niki, J. Electroanal. Chem. 147:329 (1983).

7. K. Niki, Y. Kawasaki, Y. Kimura, Y. Higuchi, and N. Yasuoka, Langmuir 3:982 (1987).

8. G. Smulevich and T.G. Spiro, J. Phys. Chem. 89:5168 (1985).

9. P. Hildebrandt and M. Stockburger, J. Phys. Chem. 90:6017 (1986).

10. I. Taniguchi, M. Iseki, H. Yamaguchi, and K. Yasukouchi, J. Electroanal. Chem. 175:341 (1984).

11. M.R. Tarasevich, in reference 2, pp 231-295.

12. J.J. Kulys and A.S. Samalius, Bioelectrochem. Bioenerg. 13:163 (1984).

13. C.-W. Lee, H.B. Gray, F.C. Anson, and B.G. Malmstrom, J. Electroanal. Chem. 172:289 (1984).

14. S.D. Varfolomeev and I.V. Berezin, in "Advances in Physical Chemistry: Current Developments in Electrochemistry and Corrosion," Ya. M. Kolotyrkin, ed., Mir Press, Moscow (1982), pp 60-95.

15. J.L. Willit and E.F. Bowden, J. Electroanal. Chem. 221:265 (1987).

16. J.L. Willit and E.F. Bowden, in "Redox Chemistry and Interfacial Behavior of Biological Molecules," G. Dryhurst and K. Niki, eds., Plenum Press, New York (1988), pp 63-76.

17. V. Senaratne and E.F. Bowden, Biochem. Biophys. Res. Commun. 157:1021 (1988).

18. R.M. Paddock and E.F. Bowden, J. Electroanal. Chem. 260:487 (1989)

19. F.R. Salemme, Ann. Rev. Biochem. 46:299 (1977).

20. S.E. Peterson-Kennedy, J.L. McGourty, P.S. Ho, C.J. Sutoris, N. Liang, H. Zemel, N.V. Blough, E. Margoliash, and B.M. Hoffman, Coord. Chem. Rev. 64:125 (1985).

21. E. Cheung, K. Taylor, J.A. Kornblatt, A. English, G. McLendon, and J.R. Miller, Proc. Natl. Acad. Sci. USA 25:4804 (1986).

22. J.T. Hupp and M.J. Weaver, J. Electroanal. Chem. 145:43 (1983).

23. T.A. Horbett and J.L. Brash, in "Proteins at Interfaces: Physicochemical and Biochemical Studies", ACS Symposium Series 343, J.L. Brash and T.A. Horbett, eds., American Chemical Society, Washington, D.C. (1987), pp 1-33.

24. G.S. Wilson, in "Biosensors: Fundamentals and Applications", A.P.F. Turner, I. Karube, and G.S. Wilson, eds., Oxford University Press, Oxford, (1987), pp 165-179.

25. W.J. Albery and P.N. Bartlett, J. Electroanal. Chem. 194:211 (1985).

26. R.E. Dickerson and R. Timkovich, in "The Enzymes", Volume XI-A, P.D. Boyer, ed., Academic Press, New York (1975), pp 397-547.

27. S. Ferguson-Miller, D.L. Brautigan, and E. Margoliash, in "The Porphyrins", Volume VII-B, D. Dolphin, ed., Academic Press, New York (1979), pp 149-240.

28. J. Zak, M.D. Porter, and T. Kuwana, Anal. Chem. 55:2219 (1983).

29. Y. Gui and T. Kuwana, J. Electroanal. Chem. 222:321 (1987).

30. E. Margoliash and N. Frohwirt, Biochem. J. 71:570 (1959).

31. D.L. Brautigan, S. Ferguson-Miller, and E. Margoliash, Methods Enzymol. 53D:131 (1978).

32. A.M. English, M. Laberge, and M. Walsh, _Inorg. Chim. Acta_ 123:113 (1986).
33. T. Yonetani and H. Anni, _J. Biol. Chem._ 262:9547 (1987).
34. K.B. Koller and F.M. Hawkridge, _J. Am. Chem. Soc._ 107:7412 (1985).
35. M. Collinson and E.F. Bowden, to be published.
36. P. Nicholls, _Biochim. Biophys. Acta_ 346:261 (1974).
37. S.M. Ahmed, in "Oxides and Oxide Films," Volume 1, J.W. Diggle, ed., Marcel Dekker, New York (1972), p. 319.
38. M. Collinson, J.L. Willit, R.M. Paddock, and E.F. Bowden, _J. Am. Chem. Soc._, submitted.
39. E. Laviron, _J. Electroanal. Chem._ 101:19 (1979).
40. J.L. Willit and E.F. Bowden, to be published.
41. T.L. Poulos and J. Kraut, _J. Biol. Chem._ 225:10322 (1980).
42. A.J. Bard and L.R. Faulkner, "Electrochemical Methods", John Wiley, New York (1980), pp 457-459.
43. Present uncertainties in determining quantitative surface coverage values include adsorption on non-electrode surfaces, edge diffusion effects, the lack of truly steady-state conditions (see Figure 4), the inability to follow absorbance changes during sample injection (see text), some non-reproducibility in quantitatively displacing solution, and some light throughput limitation that may affect absorbance readings by, at most, 8%. Work is in progress to eliminate or minimize these sources of error, thereby improving the accuracy and precision of the measurement.
44. J.D. Rush, J. Lan, and W.H. Koppenol, _J. Am. Chem. Soc._ 109:2679 (1987).

INTERFACIAL ELECTRON TRANSFER REACTIONS OF HEME PROTEINS

Yuan Xiaoling, John K. Cullison, Songcheng
Sun and Fred M. Hawkridge

Department of Chemistry
Virginia Commonwealth University
Richmond, Virginia 23284

INTRODUCTION

The kinetics and thermodynamics of electron transfer
reactions of heme proteins can be studied directly at metal
and semiconductor electrodes. We have been interested in
characterizing the heterogeneous electron transfer properties
of biological electron transfer proteins since the initial
report of electrocatalysis in the reaction of spinach
ferredoxin at a polymer modified gold electrode (Landrum et
al., 1977). This 11,000 dalton iron-sulfur protein, which
participates in green plant photosynthesis, was shown to
undergo direct, heterogeneous, electron transfer at quasi-
reverisble rates at this modified electrode. Later that year
two communications described quasi-reversible electron
transfer rates by cytochrome c at modified gold (Eddowes and
Hill, 1977) and at bare indium oxide (Yeh and Kuwana, 1977)
electrodes.

Early in our work purely electrochemical methods did not
yield voltammetric results that were useful for kinetic
analysis of heme protein heterogeneous electron transfer
reactions. Spectroelectrochemical methods that permitted the
elucidation of the heterogeneous electron transfer kinetics
of these reactions were developed collaboratively (Albertson
et al., 1979; Bancroft et al., 1981a, 1986) and by Bancroft
et al. (1981b). These spectroelectrochemical methods were
subsequently used to study the heterogeneous electron
transfer reactions of myoglobin (Bowden et al., 1981; King
and Hawkridge, 1988) and cytochrome c (Bowden et al., 1982a).

Results from these studies led to the recognition that
small concentrations of oligomeric forms of cytochrome c
demonstrably alter the kinetics of electron transfer for
solution resident molecules at indium oxide electrodes
(Bowden et al., 1982b). Moreover, it was subsequently
appreciated that lyophilization of purified samples of
cytochrome c reintroduces small, but deleterious, concentrations of oligomeric cytochrome whose effects on heterogeneous
electron transfer kinetics are manifested only at metal

surfaces such as silver (Reed and Hawkridge, 1988) and other metal electrodes (Sun et al., 1988).

The consequence of the above work is that direct electrochemical methods can be used to study the electron transfer reactions of many different heme proteins. This avoids the need to interpret effects arising from chemical reductants or oxidants as might be used in stopped flow kinetic studies and affords the opportunity to quantitatively control the rate of electron transfer. As an example, the temperature dependence of the reaction of cytochrome c at indium oxide electrodes revealed that there is a particular temperature at which a maximum rate of electron transfer occurs (Koller and Hawkridge, 1985). When the buffer media is chosen to exhibit different degrees of ion binding to the front, or reaction surface, of the cytochrome c molecule this optimum temperature differs. However, the optimum temperature corresponds to the same formal potential suggesting that the same relative conformational stabilities of the oxidized to reduced forms is achieved in different media at different temperatures. This study has been extended in the present work to include cytochrome molecules from a variety of different sources.

Recently our group also has been studying the electrochemistry of Co-enzyme Q at and in assembled monolayers and bilayers at electrode surfaces. Although Co-enzyme Q is not a heme protein it serves the function of transporting electrons and protons within the inner mitochondrial membrane. One of Co-enzyme Q's redox partners in-vivo is cytochrome c reductase which is a heme protein (Lenaz, 1985). The monolayers are built by using the spontaneous adsorption technique. The formation of spontaneously adsorbed monolayers has received considerable attention lately (Nuzzo and Allara, 1983; Netzer and Sagiv, 1983), and the ability to form well organized monolayers by this method has been realized (Porter et al., 1987). The second half of the bilayer is formed using the Langmuir-Blodgett method (Langmuir, 1939; Blodgett, 1935). The cyclic voltammetric cathodic and anodic peaks are evident but have large peak separations.

MATERIALS AND METHODS

Cytochrome c from horse heart, porcine heart, tuna heart, chicken heart and pigeon breast muscle were purchased from Sigma Chemical Co., and were purified by chromatography on carboxylmethylcellulose (CM-52, Whatman) according to a published procedure (Brautigan et al., 1978). Tris (hydroxymethy) amino methane was used as received from Sigma Chemical Co. (Trizma Base, reagent grade). Cacodylic acid (hydroxy dimethylarsine oxide, Sigma Chemical Co., 98% pure) was recrystallized twice from 2-propanol. Water used in this work was purified with a Milli RO-4/Milli-Q system (Millipore Corp.) and exhibited a resistivity of 18 MΩcm^{-1} on delivery. Cytochrome c concentrations were determined by the reduced-minus - oxidized difference molar absorptivity, $\Delta\epsilon$=21,100 M^{-1}cm^{-1} at 550 nm (Van Buuren et al., 1974), on a Hewlett Packard 8452A diode Array spectrophotometer. Ferricytochrome c was reduced by dithionite to form ferrocytochrome c. All other chemicals used in this work were ACS reagent grade.

Buffers were prepared by mixing the acid and base components to achieve the desired pH at an ionic strength of 0.2M.

The non-isothermal electrochemical cell used for the cyclic voltammetry temperature studies has been described in previously (Koller and Hawkridge, 1985). Tin doped indium oxide optical transparent electrode (OTE) materials were obtained from Donnelly Corp. The OTE electrode were cleaned by successive 5 min. sonication in Alconox solution, in 95% ethanol and twice in purified water (Armstrong et al., 1976). An IBM PC AT microcomputer system, containing a 12 μs conversion time, 12 bit , HI-674A analog/digital and digital-/analog interface board (Dash-16), was used to control and collect data for the cyclic voltammetric experiments. The high frequency noise showed in data was digitally smoothed by a five point smoothing routine (Savitzky and Golay, 1964). Slow potential scan rate experiments were also smoothed by an analog fourth-order Butterworth filter with a cut off frequency of Ca. 50 Hz. Formal potential were determined from the peak potential values from reversibly cyclic voltam-mogrames acquired at 20 mv/sec scan rate. The difference in the computer - accursed background (electrolyte alone) and total (electroactive species present) cyclic voltammograms was used in these determinations. A program then determined the potential at which the maximum cathodic and anodic currents occurred from each digitally stored cyclic voltam-mogram, providing precise values for formal potentials.

The monolayers were spontaneously assembled at a gold foil electrode (Aldrich, 99.99%) using a 1 mM solution of oc-tadecyl mercaptan (Aldrich) in absolute ethanol (USI Chem. Co.). The deposition times were on the order of ten minutes. Prior to modification the gold foil was polished using Alumina micropolish (Buehler) starting with 1 micron particle size, and then moving to 0.3 microns and finally to 0.05 microns.

Following this procedure the Langmuir Trough (Lauda filmbalance) was used a finish building the bilayer. A mixture of 10% Co-enzyme Q (Sigma Chem. Co.) and 90% phos-phatidyl choline (Sigma Chem. Co.) in hexane (Mallinckrodt, HPLC grade) was prepared. This solution was added to the Langmuir trough and the film was deposited onto the octadecyl mercaptan modified electrode. The film was deposited at a pressure of 30 mN/m and at a speed of 1 cm/min. Phosphate buffer (pH = 7.4, ionic strength = 0.15) was used as a subphase and all experiments were performed at room tempera-ture.

The cyclic voltammetry was performed using a standard three electrode cell configuration. The buffer used was a phosphate (pH = 7.4, ionic strength = 0.15). The reference electrode used was a 1M KCl Ag/AgCl electrode.

RESULTS AND DISCUSSION

Figure 1 shows the temperature dependence for the formal potential of cytochrome c from tuna heart in Tris/cacodylic acid buffers at different pH (μ = 0.2). At pH 5.3, the formal potential of cytochrome c from tuna heart linearly decreases as temperature increases in the range 5 - 60 °C. However, at

pH 7.0, a discontinuity in formal potential versus temperature relationship occurs at 50 °C. at pH 8.0, the break
occurs at an even lower temperature (30 °C). It is well known
that ferricytochrome c undergoes conformational change in
different pH media (Theorell and Akesson, 1941). The reversible transition from the neutral to the alkaline state is
governed by a pKa value between 8.5 and 9.5, depending on the
origin of the cytochrome c (Osheroff et al., 1980, and Davis
et al., 1974). The alkaline state differs from the native
conformation of cytochrome c at neutral pH by a more open
protein crevice and the presumed loss of the Met 80 Fe - S
bond (Myer, 1968). Previous studies have also indicated that
in ferricytochrome c , the Fe - S bond gradually and continuously weakens with the temperature increases (Moore et
al., 1980, and 1982), which causes the conformation transitions. The biphasic behavior of formal potential of cytochrome c as a function of temperature is due to a pronounced
change in protein structure of ferricytochrome c (Tanigushi
et al., 1984). The break temperature represents a particular
condition at which the neutral conformation of cytochrome c
changes into its alkaline state. The effects of pH and
temperature on the conformational change of cytochrome c are
cooperative, higher pH corresponds to a lower break temperature in the formal potential - temperature relationship.

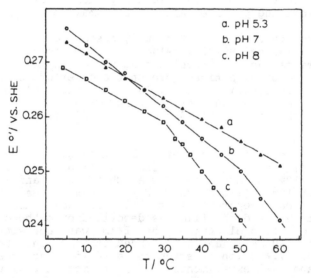

Fig. 1. Temperature dependence of the E°' of cytochrome c from
tuna heart in Tris/cacodylic acid buffer at different
pH. Cytochrome c concentration: 30 um.

The formal potential of cytochrome c from tuna heart in
pH 8.0 Tris/cacodylic acid (5 - 50 °C) and phosphate (5 - 45
°C) buffers are shown in Figure 2. The negative shift of the
formal potential in phosphate buffer was attributed to anion
binding (3,13). Phosphate anions bind to the positively
charged lysine residues surrounding the solvent exposed heme
edge (Osheroff et al., 1980) and result in an opening of the
heme crevice (Robinson et al., 1983, Pettigrew et al., 1976,

Eley rt al., 1983, and Osheroff et al., 1980), which stabilize the positively charged ferrichrome c, cause the negative
shift of the formal potential. An important phenomenon should
be pointed out here is that the binding of phosphate anions
caused the further negative shift at every temperature
studied in this work , but it did not change the break
temperature at which the biphasic behaviors. This fact
indicated that the conformational perturbation induced by
anion binding is not cooperative with the perturbation
induced by temperature and pH change. The possible reason
could be that the phosphate anions bind to both oxidation
states of cytochrome c (Osheroff et al., 1980), which results
equal perturbations for both states, but the perturbations of
temperature and pH cause much more pronounced change in the
protein structure of ferricytochrome c than that of ferrocytochrome c. The further negative shift of formal potential observed in phosphate buffer is not caused by the
further difference in protein structure between ferri- and
ferrocytochrome c, but the exist of anions in the heme
environment, which stabilize positively charged ferricytochrome c more than zero charged ferrocytochrome c.

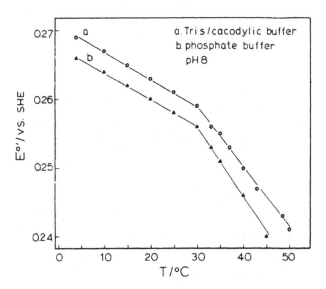

Fig. 2. Temperature dependence of E°' of cytochrome c from tuna
fish in different buffer at pH 8.0. Cytochrome c
concentration: 32 um.

Figure 3 shows the formal potential - temperature
dependence of cytochrome c from different vertebrate species
(chicken, pigeon, porcine, horse and tuna) as a function of
temperature in pH 8.0 Tris/cacodylic acid buffer (μ=0.2).
Every studied cytochromes c showed the biphasic behavior in
this condition, but the break temperatures are quite different. The break temperature for cytochrome c from chicken
is 50 °C, for pigeon is 45 °C, both of them are higher than
that of cytochrome c from horse and porcine (40 °C), cytochrome c from tuna heart has the lowest break temperature (30
°C) among the studied species.

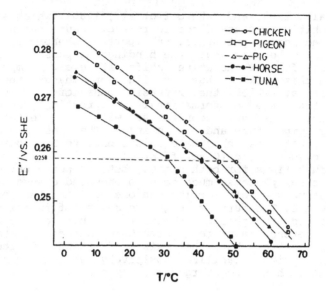

Fig. 3. Temperature dependence of E°' of cytochrome c from
different species in pH 8.0 Tris/cacodylic buffer.
Cytochrome c concentration: 30 - 40 um.

As mentioned earlier, the break temperature is an
indication of the conformational states transition, which
represents the stability of the native conformation of the
heme crevice of ferricytochrome c, as the pKa and transition
temperature values for the 695 nm absorption band studied by
Osheroff (Osheroff et al., 1980). Higher values of pKa and
transition temperature represent a more stable closed crevice
structure. It has shown that the heme crevice stability of
cytochromes c very greatly between the cytochromes c from
different mammalian taxonomic groups and is related to the
surface interactions such as the salt bridge between the ε-
amino group of lysine 13 and the γ - carboxyl group of
glutamic acid 90 and the hydrogen bond between the ε - amino
group of lysine 79 and the backbone carbonyl of residue 47.
These interactions on the surface of the molecule are
important in maintaining the closed structure of the heme
crevice. Our study also showed the importance of these two
surface bonds. Around Lys 79, the amino acid residue 60 for
cytochrome c from tuna is AspNH₂, for chicken is Gly and for
horse is Lys. Both AspNH₂ and Lys at this position can offer
a NH₂ group to compete with Lys 79 for the Ser or Thr 47 for
hydrogen bonding, moreover, for cytochrome c from tuna heart,
the residue 46 is tyr instead of Phe in both chicken and
horse, tyr 46 can also compete with Ser or Thr 47 for the ε-
amino group of Lys 79. On the other hand, the residues
around Glu 90 include 92, at this position, Val (which is
hydrophobic amino acid residue) substituted Glu (which is
charged) in cytochrome c from chicken, and around Lys 13, the
residue 2 in chicken cytochrome c substitute by Ileu, which
is more hydrophobic than Val in cytochrome c from horse and
tuna. In both cases, a more hydrophobic environment is
provided in cytochrome c from chicken, which makes the salt

bridge between Lys 13 and Glu 90 stronger than that in cytochrome c from horse and tuna. This, together with the hydrogen bond mention earlier, explain why the closed heme crevice for cytochrome c from chicken is more stable than that from horse, and the latter is more stable than that from tuna. An interesting point should be made here is the stability series obtained in this work is corresponded to the body temperatures of studied species. The body temperature of birds (chicken and pigeon) is higher than that of mammalian animals (horse and porcine), and the body temperature of tuna fish changes as surrounding water, but is much lower than that of homothermal animals.

The other information we can get from Figure 3 is that although the break temperatures for cytochromes c from different species are different, the formal potentials at this point are almost the same for the studied species, which indicated the relative conformations of ferricytochrome c and ferrocytochrome c at this point are the same. Considering the previous study (Koller and Hawkridge, 1985) has shown that there is an optimum conformation of ferricytochrome c for heterogeneous electron transfer, which occurs at 40 °C in Tris/cacodylic acid buffer for horse heart cytochrome c, the temperature dependence of the formal heterogeneous electron-transfer rate constant of cytochromes c from different species will be examined.

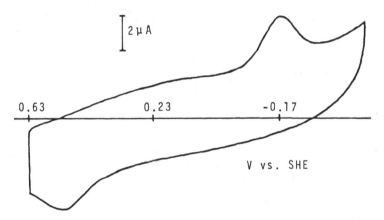

Fig. 4. Cyclic voltammogram of Coenzyme Q within the bilayer electrode. Phosphate buffer (pH = 7.4, ionic strength 0.15), scan rate = 100 mv/s .

Figure 4 shows the cyclic voltammogram of the Coenzyme Q within the bilayer. The voltammetric cathodic and anodic peaks are clear, however, the peak separation is very large. This is because the redox reactions of such quinones are complicated due to the fact that the electron transfer and the protonation steps are separate. Reversible voltammetric waves are usually not observed and peak separations of hundreds of millivolts are not unusual (Rich, 1982). The ability to turn over the Co-enzyme Q within the bilayer has been shown. In the future the ability of the membrane resident Co-enzyme Q to communicate with solution resident species as a function of temperature will be examined. The incorporation of Co-enzyme Q and cytochrome c reductase into these bilayers will be attempted. It is anticipated that if Co-enzyme Q can be turned over then it may reduce the cytochrome c reductase and this system could then be coupled to solution resident cytochrome c.

ACKNOWLEDGEMENTS

The authors gratefully acknowledge the financial support of this work by a grant from the National Science Foundation, CHE-8520270.

REFERENCES

Albertson, D. E., Blount, H. N., and Hawkridge, F. M., 1979, Spectroelectrochemical Determination of Heterogeneous Electron Transfer Rate Constants, Anal. Chem., 51:556.

Armstrong, N.R., Lin, A.W.C., Fujihira, M. and Kuwana, T., 1976, Electrochemical and Surface Characteristics of Tin Oxide and Indium Oxide Electrodes, Anal.Chem., 48:741.

Bancroft, E. E., Blount, H. N., and Hawkridge, F. M., 1981a, Single Potential Step Chronoabsorptometric Determination of Heterogeneous Electron Transfer Kientic Parameters of Quasi-Reversible Processes, Anal. Chem., 53:1862.

Bancroft, E. E., Sidwell, J. S., and Blount, H. N., 1981b, Derivative Linear Sweep and Derivative Cyclic Voltabsorptometry, Anal. Chem., 53:1390.

Blodgett, K. B., 1935, Films Built by Depositing Successive Monomolecular Layers on a Solid Surface, J. Am. Chem. Soc., 57:1007.

Bowden, E. F., Hawkridge, F. M., and Blount, H. N., 1980, Heterogeneous Electron Transfer Kinetics of Sperm Whale Myoglobin, Bioelectrochem. Bioenerg., 7:447.

Bowden, E. F., Hawkridge, F. M., and Blount, H. N., 1982a, The Heterogeneous Electron Transfer Properties of Cytochrome c, Adv. Chem. Ser., 201:159.

Bowden, E. F., Hawkridge, F. M., Chlebowski, J. F., Bancroft, E. E., Thorpe, C., and Blount H. N., 1982, Cyclic Voltammetry and Derivative Cyclic Voltabsorptometry of Purified Horse Heart Cytochrome c at Tin-Dopped Indium Oxide Optically Transparent Electrodes, J. Am. Chem. Soc., 104:7641.

Brautigan, D.L., Ferguson-Miller, S. and Margoliash, E., 1978, Mitochondrial Cytochrome c Preparation and Activity of Native and Chemically Modified Cytochromes c, Methods Enzymol., 53D:131.

Davis, L.A., Schejter, A., Hess, G.P., 1974, Alkaline Isomerization of Oxidized Cytochrome c - Equilbrium and Kinetic Measurements, J.Biol.Chem., 249:2624.

Eddowes, M. J. and Hill, H. A. O., 1977, Novel Method for the Investigation of the Electrochemistry of Metallopro teins: Cytochrome c, J. C. S. Chem. Commun., 771.

Eley, C.G.S. and Moore, G.R., 1983, H - 1 NMR Investigation of the Interaction Between Cytochrome c and Cytochrome B$_5$, Biochem.J., 215:11.

Koller, K. B. and Hawkridge, F. M., 1985, Temperature and Electrolyte Effects on the Electron-Transfer Reactions of Cytochrome c, J. Am. Chem. Soc., 107: 7412

Langmuir, I., 1939, Molecular Layers, Proc. Roy. Soc. London Ser., 170:1.

Landrum, H. L., Salmon, R. T., and Hawkridge, F. M., 1977, A

Surface-Modified Gold Minigrid Electrode Which Hetero-
geneously Reduces Spinach Ferredoxin, <u>J. Am. Chem. Soc.</u>,
99:3154.

Lenaz, G., ed., 1985, "Coenzyme Q: Biochemistry, Bioenergetics and Clinical Applications of Ubiquinone," John
Wiley and Sons, New York.

Moore, G.R., Williams, R.J.P., 1980, Stability of Ferri -
cytochrome <u>c</u> - temperature Dependence if Its NMR -
Spectrum. <u>Eur.J.Biochem.</u> 103:523.

Moore, G.R., Huang, Z.-X., Eley, C.G.S., Barker, H.A.,
Williams,G., Robinson, M.N., Williams, R.J.P., 1982,
Electron transfer in biology; The function of cytochrome
<u>c</u>. <u>Faraday Disc.</u> Faraday Discuss. Chem. Soc., 74:311

Myer, Y.P., 1968, Conformation of Cytochromes. III. Effect of
Urea, Temperature, Extrinsic Ligands, and pH Variation on
the Conformation of Horse Heart Ferricytochrome <u>c</u>.
<u>Biochemistry</u> 7:765.

Netzer, L., and Sagiv, J., 1983, A New Approach to Construction of Artificial Monolayer Assemblies, <u>J. Am. Chem.
Soc.</u>,105:674.

Nuzzo, R. G., and Allara, D. L., 1983, Adsorption of Bifunc-
tional Organic Disulfides on Gold Surfaces, <u>J. Am. Chem.
Soc.</u>, 05:4481.

Osheroff, N., Borden, D., Koppenol, W.H. and Margoliash, E.,
1980a, Electrostatic Interations in Cytochrome <u>c</u>. <u>J.
Biol. Chem.</u>, 255:1689.

Osheroff, N., Brautigan, D.L. and Margoliash, E., 1980b,
Mapping of Anion Binding Sites on Cytochrome <u>c</u> by Dif
ferential Modification of Lysine Residues, <u>Proc. Natl.
Acad. Sci. U.S.A.</u>, 77:4439

Osheroff, N., Brautigan, D.L, and Margoliash, E., 1980c,
Definition of Enzymatic Interation Domains on Cytochrome <u>c</u>
- Purification and Activity of Singly Substituted, <u>J.
Biol. Chem.</u>, 255:8245.

Pettigrew, G.W., Ariram, I. and Schejter, A., 1976, Role of
Lisines in Alkaline Heme - linked Ioniaztion of Ferric
Cytochrome <u>c</u>, <u>Biochem. Biophys. Res. Commun.</u>, 68:807.

Porter, M. D., Bright, T. B., Allara, D. L., and Chidsey, C.
E. D., 1987, Spontaneously Organized Molecular As sembli
es: Structural Characterization of n-Alkyl Thiol
Monolayers on Gold by Optical Ellipsometry, Infrared
Spectroscopy, and Electrochemistry, <u>J. Am. Chem. Soc.</u>,
109:3559.

Reed, D. E. and Hawkridge, F. M., 1988, Direct Electron
Transfer Reactions of Cytochrome <u>c</u> at Silver Electrodes,
<u>Anal. Chem.</u>, 59:2334.

Rich, P. R., 1982, Electron and Proton Transfers in Chemical
and Biological Quinone Systems, <u>Faraday Discuss. Chem.
Soc.</u>, 74:349.

Robinson, M.N., Boswell, A.P., Huang, Z.-X., Eley, C.G.S. and
Moore, G.R., 1983, The conformation of eukaryotic
cytochrome <u>c</u> around residues 39,57,59 and 74,
<u>Biochem. J.</u>, 213:687.

Savitzky, A. and Golay, M.J.E., 1964, Smoothing and
Differen tiation of Data by Simplified Least Squares
Procedures, <u>Anal.Chem.</u>, 36:1627.

Schejter, A. and George, P., 1964, 695 - nm Band of Fer
ricytochrome <u>c</u> and Tis Relationship to Protein Conforma
tion, <u>Biochemistry</u>, 3:1045.

Sun, S., Reed, D. E., Cullison, J. K., Rickard, L. H., and
Hawkridge, F. M., 1988, Electron Transfer Reactions of

Cytochrome c at Metal Electrodes, Mikrochim. Acta, III: 97.

Taniguchi, I., Iseki, M., Eto, T., Toyosawa, K., Yamaguchi, H. and Yasukouchi, K. , 1984, The Effect of pH on the Temperature Dependence of the Redox Potential of Horse Heart Cytochrome c at a Bis(4-pyridyl)disultide-modified Gold Electrode, Bioelectrochem. Bioenerg., 13:373.

Theorell, H. and Akesson, A., 1941, Studies on Cytochrome c. I. Electrophoretic Purification of Cytochrome c and its Amino Acid Composition, J. Am. Chem. Soc., 63:1804.

Vanbuuren, K.J.H., Van Gelder, B.F., Wilting, J. and Braams, R., 1974, Biochemical and Biophysical Studies on Cytochrome c Oxidase. 14. Reaction with Cytochrome c as Studied by Pluse - radiolysis, Biochim. Biophys. Acta, 333:421.

Yeh, P. and Kuwana, T., 1977, Reversible Electrode Reaction of Cytochrome c, Chem. Lett., 1145.

USE OF PROMOTER MODIFIED ELECTRODES FOR

HEME PROTEIN ELECTROCHEMISTRY

Isao Taniguchi

Department of Applied Chemistry
Kumamoto University
Kurokami, Kumamoto 860 (Japan)

INTRODUCTION

In recent years, electrochemical studies of proteins such as cytochrome c have become possible by conventional electrochemical techniques such as cyclic voltammetry by using several types of electrodes: For example, promoter modified electrodes, of which surfaces are modified by suitable compounds, so-called promoters, for the rapid electron transfer of proteins, have been developed by the author's group (Taniguchi et al, 1982; Taniguchi et al., 1986b; Taniguchi, 1988); some metal oxides such as indium oxide are also useful (Yeh and Kuwana, 1977; Bowden et al, 1982; Harmer and Hill, 1984); also, in some cases, bare metal electrodes without any surface modification are reported to be used (Bowden et al, 1984; Reed and Hawkridge, 1987; Sun et al., 1988). Among these electrodes, promoter modified electrodes are sometimes more convenient to use than other electrodes, because they can be prepared so easily (e.g., just dipping an electrode into a promoter solution), show very reproducible results, and are applicable under various experimental conditions, when we choose a suitable promoter. In the present paper, some typical examples using promoter modified electrodes for cytochrome c electrochemistry are given to demonstrate how these electrodes can be widely used.

EXPERIMENTAL

Cyclic voltammetry was mainly used to examine electrochemical behavior of cytochrome c. Bis(4-pyridyl)disulfide (PySSPy), 6-mercaptopurine (PuSH) and L- and D-cysteine were used as promoters to prepare effective modified electrodes. The electron transfer reaction between cytochrome c and cytochrome c oxidase was estimated from the cathodic catalytic current using the electrochemical model of the terminal of the respiratory chain: promoter modified electrode/ cytochrome c/ cytochrome c oxidase/ dioxygen (Taniguchi, 1987).

Horse heart cytochrome c (Sigma, type VI) was purified by chromatographically using a Whatman CM-32 column at 4 °C, according to the procedure as described elsewhere (Taniguchi et al., 1984a; Bowden et al., 1982) to obtain deamidated, oligomeric and native purified cytochrome c; without specified, cytochrome c used was in the native form.

To obtain amino acid residue modified cytochrome c, cytochrome c was treated with 2,4-dinitrofluorobenzene or 4-chloro-3,5-dinitrobenzoic acid in a phosphate buffer (PB, pH=7.8) solution at room temperature, giving DNP- and CDNP-modified cytochrome c, respectively. After DNP (or CDNP) remained unreacted was removed by gel filtration using a Sephadex G-25 column, the modified cytochrome c was separated into several fractions by cation exchange chromatography using again a Whatman CM-32 column (Hill and Whitford, 1987). The numbers of DNP (or CDNP) immobilized per a cytochrome c molecule were estimated from the difference in absorbance at 360 nm (for DNP) or at 450 nm (for CDNP) between the modified cytochrome c and native one.

RESULTS AND DISCUSSION

Functional electrodes modified with sulfur containing compounds

When a gold electrode was modified with such compounds as those shown in Fig. 1, where the sulfur atom of the modifier would interact strongly with the gold electrode surface, a suitable promoter modified electrode was obtained. The surface profiles of the modified electrodes were estimated to be those as shown schematically in Fig. 1 by using surface enhanced Raman scattering (SERS) technique (Taniguchi et al., 1984b; Taniguchi et al., 1985) as well as electrochemical behavior of cytochrome c on these electrodes (Taniguchi et al., 1988): a PySSPy modified gold (PySSPy-Au) electrode (Fig. 1a), where PySSPy adsorbs at the S atom with the cleaved S-S bond (Taniguchi et al., 1985) works excellently for the rapid electron transfer of cytochrome c, but, unfortunately, this electrode is not useful in acid solutions of pH <5, while a PuSH-Au electrode (Fig. 1b), where PuSH would adsorb in its alkaline form through the S atom, is used even in acid solutions of pH >3, probably because of smaller pKa value of the purine N atom than the pyridine N atom (Taniguchi et al., 1986b; Taniguchi, 1988). L- and D-cysteine also adsorb onto a gold electrode to give promoter modified electrodes (Fig. 1c), which can be used even in acid solutions (Taniguchi, 1988), although the

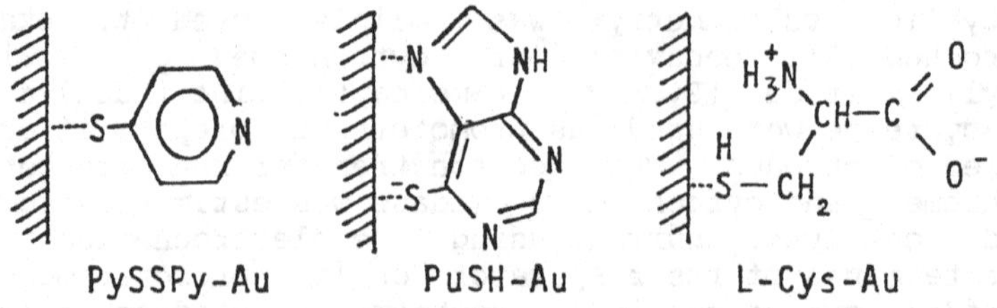

PySSPy-Au   PuSH-Au   L-Cys-Au

Fig. 1.   Possible surface structures of promoter modified electrodes.

durability of these electrodes were less than PySSPy and PuSH modified electrodes. Also, on L- and D-cysteine modified electrodes, electrochemical response of cytochrome c was interfered when poly-L-lysine was added to the solution (Taniguchi, 1988), while no significant influence of poly-L-lysine addition on the cyclic voltammograms of cytochrome c was observed at PySSPy-Au and PuSH-Au electrodes (Taniguchi et al., 1986a; Taniguchi, 1987).

On these electrodes, well-developed voltammograms of cytochrome c under various conditions are easily obtained. The heterogeneous electron transfer rate constants of cytochrome c observed were ca. $5-10 \times 10^{-3}$ cm s^{-1}.

Dependence of E°' of cytochrome c on experimental conditions

Using a promoter modified electrode, a redox wave of cytochrome c under various conditions was obtained: The formal redox potential, E°', estimated from the midpoints of anodic and cathodic peak potentials of cyclic voltammograms changed rather sensitively with changes in experimental conditions such as pH, temperature, type of the buffer solution and ionic strength.

The E°' values shifted toward more negative values in alkaline and acid solutions at a given temperature. Also, in acid and alkaline solutions, the E°' values shifted biphasically toward more negative potentials with increasing temperature (Taniguchi et al., 1984c; Ikeshoji et al., 1989). On the basis of the E°' values, the transition (inflection) points of E°' observed with changes in pH and temperature showed that at least four distinguishable states of ferricytochrome c exist as a function of pH and temperature, which are in good agreement with those suggested by other methods such as spectrophotometry (Theorell and Akesson, 1941) and resonance Raman spectroscopy (Myer et al., 1983); the pH-Temperature diagram for ferricytochrome c with thermodynamic parameters, such as ΔH, ΔS and an apparent protonation number coupled with the electron transfer, of these four states were obtained (Ikeshoji et al., 1989).

Also, influences of solution conditions such as supporting electrolyte added, type of the buffer solution used and ionic strength were clearly observed: For example, by adding 0.1 M NaClO$_4$ instead of 0.1 M NaCl in a phosphate buffer solution, the E°' value of cytochrome c shifted

Table 1. The E°' values of cytochrome c observed at a PuSH-Au electrode at 25 °C in various buffer solutions.

Buffer	HEPES	Tris/cacodylic	B&R	PB
E°' (mV vs. SCE)	41 ± 2	36 ± 1	29 ± 1	25 ± 1

Buffer solutions (pH 7) in ionic strength of 0.1 M with no additional electrolyte was used. HEPES=Good's buffer; B&R=Briton-Robinson buffer; PB=Phosphate buffer.

toward more negative potentials by ca. 8 mV. The decrease in ionic strength from 0.2 M to 0.01 M in a phosphate buffer solution (pH 7) caused a positive shift of the $E°'$ value by ca. 15 mV. The effect of type of the buffer solution was also clearly seen as shown in Table 1, where so-called the non-binding buffer, such as HEPES and Tris-cacodylic acid buffers, gave more positive $E°'$ values than the so-called the binding buffer, such as phosphate buffer, indicating an anion of the binding buffer interacted with positively charged cytochrome c to relatively stabilize the oxidized form more than reduced form of cytochrome c. Similar results to those described above were observed at any promoter modified electrodes used.

Electrochemistry of cytochrome c components

At indium oxide and bare metal electrodes, it has been reported (Bowden et al., 1982; Reed and Hawkridge, 1987) that purification of cytochrome c is essential to obtain the well-developed voltammograms, and the oligomeric cytochrome c, which is one of the components included in the commercially available sample, has been suggested to interfere the rapid electron transfer of native cytochrome c due to its strong adsorption onto these electrodes (Reed and Hawkridge, 1987). The promoter modified electrodes were thus expected to be effective for such a component of cytochrome c, because, when promoter molecules adsorbed so strongly onto the electrode, strong adsorption of a cytochrome c component itself would be excluded, and as was expected, this really happened on promoter modified electrodes. Figure 2 shows the typical voltammograms of the cytochrome c components that obtained in the purification procedure of the commercially available sample: The redox potentials of deamidated (ca. 10 mV vs. SCE) and oligomeric (-29 mV vs. SCE) cytochrome c in a phosphate buffer solution with 0.1 M $NaClO_4$ at 25 °C were more negative than the native one (15 mV vs. SCE). The circular dichroism (CD) spectrum of the deamidated ferricytochrome c was very similar to that of native one, but a little more negative

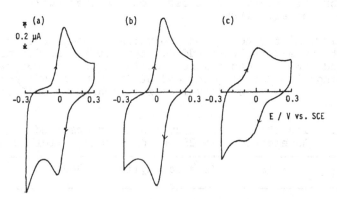

Fig. 2. Typical cyclic voltammograms of three components of ca. 0.2 mM cytochrome c (a: deamidated; b: purified native; c: oligomeric) in a phosphate buffer solution (pH 7) with 0.1 M $NaClO_4$ at 25 °C at a PySSPy-Au electrode. Scan rate used was 20 mV/s.

value of E°' and a somewhat smaller peak current of deamidated cytochrome c than those of native one may be due to a little change in protein structure: The E°' value may be used as a very sensitive measure of a change in the protein structure in some cases. For oligomeric cytochrome c, the E°' value shifted largely toward negative direction and the peak current became about a half of that of native one. The CD spectrum of oligomeric cytochrome c showed a much smaller peak at 220 nm than that of native one, indicating structure of oligomeric cytochrome c is less ordered than the native one, which may induce the more exposed heme edge to the solution. Interestingly, after the temperature of the solution of oligomeric cytochrome c was raised up to 40 °C and returned down to 25 °C, the E°' value (initially, -29 mV at 25 °C) shifted irreversibly to be ca. 8 mV (vs. SCE) with an increase in the peak current. This is probably because at higher temperatures oligomeric cytochrome c irreversibly dissociated to be monomeric one.

These results indicate that electrochemical study of each component of cytochrome c would be carried out by using suitable promoter modified electrodes and also that exclusion of strong adsorption of cytochrome c onto the electrode surface would be one of the most important points to obtain electrochemical response of cytochrome c at an electrode.

Electrochemistry of modified cytochrome c

Use of promoter modified electrodes makes possible to monitor electrochemistry of cytochrome c at the electrode as well as the reaction between cytochrome c and cytochrome c oxidase, which is a physiological redox partner of cytochrome c (Taniguchi, 1987). In the present study, cytochrome c was modified by DNP or CDNP; the numbers of the immobilized DNP (or CDNP) used were less than four, because the modification of amino acid residues of cytochrome c in

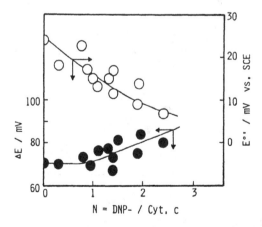

N = DNP- / Cyt. c

Fig. 3.   Changes in peak separation, ΔE, and formal redox potential, E°', as a function of the number of DNP introduced per a molecule of cytochrome c at 25 °C in a phosphate buffer solution with 0.1 M NaCl (pH 7). Scan rate of 20 mV/s was used. A PuSH-Au electrode was used.

large numbers may change the structure of cytochrome c so much. As the number of modified DNP increased, the $E°'$ value of the modified cytochrome c shifted negatively as shown in Fig. 3, indicating that the modified ferricytochrome c was relatively stabilized to the corresponding ferrocytochrome c. This is probably because the modification of cytochrome c with DNP caused a somewhat change in the cytochrome c structure; in fact the modified cytochrome c showed a change in its CD spectrum.

The cathodic current of modified cytochrome c changed gradually with an increase in the number of introduced DNP as shown in Fig. 4, and also the peak separation of the redox wave changed in a similar manner to that of the cathodic current (see Figs. 3 and 4), indicating cytochrome c molecule became inactive at the electrode gradually with the modification. Similar results were again observed when cytochrome c was modified with CDNP, where the negative charge due to the dissociated carboxylic group was introduced. This indicates that the charge of cytochrome c does not affect so much to the electron transfer of cytochrome c at promoter modified electrodes used under the present experimental conditions. Although the interfacial structure of the electrode surface and the mechanism of electron transfer of modified cytochrome c at the electrodes have not yet been clear at present, these results show that promoter modified electrodes are applicable to electrochemical studies of modified cytochrome c as well.

Using the electrochemical model of the terminal of the respiratory chain (Taniguchi, 1987), the electron transfer reaction between modified cytochrome c and native cytochrome

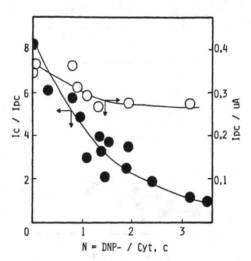

Fig. 4.   Changes in catalytic current, i_c, as in i_c/i_{pc} and peak current, i_{pc}, as a function of the number of DNP introduced per a molecule of cytochrome c at 25 °C in a phosphate buffer solution with 0.1 M NaCl (pH 7). Scan rates used were: 1 mV/s for solid circles and 20 mV/s for open circles.

c oxidase (from bovine heart, Sigma) in the solution was demonstrated to be inhibited significantly even at the beginning of the modification of amino acid residues (mainly lysine, but not specified which amino acid was modified) of cytochrome c: The catalytic current decreased more significantly than the cathodic current of cytochrome c at an electrode when modified cytochrome c was used as shown in Fig. 4 for the case of DNP modified cytochrome c. Similar results were also obtained when CDNP modified cytochrome c was used. These results indicate that molecular recognition of cytochrome c is more rigorous for the protein-protein reaction with cytochrome c oxidase than for the electron transfer reaction at an electrode.

Optically transparent thin layer cell with promoter modified electrodes

Optically transparent thin layer electrode (OTTLE) cells have been widely used in bioelectrochemical studies (Heineman et al., 1984). A usual OTTLE cell with a mediator is useful to measure spectra in the visible region. However, mediator itself has usually large absorption peaks in the UV region, and this often makes difficult to obtain UV spectra of proteins. When a promoter modified electrode with no mediator in the solution was used, no influence due to either mediator or promoter immobilized on the electrode surface was observed, because promoter is only on the

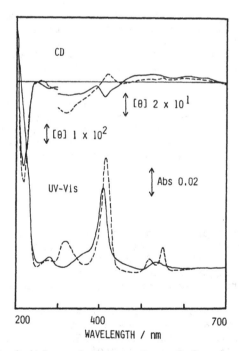

Fig. 5.   UV-visible and CD spectra of ferri- (——) and ferro- (---) cytochrome c (ca. 40 uM) in Briton-Robinson buffer solution (pH 7) with 0.1 M NaCl. An OTTLE cell with a PuSH-Au electrode was used.

electrode surface in about monolayer. Thus, pure spectra of proteins can be obtained using a high quality quartz cell even in the UV region, where sometimes large absorption peaks, like the Soret band, of a protein are located. In the present study, a simple OTTLE cell, in which a promoter modified gold mesh (333 or 200 lines/inch) working electrode was sandwiched with two quartz plates, was used. Using this cell, in-situ spectra of cytochrome c in the whole UV-visible region were obtained as shown in Fig. 5. Peaks in the UV region provide more sensitive measurements for cytochrome c because the absorption coefficients of the UV peaks and their differences between oxidized and reduced forms are much larger than those in the visible region (see Fig. 5), meaning the concentration of protein to be measured can be largely reduced; this is often very advantageous for spectroelectrochemical studies of proteins. The E°' values estimated from the change in absorbance at the Soret band by using the present OTTLE cell with promoter modified electrode under various experimental conditions for a 20 μM cytochrome c solution were in good agreement with those obtained by cyclic voltammetry using promoter modified electrodes for ca. 200 μM cytochrome c solutions.

The CD spectra of cytochrome c were also obtained by using the present OTTLE cell (see Fig. 5), and in-situ CD spectra of cytochrome c in the whole UV-visible region as a function of the applied potential were clearly monitored. This is a very convenient and rapid method to obtain the CD spectrum of reduced form of cytochrome c, because no additional reductant is required: To obtain spectral data, since a usual reductant used to reduce ferricytochrome c to ferrocytochrome c has absorption and/or CD signals in the UV region, the following time consuming purification procedure to remove the reductant remained unreacted is usually required. By using this OTTLE cell, CD measurements under various conditions can be carried out with no significant difficulty.

In conclusion, promoter modified electrodes are applicable in various ways for electrochemical and spectroelectrochemical studies of proteins, and such electrodes make conventional electrochemical techniques very useful as one of the most powerful methods for protein chemistry.

ACKNOWLEDGEMENTS

The author appreciates Professor Fred M. Hawkridge of Virginia Commonwealth University for his helpful suggestions for the present work, and Messrs. Kenji Yoshida, Masato Tominaga and Hirofumi Kurihara, graduate students of the author's research group, for their experimental assistance. The financial support of this work by a Grant-in-Aid for Scientific Research on Priority Areas of "Macromolecular" (#01612003), by a Grant-in-Aid for Co-operative Research (#62303011), and by the CIBA-GEIGY Foundation for the Promotion of Science is also gratefully acknowledged.

REFERENCES

Bowden, E.F., Hawkridge, F.M., Chlebowski, J.F., Bancroft, E.E., Thorpe, C., and Blount, H.N., 1982, Cyclic Voltammetry and Derivative Cyclic Voltabsorptometry of Purified Horse Heart Cytochrome c at Tin-Doped Indium Oxide Optically Transparent Electrodes, J. Am. Chem. Soc., 104:7461.

Bowden, E.F., Hawkridge, F.M., and Blount, H.N., 1984, Interfacial Electrochemistry of Cytochrome c at Tin Oxide, Indium Oxide, Gold and Platinum Electrodes, J. Electroanal. Chem., 161:315.

Harmer, M.A. and Hill, H.A.O., 1984, The Direct Electrochemistry of Horse Heart Cytochrome c Ferredoxin and Rubredoxin at Ruthenium Dioxide and Iridium Dioxide Electrodes, J. Electroanal. Chem., 170:369.

Heineman, W.R., Hawkridge, F.M., and Blount, H.N., 1984, Spectroelectrochemistry at Optically Transparent Electrodes. II. Electrodes Under Thin-Layer and Semi-infinite Diffusion Conditions and Indirect Coulometric Titrations, in: "Electroanalytical Chemistry," Vol. 13, A.J. Bard, ed., Marcel Dekker, New York, 1.

Hill, H.A.O. and Whitford, D., 1987, Direct Electrochemistry of Native and 4-Chloro-3,5 dinitrophenyl (CDNP)-substituted Cytochrome c at Surface-modified Gold and Pyrolytic Graphite Electrodes, J. Electroanal. Chem., 235:153.

Ikeshoji, T. Taniguchi, I., and Hawkridge, F.M., 1989, Electrochemically Distinguishable States of Ferricytochrome c and Their Transition with Changes in Temperature and pH, J. Electroanal. Chem., in press.

Myer, Y.P., Srivastava, R.B., Kumar, S., and Raghavendra, K., 1983, State of Heme in Heme c Systems: Cytochrome c and Heme c Models, J. Protein Chem., 2:13.

Reed, D.E. and Hawkridge, F.M., 1987, Direct Electron Transfer Reactions of Cytochrome c at Silver Electrodes, Anal. Chem., 59:2334.

Sun, S. Reed, D.E., Cullison, J.K., Rickerd, L.H., and Hawkridge, F.M., 1988, Electron Transfer Reactions of Cytochrome c at Metal Electrodes, Makrochim. Acta, III:97.

Taniguchi, I., Toyosawa, K., Yamaguchi, H., and Yasukouchi, K., 1982, Reversible Electrochemical Reduction and Oxidation of Cytochrome c at a Bis(4-pyridyl)disulfide Modified Gold Electrode, J. Chem. Soc. Chem. Commun., 1032; Voltammetric Response of Horse Heart Cytochrome c at a Gold Electrode in the Presence of Sulfur Bridged Bipyridines, J. Electroanal. Chem., 140:187.

Taniguchi, I., Iseki, M., Toyosawa, K., Yamaguchi, H., and Yasukouchi, K., 1984a, Purines as New Promoters for the Voltammetric Response of Horse Heart Cytochrome c at a Gold Electrode, J. Electroanal. Chem., 164:385.

Taniguchi, I., Iseki, M., Yamaguchi, H., and
    Yasukouchi, K., 1984b, Surface Enhanced Raman
    Scattering Study of Horse Heart Cytochrome c at a
    Silver Electrode in the Presence of Bis(4-
    pyridyl)disulfide and Purine, J. Electroanal.
    Chem., 175:341.
Taniguchi, I., Iseki, M., Eto, T., Toyosawa K.,
    Yamaguchi, H., and Yasukouchi, K., 1984,
    The Effect of pH on the Temperature Dependence of
    the Redox Potential of Horse Heart Cytochrome c at
    a Bis(4-pyridyl)disulfide-modified Gold Electrode,
    Bioelectrochem. Bioenerg., 13:373.
Taniguchi, I., Iseki, M., Yamaguchi, H., and
    Yasukouchi, K., 1985, Surface Enhanced Raman
    Scattering from Bis(4-pyridyl)disulfide and 4,4'-
    Bipyridine Modified Gold Electrodes, J.
    Electroanal. Chem., 186:299.
Taniguchi, I., Funatsu,T., Umekita, K., Yamaguchi, H.,
    Yasukouchi, K., 1986a, Effect of Poly-L-lysine
    Addition on the Redox Behavior of Horse Heart
    Cytochrome c at Functional Electrodes,
    J. Electroanal. Chem., 199:455.
Taniguchi, I., Higo, N., Umekita, K., and Yasukouchi,
    K., 1986, Electrochemical Behavior of Horse Heart
    Cytochrome c in Acid Solutions Using a 6-
    Mercaptopurine Modified Gold Electrode,
    J. Electroanal. Chem., 206:341.
Taniguchi, I., 1987, Promoter Modified Electrodes as
    Probes for Analysis of the Interaction Between
    Cytochrome c and Cytochrome c Oxidase,in:
    "Proceedings of the 2nd Japan-Korea Joint Symposium
    on Analytical Chemistry," Y. Ohkura ed., Society of
    Analytical Chemistry, Japan, Tokyo, 2:55.
Taniguchi, I., 1988, Interfacial Electrochemistry of
    Promoter Modified Electrodes for the Rapid
    Electron-Transfer of Cytochrome c, in:"Redox
    Chemistry and Interfacial Behavior of Biological
    Molecules," G. Dryhurst and K. Niki eds., Plenum,
    New York, 113.
Theorell, H. and Akesson, A., 1941, Studies on
    Cytochrome c. I. Electrophoretic Purification of
    Cytochrome c and its Amino Acid Composition,
    J. Am. Chem. Soc., 63:1804.
Yeh, P. and Kuwana, T., 1977, Reversible Electrode
    Reaction of Cytochrome c, Chem. Lett., 1145.

ION AND ELECTRON TRANSPORT PROPERTIES OF

BIOLOGICAL AND ARTIFICIAL MEMBRANES

INTERACTION OF COMBINED STATIC AND EXTREMELY LOW FREQUENCY MAGNETIC FIELDS WITH CALCIUM ION TRANSPORT IN NORMAL AND TRANSFORMED HUMAN LYMPHOCYTES AND RAT THYMIC CELLS

Bertil R.R. Persson[1], Magnus Lindvall[2], Lars Malmgren[1], and Leif G. Salford[3]

Departments of Radiation Physics[1], Tumour Immunology[2] and Neurosurgery[3]
Lund University, S-221 85 Lund Sweden

INTRODUCTION

Liboff (1985) proposed a model which might explain the interaction of undulating electromagnetic fields with ionic species at geomagnetic flux densities. A charged ion moving in a plane normal to the Earth's magnetic field will experience a radial force (Lorentz's force):

$$q \cdot v \cdot B = \frac{m \cdot v^2}{R}$$

where: q is the charge of the ion,
           m the mass of the ion,
           v its velocity,
           R the radius of the curvature of the path.

Because of this force, the ion will execute a circular or a helical path. The velocity can be simply expressed as the product of the frequency of rotation f, and the pathlength, leading to a unique frequency corresponding to the the geomagnetic field B:

$$f = \frac{q \cdot B}{2\pi \cdot m}$$

This is the same condition for accelerating charged particles in a cyclotron and the phenomenon is therefore named "Ion Cyclotron Resonance".

The Earth's geomagnetic field varies from about 70 micro-tesla (μT) at the poles to 25 micro-tesla (μT) at the geomagnetic equator and averages about 50 micro-tesla (μT) at mid latitues. For such fields, frequencies in the range of 10-100 Hz correspond approximately to charge/mass-ratios of 0.01 to 0.1 electronic charge per atomic mass unit indicating that biologically important ions, heavier than protons but lighter than enzymes and proteins appear to have geomagnetic cyclotron resonance frequencies.

The "Ion Cyclotron Resonance Hypothesis" was explored by Liboff et al., (1987) in an experiment of incorporation of calcium-45 ($^{45}Ca^{2+}$) in mixed human lymphocytes. The geomagnetic horizontal field component was adjusted to 21 micro-tesla (μT). The experiment was first performed at an amplitude of 150 micro-tesla(μT) and a sharp minimum was obtained at the frequency of 14.3 Hz which corresponds to the ion-cyclotron-frequency at 21 micro-tesla (μT). The experiment was then repeated at an amplitude of 21 micro-tesla (μT) and now a sharp narrow maximum occured at 14.3 Hz.

We found this remarkable interaction mechanism very interesting and have since March 1988 set up experiments to explore the presence of "Ion Cyclotron Resonance" for calcium ions in human normal and transformed lymphocytes and in rat thymocytes.

MATERIALS AND METODS

Magnet coils. The apparatus used for exposure consisted of two pairs of Helmholtz-coils placed orthogonally to each other. The axis of the vertical coils was oriented in the North-South direction and the axis of the other pair in the horizontal plane. The diameter of the coils was 230 mm winded with 100 turns of 1.5 mm diameter enameled copper wire. The horizontal coils were coupled in series at a distance of 230 mm and used for compensating the vertical component of the Earth's geomagnetic field. A fluxgate magnetometer was used as an indicator and was balanced to zero field in the vertical direction. The vertical component of the geomagnetic field was balanced to 21.0 μT using the bias voltage from the pulse generator. A sinusoidal time varying field with adjustable frequency and amplitude was also applied to the vertical coils. The frequency was monitored by using a frequence meter. The amplitude of the undulating field was checked at the center of the coils using a pickup coil. The induced electromotoric force was recorded on the oscilloscope.

Calcium tracer and radioactivity measurements. Radioactive calcium-45 with a radioactivity concentration of 370 MBq/ml and low stable calcium concentration was used. About 0.2 μl was added to the stock solution of mixed media used in each experimental series. The cells were collected on filter Whatman GFA and washed 7 times with inactive media of the same composition as the one in which the cells were exposed. The filters were mounted on the glasses of diaframes and slided in a reproducible position under an endwindow GM-tube counter.

Cells and media were specially prepared as described for each experiment.

EXPERIMENTS AND RESULTS

Uptake of ^{45}Ca in normal human lymphocytes. Normal human lymphocytes were prepared to $6 \cdot 10^6$ cells per ml in calcium free buffer solution (Hank). Calcium-45 tracer solution was prepared in 0.02 mM Ca to an activity concentration of 40 kBq/ml. Triplicate control and experimental round bottomed microtiter plates were prepared immediately prior to magnetic field exposure by combining 50 μl ^{45}Ca-tracer solution and

50 μl cell suspension. The horizontal magnetic field was adjusted to zero and the vertical field to 21-22 μT. The applied oscillating vertical field had an amplitude of 29.7 μT peak to peak. The frequency of the vertical magnetic field was 14.27 Hz. The experiment was first performed at 60 min exposure-time and the repeted several months later at 15 and 60 min exposure-time.

The ratio of activity measurements of exposed and control cells was 1.5 ± 0.4 (SD) in the first experiment after 60 min exposure. In the second experiment the corresponding ratio at 15 min was 1.5 ± 0.6 and at 60 min 0.6 ± 0.2. Thus these seems to be no significant difference in calcium uptake between exposed and control cells in those two experiments.

Study of the ELF frequency on the uptake of ^{45}Ca in human lymphocytes and transformed lymphoma cells. Normal human lymphocytes and transformed lymphoma (YAG) cells were adjusted to $2 \cdot 10^6$ cells per ml in calcium free buffer solution (Hank). Calcium-45 tracer solution was prepared in 0.23 mM Ca to an activity concentration of 74 kBq/ml. Triplicate control and experimental round bottomed microtiter plates were prepared immediately prior to magnetic field exposure by combining 50 μl ^{45}Ca-tracer solution and 50 μl cell suspension. The horizontal magnetic field was adjusted to zero and the vertical field to 21-22 μT. The applied oscillating vertical field had an amplitude of 29.7 μT peak to peak and the exposure was performed at the three different frequencies: 14.27, 14.50 and 15.00 Hz.

The quotient of calcium-45 activity was measured after 60 min in exposed cells and control cells are given in Table 1. The frequency was varied at a constant vertical magnetic flux density of 21 μT.

Table 1. The ratio of calcium-45 activity uptake in
exposed cells and control

Frequency Hz	Normal Lymphocytes ratio to control	Lymphoma YAG ratio to control
14.27	0.94	0.83
14.50	1.32	1.25
15.00	0.80	1.10

Uptake of and efflux of ^{45}Ca in rat thymus lymphocytes. Lymphocytes from thymus of Wistar W/FU rats were prepared either in RPMI (10% fetal calf serum) or in buffer solution to $5 \cdot 10^6$ cells per ml with calcium concentration of 2 mM. Calcium-45 tracer solution was prepared either in normal medium or buffer solution with calcium concentration of 2 mM and with the activity concentration of 150 kBq/ml. Triplicate control and experimental vials with 0.5 ml cell suspension and 0.5 ml calcium-45 solution were prepared immediately prior to magnetic field exposure. The vials were exposed in the center of the combined pairs of coils where the horizontal magnetic field was adjusted to zero and the vertical field to 21-22 μT. The applied oscillating vertical field had an amplitude of 29.7 μT peak to peak. The frequency of the vertical magnetic field was 14.27 Hz. Aliquots were taken from the vials every 15 minutes. The cells were immediately separated and measured for radioactivity. After 1 hour in the magnet the cells were spun down and washed with activity free medium 7 times. The same procedure was performed with the control. Then the exposure was continued in the magnet for another hour and samples were taken every 15 minutes.

The ^{45}Ca-activity was measured in each triplicate preparation and the mean values were calculated. In Table 2 are given the ratios between exposed/control rat thymus cells in plasma and buffer solution respectively. Corresponding results for human lymphocytes in buffer soultion was a ratio of 1.0 ± 0.3 at 15 min and 2.7 ± 0.6 at 60 min.

Table 2. Ratio of ^{45}Ca-activity between exposed and control cells incubated in either buffer or plasma·

Time of exposure min	Cells in Buffer	Cells in Plasma
15	2.01	1.03
30	1.12	0.66
45	0.88	1.71
60	0.69	0.97
washed		
15	0.30	0.38
30	0.18	0.69
45	0.49	0.56
60	0.70	1.16

Uptake of ^{45}Ca in human lymphocytes and leukemia cells. Premyelocytic leukemia cells (U 937 from Uppsala) and fresh human lymphocytes were prepared in RPMI (10% fetal calf serum from Flow, Edinburgh, Scotland). The cells were spun down at 12 000 rpm for 5 minutes and washed twice in Ca-free trypsin buffer solution, PBS without Ca and Mg. Fifty (50) μl of the cell suspension ($2 \cdot 10^5$ cells) was mixed with 50 μl ^{45}Ca = "carrier free" in micro-titer plates (NUNC Odense Denmark). Triplicate control cells and experimental cells were prepared. The plate was exposed in the center of the combined Helmholtz coil arrangement. The vertical component of the geomagnetic field was adjusted to zero and the horizontal to 21 - 20 μT. The applied oscillating field had an amplitude of 29.7 μT peak to peak and a frequency of 14.27 Hz. Exposure times of leukemia cells were 5, 10, 15, 30 and 60 minutes and for human lymphocytes 15 and 60 min. After exposure the exposed cells and their controls were separated from the buffer and collected on Millipore filter type AP-20 (Cat. No. AP 200 2200), Millipore, Ireland. One fraction of the cells exposed for 60 minutes was washed cooled on ice to 4 °C and washed in ice cooled calcium-free buffer. It was then exposed for another 15 minutes in the magnet.

The radioactivity was recorded in each triplicate preparation of leukemia cells and the mean value was calculated. The quotient between exposed and control samples as given in Table 3 varied from 0.89 - 3.24. There was a significant difference at 15 minutes with a quotient of 3.24 ±

1.00 (SD). For human lymphocytes the quotient between magnet exposed and controls were 0.83 ± 0.45 at 15 minutes and 0.54 ± 0.62 at 60 minutes. We can observe the tendency of a higher quotient at 15 minutes although the values are lower than for leukemia cells where the ratios were 3.2 at 15 min and 1.32 at 60 min.

Efflux of ^{45}Ca in leukemia cells. Premyelocytic leukemia cells (U 937 from Uppsala) were prepared in RPMI (10% fetal calf serum from Flow, Edinburgh, Scotland). The cells were incubated during 1 h in RPMI medium with ^{45}Ca added. The cells were washed twice in cooled (4 °C) Ca- free medium and exposed in the coil arrangement as described previously for 15 minutes. After exposure the cells were filtered and measured as above.

A slight difference in the efflux of ^{45}Ca was found. But the quotient between exposed and control was 1.04 ± 0.21 (SD) as seen in Table 3.

Table 3. Ratio of ^{45}Ca-activity between exposed and control cells incubated in Ca-free medium

Time of exposure min	Leukemia Cells exp./contr. Ave. + S.D.	Human lymphocytes exp./contr. Ave. + S.D.
5	1.24 ± 0.26	
10	1.18 ± 0.32	
15	3.24 ± 1.00	0.83 ± 0.45
30	0.89 ± 0.39	
60	1.32 ± 0.22	0.54 ± 0.62
washed		
15	1.04 ± 0.18	

Temperature dependence of uptake of ^{45}Ca in leukemia cells. Premyelocytic leukemia cells (U 937 from Uppsala) were prepared in MI (10 % fetal calf serum from Flow, Edinburgh, Scotland). The cells were spun down at 12 000 rpm for 5 minutes and washed twice in Ca-free trypsin buffer solution , PBS without Ca and Mg. Fifty (50) μl of cell suspension ($2 \cdot 10^5$) was mixed with 50 μl ^{45}Ca "carrier free" solution in small conical vials. The exposure took place for 30 min at the temperatures 4, 21, 37 and 43 °C in a specially constructed waterbath placed within the coil arrangement.

The results shown in Table 4 indicate that the quotient between exposed and control was less than 1 but with no significance.

Table 4. The ratio of the ^{45}Ca-activity
at various temperatures

Temperature of exposure °C	Cells in Ca-free medium Ave. ± S.D.
4	1.01 ± 0.34
21	0.69 ± 0.22
37	0.90 ± 0.22
43	0.82 ± 0.38

CONCLUSIONS

The experimental method of studying influx and efflux of ^{45}Ca is not yet sufficient to verify the presence of the cyclotron resonance effect!

According to our experiences further studies on this subject should be performed on the following premises.

- Premyelocytic cells should be used!
- The temperature should be kept at 37 °C.
- The frequency amplitude of the oscillating field should be varied in order to find the resonance conditions.
- The Ca-concentration dependence should be carefully verified.
- Biological transformations should be studied.

REFERENCES

Liboff, A. R., 1985, Ion Cyclotron Resonance in Living Cell. Journal of Biological Physics. 13:99-102.
Liboff, A. R., Rozek, R. J., Sherman, M. L., McLeod, B. R., and Smith, S. D., 1987, Ca(2+)-45 Cyclotron Resonance in Human Lymphocytes. Journal of Bioelectricity, 6(1):13-22.

Acknowledgement. The help from Kjell Åke Carlsson with construction and building of the coil-arrangement and experimental assistance is gratefully acknowledged.

THEORETICAL MODELING OF THE ETHANOL INDUCED CHANGES IN THE LATERAL INTERACTIONS OF THE IONIC HEAD GROUPS OF PHOSPHATIDYLCHOLINE AT THE WATER-LIPID INTERFACE

George P. Kreishman

Department of Chemistry
University of Cincinnati
Cincinnati, OH 45221-0172 U.S.A.

INTRODUCTION

Because of the direct relationship between the efficacy of alcohols and other anesthetics with their oil/water partition coefficients, most research concerning the mechanism of action of ethanol has focused on the disruptive effects of the alcohol on the molecular order in the interior of neuronal and model membrane systems. The observed changes are, however, small and alone cannot explain the total pharmacological actions of alcohols[1]. Recent studies have shown that ethanol and other anesthetics not only interact with the interior of the membrane but also interact with the surface domain[2-6].

Ethanol bound to the surface and interior domains of dipalmitoylphospatidylcholine (DPPC) liposomes and rat synaptic plasma membranes have different exchange lifetimes in relation to the deuterium NMR time scale. The two binding domains can be differentiated and the partition coefficients to each can be determined[2,3]. Whereas the binding constant of ethanol to the interior was independent of concentration, the binding constant to the surface increased with increasing concentration in an apparent cooperative manner. The effects of the bound ethanol on the molecular order of the two domains are different.[7,8] The binding of ethanol to the interior has disordering effects on the fatty acid chains. In contrast, surface binding of the ethanol has an ordering effect on the choline head groups in pure DPPC liposomes at low concentrations of ethanol ($<$ 1%). Under conditions of high ethanol concentration used in fluorescence studies of lipids, ethanol has a disordering effect on the surface domain[8]. At very high concentrations of ethanol ($>$ 7% v/v), DPPC forms the interdigitated gel state[9]. This state can be induced if there is sufficient repulsion between the head groups and if favorable solvation between the head groups can be achieved. Phosphatidylethanolamine (PE) lipids do not form the interdigitated state[10]. This difference in the properties of the two lipids is a surface phenomenum and is solely due to the difference in the hydrophobicity of the head groups.

Recently, Dill and Stigter have developed a theoretical model that successfully predicts the lateral repulsive forces between the head groups of DPPC at the lipid-water interface[1]. In the current study, this theoretical approach has been modified to incorporate the binding of ethanol to the surface and interior domains. The predicted changes in the repulsive forces of the choline head groups of DPPC bilayers are compared to the experimental observations.

The bilayer is approximated by a two phase water-hexane system. The zwitterionic choline head group of DPPC is placed with the phosphate group positioned at the interface and the dipole between the charges is free to assume any position relative to the interface. The head group orientation is then simply given by the vertical displacement (Z) of the N^+ from the interface. The equilibrium position of the head is determined by the balance of two opposing forces: (1) the attraction of the N^+ charge to the water phase due to the Born electrostatic free enegy changes and (2) the attraction of the head group to the oil phase due to the hydrophobic interaction.

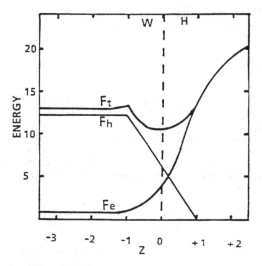

Figure 1. Free energy functions of a charged hydrophobic sphere as a function of distance, Z, from the water/heptane interface, Z is expressed in units of the radius of the sphere (3.5 Å) and energies are in arbitrary units. Fe is the electrostatic free energy, F_h is the hydrophobic energy, and F_t is the total free energy. The energies were calculated using the equations in ref 11. The dashed line is the boundary between the hexane and water phases.

The electrical free energy, F_e, is approximated as a continuous function across a water/hexane interface and increases with the penetration of the head group into the hexane layer (Figure 1). The free energy from the hydrophobic interaction decreases in a linear manner across the interface and reflects the proportional loss of the water of hydration upon penetration of the head group into the hexane layer (Figure 1). The equilibrium position of the head group occurs at the minimum of the sum of the two forces. For DPPC, this occurs with the dipole virtually parallel to the interface (Figure 1). At any temperature, the head group is not static but can oscillate around this equilibrium position. As predicted by this model, the lateral repulsive forces betwen the head groups arise predominantly from the movement of the head into the low dielectric lipid phase. The predicted temperature dependence of these disruptive forces is in excellent agreement with experimental results for DPPC.

MODIFICATIONS TO THE THEORY

Since it has been previously demonstrated that ethanol exists in three states in the presence of DPPC, (1) in the bulk solvent, () as "solvation ethanol" at the surface, and (3) in the lipid phase, the water/hexane system of the model is modified to include a solvation region (Figure 2). The depth and position of the region of solvation are arbitrary. The measured partition coefficients for interior binding and surface binding, K_1 and K_2, respectively,[2] are used to calculate the number of moles of ethanol in each of the zones. These are given by

$$n_1 = K_1 \times n_t/(1 + K_1)$$

where n_1 is the number of moles of ethanol in the lipid phase, K_1 is the partition coefficient for the lipid phase and n_t is the total number of moles of ethanol. The number of moles of ethanol at the surface, n_2, is given by an analogous equation. The ethanol in the water phase, n_3, is given by

$$n_3 = n_t - n_1 - n_2$$

The mole fractions in the bulk and lipid phases are calculated in the normal manner. The mole fraction at the interface is approximated as the fraction of the moles bound at a given concentration relative to the number of moles when the interface is saturated with ethanol. The average dielectric constant is estimated in terms of the partial molar contribution of each solvent in the bulk and surface phases. To account for the large differences in molecular weight between ethanol and DPPC, the dielectric constant is estimated in terms of the partial molar

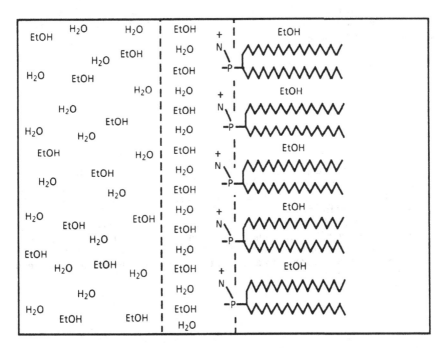

Figure 2. Graphical representation of the three regions of the interface, namely the bulk phase, the solvation region and the lipid phase.

volume contribution of each. The quadratic interpolation function of Dill and Stigter is used to estimate the electrical free energy change across each interface using the appropriate average dielectric constant of each phase. The magnitude of the hydrophobic energy is assumed to be unaffected by the presence of low concentrations of ethanol.

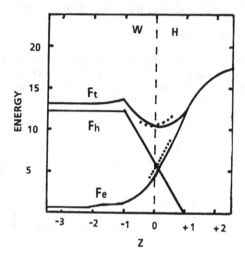

Figure 3. Free energy functions as described in Figure 2 for a 0.2% v/v solution of ethanol. The solvation region is from -1 to -2. Dotted lines are the energy functions in the absence of ethanol.

RESULTS AND DISCUSSION

The energy calculations for 0.2% and 1.0% ethanol solutions (v/v) are shown in Figures 3 and 4, respectively. As the average dielectric of the lipid phase increases and solvation region decreases with increasing ethanol concentration, the increase in the electrostatic energy of the head group in the lipid phase is attenuated. As a result, the equilibrium position of the head group moves further into the lipid bilayer with increasing ethanol concentration. Since the changes in the strength of the hydrophobic interaction with increasing ethanol solvation of the interface have not been taken into account, the degree of penetration of the head group is probably overestimated.

Strong lateral repulsive forces between head groups are generated from thermally induced vibrations in and out of the lipid phase. Because of the high dielectric constant toward the bulk water phase, the repulsive forces are relatively small if the head group protrudes above the interface. Toward the lipid phase, even small changes in the distance from the interface result in a large decrease in the dielectric constant between head groups and large repulsive forces are generated. When increasing ethanol concentration, the equilibrium position of the head group moves into the lipid phase but the dielectric constant of the lipid phase in that region increases significantly. Although the region over which the head group fluctuates is quite similar to that in the absence of ethanol, the repulsive forces will be less than in the absence of ethanol because of the decrease in the dielectric constant. As the equilibrium position of the head group moves in with increasing ethanol concentration, the dielectric constant of the overall lipid phase increases. However, the region over which the fluctuations occur has a dielectric constant which is lower than the corresponding region at the low concentrations of ethanol or in the absence of ethanol. Therefore, after an initial attenuation of the repulsive forces with added ethanol, the repulsive forces increase. At very high concentrations of ethanol, the repulsive forces between the head groups could become large enough such that the head groups move apart and the interdigitated state will form. For DPPC, this occurs when ethanol concentrations exceed 7% v/v.

These qualitative predictions are in agreement with the observed changes in molecular order of the surface domain in liposomes with increasing ethanol concentration. Although this initial theoretical treatment is qualitative and

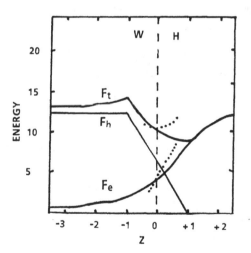

Figure 4. Free energy functions as described in Figure 1 for a 1.0% v/v solution of ethanol. The solvation region is from -1 to -2. Dotted lines are the energy functions in the absence of ethanol.

utilizes many assumptions, the possible role of subtle changes in the composition of the water-lipid interface in the determining of gross structural properties of the lipid bilayer is demonstrated and warrants further attention.

Acknowledgement: This work was supported by the National Institute of Alcohol Abuse and Alcoholism (AA 07055).

References

1. E. Rubin, "Alcohol and the Cell", Anal. N.Y. Acad. Sci., New York (1987).

2. G. P. Kreishman, C. Graham-Brittain, and R. J. Hitzemann, Determination of Ethanol Partition Coefficients to the Interior and the Surface of Dipalmityl-Phosphatidyl Choline Liposomes Using Deuterium Nuclear Magnetic Resonance Spectroscopy, Biochem. Biophys. Res. Commun. 130:301 (1985).

3. R. J. Hitzemann, H. E. Schueler, C. Grahama-Brittain and G. P. Kreishman, Ethanol-Induced Changes in Neuronal Membrane Order; An NMR Study, Biochem. Biophys. Acta 859:189 (1986).

4. J. Figueirinhas and P. W. Westerman, Direct Observation of Ethanol Binding to Phospholipid Bilayers, A ^{2}H NMR Study, Biophys. J., 55:324a (1989).

5. L. R. De Young and K. A. Dill, Solute Partitioning in Lipid Bilayer Membranes, Biochemistry 27:5281 (1988).

6. A. S. Avers, B. A. Berkowitz and D. A. d'Avignon, Correlation between the Anaesthetic Effect of Halothane and Saturable Binding in Brain, Nature 328:157 (1987)

7. R. J. Hitzemann, C. Graham-Brittain, H. E. Schueler and G. P. Kreishman, Ordering/Disordering Effects of Ethanol in Dipalmitoyl Phophatidyl Choline Liposomes, Anal. N.Y. Acad. Sci. 492:142 (1986).

8. R. A. Harris, L. M. Zaccaro, S. McQuilkin and A. McClard, Effects of Ethanol and Calcium on Lipid Order of Membranes from Mice Selected for Genetic Differences in Ethanol Intoxication, Alcohol 5:251 (1988).

9. E. Rowe, Lipid Chain Length and Temperature Dependence of Ethanol-Phosphatidylcholine Interactions, Biochemistry 22:3299 (1983).

10. E. S. Rowe, Thermodynamic Reversibility of Phase Transitions, Specific effects of alchols on phosphatidyl cholines, Biochim. Biophys. Acta 813:321 (1985).

11. K. A. Dill and D. Stigter, Lateral Interactions among Phosphatidylcholine and Phosphatidylethanolamine Head Groups in Phospholipid Monolayers and Bilayers, Bochemistry, 27:3446 (1988).

THE ELECTRONIC PROPERTIES OF DYSTROPHIC
MUSCLE MEMBRANE SYSTEMS

Milton J. Allen and Gwendolyn Geffert

Biophysical Laboratory, Department of Chemistry
Virginia Commonwealth University
Richmond, Virginia 23284

INTRODUCTION

The genetic differences between 'normal' and dystrophic muscula-ture have been well documented in recent years. In order to further elucidate these diverse characteristics, the electronic properties of muscle tissue of dystrophic mice and their phenotypically normal litter mates were examined. Utilizing various electrical techniques it was found that the electron:hole (electron receptor sites) ratio, Arrhenius plots of conductance vs $1/T$ and the activation energies obtained there-from, and the effects of pressure on the muscle membranes demonstrat-ed significant differences.

EXPERIMENTAL

Animals

The muscle tissue used in these studies was obtained from the mouse strain, dystrophia muscularis (129 ReJ, Jackson Laboratories). In order to compare the electronic behavior of the homozygote, control samples obtained from phenotypically normal litter mates were also examined. The homozygote is readily distinguished from the normal litter mate by the fact that within a relatively short time after birth (3-4 weeks) a weakness in the hindlimbs becomes apparent. This progresses to a further degeneration of the axial and forelimb muscula-ture. The skeletal muscle exhibits degenerative changes with prolifera-tion of sacrolemmal nuclei and size variation among individual muscle fibers. The diaphragm musculature was chosen because its size per-mitted multiple experiments and microscopic observations indicated minimal structural changes of the dystrophic's diaphragm compared to that of the 'normal' litter mate. The mice used were between the ages of 45-65 days with the phenotypically normal weighing 23-28 g and the dystrophic 12-14 g. In all instances studies on a dystrophic were accompanied by those on a phenotypically normal control. After dissec-tion the entire diaphragm was placed in a Hanks' balanced salt solution (Flow Laboratories, McLean, VA) and equilibrated to the media for a minimum of 6 hr at 4°C. It was found that the tissue maintained its viability as determined by electrical behavior patterns. After 36 hr under these storage conditions a degree of deterioration was observed.

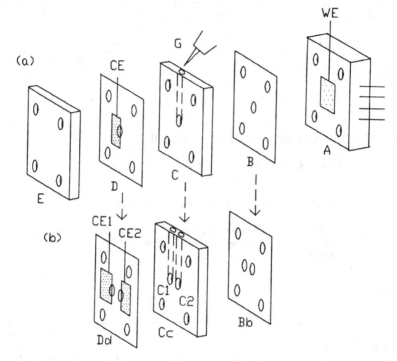

Figure 1a. Single compartment cell. (A) thermoelectric temperature block 5 x 7.5 x 0.6 cm (B,D) 0.1 cm thick silicone rubber gaskets with centrally located 0.3 cm ID window, (C) perspex compartment 5 x 7.5 x 0.4 cm with 0.3 cm ID window and channel for introduction of media and S.C.E. reference electrode salt bridge (G) containing Hanks solution, (E) perspex cover plate 5 x 7.5 x 0.6 cm (WE) membrane 'plated' platinum electrode, (CE) platinum counter electrode.

Figure 1b. Dual Compartment Cell. Bb, Cc, Dd replaced B,C,D, in Figure 1a. (Bd,Dd) contained two parallel 0.3 cm ID windows separated at the narrowest point by 0.1 cm of the silicone rubber gasket material, (Cc) consisted of two matching 0.3 cm ID windows and two channels to compartments C1 and C2.

Generally all experiments reported were performed using membranes which had been equilibrated from 12-16 hrs.

Cells

The electrical cells used in these studies are described in Figure 1a and 1b.

Instrumentation

The schematic (Figure 2) describes the general configuration of the instruments used in these studies. The potentiostat was a high sensitivity, low noise, Wenking Model LB75L and the voltage scan generator a Wenking VSG72. A high impedance Wenking PPT69 was used to determine voltage between the working (WE) and counter electrodes (CE). The high sensitivity voltage and current integrators were supplied by Time Electronics Ltd. A Eurotherm Controller/Programmer Type 818 served to control the temperature of the electronic-temperature block (A in Figure 1) over a temperature range of -20° to 60° ± 0.1°C. This unit was also programmed within the range of 20° to 44°C with the various temperature rise and dwell periods required to obtain Arrhenius plots.

Procedures

Electrokinetic Behavior (EKB). The objective in this procedure was to determine the electron:hole relationship from charge values obtained as a result of superimposing on the biological membrane under voltage-clamp conditions or potentiostasis a positive or negative going voltage pulse. Simultaneously it was also possible to obtain the conductance of the membranes.

Prior to use the platinum electrodes were treated first with 50% aqueous aqua regia (50°C) and then with concentrated nitric acid (50°C), (4 min per solution). After exposure to each of the above the electrodes were rinsed with distilled water and dried.

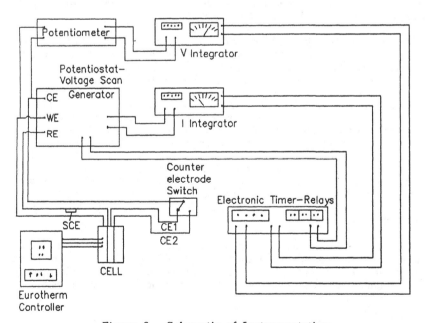

Figure 2. Schematic of Instrumentation

A section of muscle tissue was taken from the whole Hanks equilibrated diaphragm and spread on a microscope slide. Gentle pressure was then applied with another slide for a few seconds. This was sufficient to expand the section to approximately the dimensions it was in situ. The section should be of a size so as to occupy the 3 mm diameter opening with a small overlapping of the edges. With experience it was possible to obtain 4-6 of these sections from a single diaphragm.

After the diaphragm section was 'plated' on the Pt working electrode, it was placed on the temperature control block (A in Figure 1) and aligned so as to mate with the 3 mm window in the silicone rubber gasket (B in Figure 1). The perspex section followed by the second gasket (C and D respectively in Figure 1) was added to the units already in place. Before adding cover plate E (Figure 1) a 4 mm wide Pt counter electrode was inserted so as to cover approximately one-half of the 3 mm window. The total unit was bolted with four 4 mm machine head screws. A torque screwdriver (75G-M) is used so as to obtain a uniform pressure on all sections of the cover plate.

To the channel in C was added 100 μl of the Hanks' (pH 7.4) media and the compartment checked for the absence of bubbles. The cell was then mounted in an isolation cage, all electrical connections made and the Eurotherm controller set to the desired temperature.

In these studies temperatures of 20°C and 37°C were chosen. These temperatures were used because it was found in an earlier study in the course of a series of Arrhenius plots of conductance vs $1/T$, that transitions were noted at about 15°C and 30°C. These values coincide in the first instance with the change of lipidity within the membrane and at the higher temperature in the activation of calcium-ATPase as well as other enzyme systems[1].

Upon equilibration of the cell system for 10 min the potential of the membrane 'plated' working (WE) electrode vs the reference electrode was noted and then voltage clamped with the potentiostat at this potential. After a further 5 min period a 5 sec +300 mV ramp (60 mV/sec) was imposed on the system. At the end of this period the integrated current and voltage between the WE and the CE was recorded. Following a 5 min recovery period an additional five consecutive pulse/recovery periods were performed. The averaged data was used to calculate the charge (μC) and conductance (μS).

The use of six pulse periods were chosen because it was found earlier in the investigation on the EKB of plant membrane systems that this number was optimal for discrimination between varieties within the same plant species. Three pulse periods were generally found to be insufficient for resolving the differences[2].

A similar experiment was performed on a second section from the same diaphragm. However in this instance a 5 sec -300 mV ramp pulse was used. By relating the charge value obtained in this instance to similar data from the +300 mV pulsed experiment it was possible to ascertain the electron:hole (e:h) ratio.

Arrhenius plots. As indicated in the previous section, Arrhenius plots indicated a transition at about 30°C which related to the activation of the various enzyme systems within the membrane. Therefore as it was the most critical region for study the temperature range of 20° to

44°C was chosen for examination. In order to accomplish this study effectively a preliminary series was performed to determine the optimal dwell period at each temperature. It was found that a 15 min dwell period was ideal for a relatively well defined transition in plots of conductance vs $1/T$. Conductance measurements were chosen as they included both the current and voltage changes which occurred as a consequence of increasing temperature. The plots of current or charge on its own did not yield well defined transitions. This observation was also made in earlier studies with plant membranes[3]. The protocol outlining the preparation for these studies are the same as described in the EKB section.

These experiments were performed in a reproducible manner by taking advantage of the unique programmable characteristics of the Eurotherm temperature controller. The instrument was programmed to ramp 4°C over a 30 sec period and then dwell at the set temperature for 15 min. At this point the remote control relay system simultaneously signaled the Eurotherm to go on a hold state and a 5 sec +300 mV ramp at 60 mV/sec was imposed on the voltage clamped system. At completion of the 5 sec pulse, a signal from the timer initiated a second 30 s ramp (4°C) to the next dwell temperature. This sequence was continued to 44°C after which the system automatically returned to 20°C.

The conductance at each temperature was plotted as log μS vs $1/T°K$. The equivalent charge values (μC) were chosen from two relevant points on the second linear transitional rise of the plots occurring at 24 to 28°C. These values served for the rate constants k_1 and k_2 in the Arrhenius equation to obtain the activation energy (Ea).

$$\log \frac{k_2}{k_1} = \frac{Ea}{2.303R} \left(\frac{T_2 - T_1}{T_1 \times T_2}\right)$$

In the two preceding experimental studies all data accumulated were corrected for the electrical behavior of the background electrolyte, Hanks' balanced salt solution, under identical experimental conditions.

Pressure effects. In a previous study the effects of localized pressure on the charge transfer and conductance properties of leaf membrane systems were described[4]. As the effects noted were rather striking and varied with different species, it was felt, using this approach, that a further elaboration of the electronic differences between the phenotypically 'normal' and the dystrophic diaphragm muscle membrane would be of value.

In order to utilize a cell system (Fig. 1b) which would serve for the limited sized membrane available, it was necessary to develop a somewhat more elaborate control than that used with the larger dual compartment cells in the plant membrane studies. After the addition of the Hanks' solution, in the present instance, it was necessary to remove all bubbles from the compartments. This required a somewhat vigorous tapping of the assembled cell unit which could not be accomplished once it was mounted in its isolation cage. Therefore the previously used technique of performing a measurement of the membrane section covered by compartment 1 using CE1, and then adding electrolyte to compartment 2 and ultimately using CE2 in the measurement was impossible in the present studies. An alternative control was obtained by using both compartments simultaneously with an electrolyte bridge between both C-1 and C-2. This bridge was created with a gasket similar to Dd but with the separation rubber segment removed from between the windows.

The background controls were obtained using a single compartment and the bridged dual compartments of Fig. 1b. The procedure was similar to that described in the previous section. In the bridged cell containing 100 µl Hanks/compartment it was only necessary to use a single CE between Dd and E. After equilibration to temperature (37°C) for 10 min the potential between the WE and the reference electrode was determined and the voltage clamped with a potentiostat. After 5 min a 5 sec +300 mV ramp (60 mV/sec) was imposed on the system. This was followed by a dwell period (5 min) and a subsequent second +300 mV ramp. The same experiments were repeated using -300 mV ramps. The data obtained from these four different studies were used as correction factors in the next set of flip-flop (FF) experiments.

Similar experiments were performed on diaphragm membrane samples from normal and dystrophic mice. In this instance the Dd was used as shown (Fig. 1b). With this gasket in place the only electrical connection with C2 was the compressed section of the membrane between C1 and C2. After the usual experimental preamble a 5 sec +300 ramp pulse was imposed on the section in the first compartment using CE1. Immediately after this pulse the CE1 was switched out of the circuit and CE2 switched in. This then included in the system both sections of the membrane and the compressed membrane connection between both compartments. After this flip-flop (FF) and a 5 min dwell period a second pulse was imposed upon the system. The above two experiments were repeated again in a similar manner using 5 sec -300 mV (60 mV/sec) pulse exposures. All data presented for studies on FF behavior were corrected for the contribution of the Hanks' background electrolyte.

RESULTS AND DISCUSSION

Electrokinetic behavior

The averaged results obtained with studies at 20°C performed on three different sets of 129/ReJ dystrophic (dy/dy) and their phenotypically normal (?/+) litter mates and five other sets at 37°C are shown in Table 1.

There was, as expected with animal experiments, variations which might have been related to the diverse ages of the subject and/or other factors. However from examination of data for individual pairs the e:h ratios showed a greater electron availability in both the ?/+ and dy/dy with the dy/dy always greater than the phenotypically normal ?/+. In all instances the µC and µS values were lower for the dy/dy mice both with the + and - pulsed systems.

The observance of a preponderance of negative charges is to be anticipated as the muscle sarcolemma membrane, which consists of a superficially located basement membrane and a plasma membrane, contains an abundance of negatively charged free or bound mucopolysaccharide material. In addition there is also present negatively charged phosphorylated proteins superficially located in the sarcolemma membrane. Many of these may represent specific binding sites for calcium.

In order to test whether much of the change was due to the existence of Ca^{+2} binding sites experiments were run comparing the behavior of the ?/+ and dy/dy at 37°C in the presence of Hanks' and

Hanks' containing 10^{-5} M Diltiazem (Sigma) a Ca^{+2} antagonist. The function of this agent is to block specific Ca^{+2} binding sites. The results are presented in Table 2.

As will be noted there was a significant depression of the e:h ratio as compared to the controls without the antagonist. However this occurred at the expense of increased values for the h values. On an absolute basis in comparing the coulombic (μC) values as obtained in the + pulsed experiments the dy/dy demonstrates no significant change in the presence of Diltiazem whereas the control (?/+) does indicate a decrease in available electrons. The increased number of 'holes' or electron acceptor sites is noted for both ?/+ and dy/dy muscle suggesting an increased availability of anion binding sites in the Diltiazem treated membranes.

In addition to the correction of the controls, ?/+ and dy/dy, for the contribution of the Hanks background, in this series corrections were also made for the Hanks -10^{-5} M Diltiazem background behavior.

Table 1. Comparative behavior of 129/ReJ
normal (?/+) and dystrophic (dy/dy) mice

	+ Pulsed (e↑)*			- Pulsed (h = e↓)*	
	μC	μS		μC	μS
20°C ?/+	7.63	5.43		6.61	3.99
SD	(1.00)	(1.18)		(2.49)	(1.62)
			e:h = 1.15:1		
dy/dy	7.12	5.30		4.49	2.98
SD	(2.20)	(1.76)		(2.21)	(1.59)
			e:h = 1.59:1		
37°C ?/+	12.49	8.38		10.58	5.18
SD	(3.28)	(2.28)		(3.50)	(2.22)
			e:h = 1.18:1		
dy/dy	8.86	6.49		5.90	3.23
SD	(2.86)	(2.02)		(2.27)	(1.78)
			e:h = 1.50:1		

*e↑ signifies electron withdrawal and h = e↓ the holes available for the acceptance of electrons. The SD values in parenthesis represents the standard deviation for the mean values listed.**

**The SD values listed in this manuscript do not in any manner imply a truly valid statistical evaluation of the data presented. These are merely introduced to show the deviations from the mean values obtained with the limited number of animals available for this study.

Table 2. Effects of Diltiazem on a 129/ReJ normal (?/+)
and dystrophic (dy/dy) mouse diaphragm at 37°C

	+ pulsed (e↑)			– pulsed (h=e↓)	
	μC	μS		μC	μS
?/+ Hanks	14.76	9.64		10.96	5.62
SD	(0.55)	(0.28)		(0.05)	(0.04)
			e:h = 1.35:1		
?/+ Hanks 10⁻⁵ M Diltiazem	12.52	8.11		12.53	7.54
SD	(0.57)	(0.11)		(0.32)	(0.10)
			e:h = 1.00:1		
dy/dy Hanks	11.37	8.00		5.00	2.80
SD	(0.72)	(0.02)		(0.02)	(0.05)
			e:h = 2.3:1		
dy/dy Hanks 10⁻⁵ M Diltiazem	11.45	8.28		8.90	4.60
SD	(0.73)	(0.26)		(0.10)	(0.38)
			e:h = 1.4:1		

e↑ and e↓ as in Table 1. The above results represent experiments performed in duplicate from sections of a diaphragm from a ?/+ mouse and its dystrophic (dy/dy) litter mate. The SD values are in parenthesis.

Arrhenius plots

These plots were made over the temperature range of 20°-44°C. The plots of conductance (μS) vs 1/T°K indicated a transition at 28°C for the ?/+ and at 24°C for the dystrophic (dy/dy). This in itself, at present, cannot be considered of much significance. However the activation energies obtained (Table 3) demonstrate meaningful differences between the normal (?/+) and the dystrophic (dy/dy) musculatures. These differences were also manifested in the presence of Diltiazem.

Pressure effects

The data presented in Table 4 represents the results from duplicate experiments on diaphragm membranes obtained from two different maternal litters (A and B).

In an earlier report[4], in which a dual compartment cell with an inert membrane (Visking) plated electrode, it was found that the charge obtained with the dual compartment cell gave approximately twice the charge of the single compartment. A charge significantly greater than two-fold was considered to be due to the presence of a biological membrane.

Table 3. Activation Energies of Normal and Dystrophic
Diaphragm Muscle in Hanks and in Hanks -10^{-5} M Diltiazem

$$Ea = kJ/mole$$

?/+	20.42	(0.15)	
dy/dy	29.66	(0.49)	Hanks
?/+	39.49	(0.17)	
dy/dy	42.04	(0.19)	+ Diltiazem

Ea values obtained from a duplicate sampling from same
diaphragm of a phenotypically normal (?/+) and a dystrophic (dy/dy) litter mate. Parenthesis encloses standard
deviation values.

Table 4. Effect of pressure on normal and dystrophic membrane
(?/+ and dy/dy respectively)

			+ Pulsed (e+)			- Pulsed (e+)		
			μC		μS	μC		μS
A	?/+	P-1	17.80		6.31	14.03		3.70
	FF	P-2	63.07	(3.54)*	19.59	99.31	(7.08)*	25.72
	dy/dy	P-1	7.01		3.21	7.34		1.71
	FF	P-2	28.28	(4.03)*	10.93	72.33	(9.85)*	21.06
B	?/+	P-1	13.17		4.75	14.15		3.17
	FF	P-2	56.73	(4.31)	17.01	122.90	(8.69)	30.03
	dy/dy	P-1	9.10		3.43	8.94		1.15
	FF	P-2	43.98	(4.83)	13.93	105.5	(11.80)	25.40

*These values represent the potentiation of charge when flip-flop (FF)
is performed by changing from CE1 which is used during the first pulse
period (P-1) to CE2 for the second pulse period (P-2). A and B
represent two different pairs of donors.

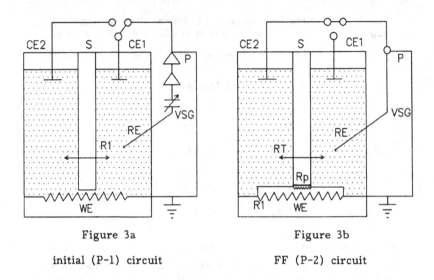

Figure 3a   Figure 3b

initial (P-1) circuit   FF (P-2) circuit

VSG-voltage scan generator, P-potentiostat, WE-membrane 'plated' electrode, CE1, CE2-counter electrodes, R1-resistance of total membrane-electrode in Fig. 3a, S-perspex separator, Rp is resistance of membrane introduced in circuit Fig. 3b.

In the present flip-flop experiments the background data was obtained with Hanks on its own as per the protocol described in the experimental section with the dual compartment cell. These data were then used to correct for the contribution of the electrolyte to the raw data acquired using the membrane 'plated' electrode.

With a system wherein both compartments contain a continuum of membranous 'plated' electrode, it is conceivable that series resistors are being dealt with when a voltage is imposed on the system (Figure 3a). However when a flip-flop is made from CE1 to CE2 the membrane resistance at the site of pressure appears to assume a configuration comparable to a parallel circuit[4].

An equivalent circuit (ignoring the capacitive effects) as a first approximation is described in Figure 3a and b. Further validity of the parallel circuit comes from the fact that all charge values obtained from the F-F data demonstrated a significantly greater amount of charge than observed in the systems in their pre-FF modes.

From the results can be noted the significant differences between the normal (?/+) and dystrophic musculature (dy/dy) systems. With the resistance values (reciprocal of the conductance values) listed in Table 5 it was possible to approximate the resistance of the pressurized portion of the membrane. For this it was assumed that during P-1 a series circuit existed. This included the sections of membrane in C1, C2 and under S. When switching to CE2, electrically it appeared that a parallel resistance was introduced into the circuit. This parallel resistive element was undoubtedly the pressurized membrane section. The P-1 values represent the total series resistance (R1) and the P-2 values were equivalent to the total resistivity (RT) of the parallel circuit. The resistivity (Rp) of this pressurized section (Table 5) was calculated from:

$$R_p = \frac{R1 \times RT}{R1 - RT}$$

Undoubtedly other factors are involved in the observed potentiation of charge in going from CE1 to CE2 and will be investigated further. However regardless of the total validity of the assumptions made there is again presented the significant differences in the electronic behavior of the phenotypically normal vs the dystrophic musculature.

The objective of demonstrating the existence of electronic differences between musculature from phenotypically normal and dystrophic mice has been achieved. This was shown with three entirely different technical approaches. The diversity of responses between the normal controls and the dystrophic may be due in part to the absence or a diminished quantity of particular charge transfer proteins, e.g. dystrophin[5], and degenerative changes in the metabolic viability of the dystrophic muscle tissue. The last mentioned possibility is not an unlikely one for it was shown some years ago[6] that the metabolic viability of the erythrocyte decreases with the progress of muscular dystrophy in young children. Studies in progress will undoubtedly further define the significance of the electronic behavioral patterns observed.

Table 5. Resistivity of normal (?/+) and dystrophic (dy/dy) muscle to localized pressure

			+ pulsed (e↑)		− pulsed (e↓)	
			R/nΩ	Rp/nΩ	R/nΩ	Rp/nΩ
A	(?/+)	P-1	158.48		270.27	
	FF	P-2	51.05	75.31	38.88	45.41
	(dy/dy)	P-1	311.53		584.80	
	FF	P-2	91.49	129.53	47.48	51.68
B	(?/+)	P-1	210.53		315.46	
	FF	P-2	58.79	81.57	33.30	37.23
	(dy/dy)	P-1	291.55		869.57	
	FF	P-2	71.79	95.24	39.37	41.24

P-1 = R1, P-2 = RT, Rp represents resistance of pressurized membrane section. A and B represent two different pairs of donors.

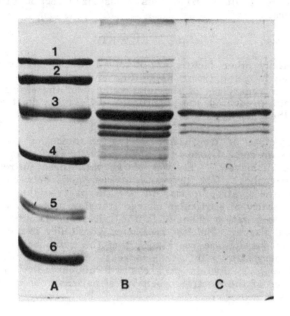

Figure 4.   Polyacrylamide gel electrophoresis

(A) markers. (1) phophorylase b M.W. 94,000, (2) albumin M.W. 67,000, (3) ovalbumin M.W. 43,000, (4) carbonic anhydrase M.W. 30,000, (5) Trypsin inhibitor M.W. 20,100, (6) α lactalbumin, M.W. 14,400. (B) phenotypically normal, (C) dystrophic mouse diaphragm muscle proteins.

The gel electrophoresis patterns shown in Fig. 4 represents a preliminary separation of the proteins obtained by extracting the homogenised diaphragm muscles from the normal (B) and dystrophic (C) mice. The significant differences between them suggest that the assumption made in the concluding statement, regarding the absence or the diminished amount of charge transfer proteins may be in-part, a valid explanation for the differences noted in electronic behavior.

References

1. M.J. Allen, Bioelectrochemistry - Before and After in "Electrochemistry Past and Present", John T. Stock and Mary V. Orna, eds., ACS Symposium Series No. 390, Washington, D.C. (1989).
2. M.J. Allen, The electrogenetic characteristics of plant membranes in "Charge and Field Effects in Biosystems", M.J. Allen and P.N.R. Usherwood, Eds. (1984).
3. M.J. Allen, The temperature-dependent behaviour of a semiconductive property of plant membranes, J. Exp. Bot., 32:855 (1981).
4. M.J. Allen, Effects of localized pressure on charge transfer and conductance properties of leaf membranes, Studia biophys. 90:19 (1982).

5. E.P. Hoffman, B.H. Brown, Jr. and L.M. Kunkel, Dystrophin: the protein product of the Duchenne muscular dystrophy gene, Cell, 51:919 (1987).
6. M.J. Allen, A voltammetric study of metabolising erythrocytes, Coll. Czech. Chem. Comm. 36:658 (1971).

ACKNOWLEDGEMENT

This study was supported by grants from the Department of Chemistry and the University Biomedical Research Fund. We also appreciate the helpful comments of Prof. E. Neumann, Drs. J.V. Quagliano and J. Stewart in the preparation of this manuscript. Our sincere thanks to Dr. M.G. Garcia for performing the gel electrophoresis.

HEMOGLOBIN ELECTRON TRANSFER REACTIONS

William T. Grubbs and Lyman H. Rickard

Physical Science Department
High Point College
High Point, NC

Electrochemical techniques have become established tools for the study of the electron transfer reactions of biological molecules (1,2,3). These reactions are the basis of life processes within the cell. The mechanisms by which these molecules transport electrons have been the subject of intense investigation. Electron transfer kinetic data of these molecules in solution and at electrode surfaces have been used to help establish the mechanisms of the physiological reactions of these molecules. Early electrochemical investigations of biological redox molecules involved the electron transfer between two biological molecules or between a biological molecule and a mediator used to couple the molecule to an electrode. These studies have been reviewed by Heineman et al. (4) and Szentrimay et al. (5). Later studies have focused on heterogeneous electron transfer reactions at modified metal electrodes (6-10), semiconductor electrodes (11-16), and at bare metal electrodes (10, 17-18).

Heterogeneous electron transfer reactions are of particular importance in understanding the physiological electron transfer reactions of biological molecules at membrane surfaces in processes such as oxidative phosphorylation. Both biological membrane surfaces and solid electrode surfaces are electrically charged and have structured water layers at their solution interface. These two common interfacial properties provide a means of understanding heterogeneous electron transfer at biological membranes by studying the reactions at solid electrodes (1).

The focus of this study has been the heterogeneous electron transfer of hemoglobin at an unmodified indium oxide electrode. Although hemoglobin has been intensely studied, there have been very few electrochemical studies. Early electrochemical studies of hemoglobin were done at mercury (19-22). A recent investigation has reported the reduction of hemoglobin at a methylene blue modified platinum electrode and at a bare platinum electrode (10).

129

TABLE I. HETEROGENEOUS ELECTRON TRANSFER

MOLECULE	MW	KINETICS	
Cytochrome c_7	9800	Fast	(25)
Cytochrome c_3	13000	Fast	(26, 27)
Cytochrome c_{553}	9000	7.5×10^{-3} cm/sec	(12)
Cytochrome c	12400	12.5×10^{3} cm/sec	(16)
Myoglobin	17800	2.6×10^{-5} cm/sec	(13)
Hemoglobin	68000		

Although hemoglobin does not function physiologically as an electron carrier it is an ideal molecule for the study of heterogeneous electron transfer because it is readily available, is stable, has a well documented structure (23-24), and is a size which is readily comparable to molecules presently found in studies in the literature. It should be pointed out that a small percentage of hemoglobin exists in vivo in the inactive oxidized form and that mechanisms for reducing this hemoglobin to its oxygen binding form also exist in vivo. Stability is important in selection of a molecule for study because of the ease with which many biological redox molecules denature in aqueous solution. A documented structure will allow for postulation of the relationship between structure and the kinetic and thermodynamic results obtained. The size of the hemoglobin molecule is important in order to allow comparisons to other studies in the literature. Table I shows the relationship between molecular weight and rate constant for heterogeneous electron transfer of several well studied molecules. Cytochrome c_7 and cytochrome c_3 both exhibit fast electron transfer kinetics.   Both of these molecules contain multiple heme groups.   Cytochrome c_7 contains three hemes and cytochrome c_3 contains four. Cytochrome c_{553}, cytochrome c and myoglobin all exhibit quasi-reversible electron transfer kinetics.   The size of cytochrome c_7 is comparable to cytochrome c_{553} but the three heme groups of cytochrome c_7 give it much faster kinetics. The same is true for cytochrome c_3 as compared to cytochrome c.   Both molecules are similar in molecular weight but the electron transfer kinetics of cytochrome c_3 are much faster since it contains four heme groups. Hemoglobin's molecular weight is much larger than the other molecules listed in Table I. However, hemoglobin contains four heme groups. Therefore the molecular weight per heme group gives a value that is comparable to the molecular weight of myoglobin.   It is postulated that the electrochemical behavior of hemoglobin might then be comparable to that of myoglobin.

The electron transfer reaction of hemoglobin is given by the following equation:

$$Fe(III)Hb + e- \longleftrightarrow Fe(II)Hb$$

where Fe(III)Hb is the oxidized form of the protein also known as ferrihemoglobin or methemoglobin. Fe(II)Hb is the reduced form of the hemoglobin also known as ferrohemoglobin or deoxyhemoglobin. Only the reduced form of hemoglobin is physiologically active as an oxygen carrier as shown by the following equation:

$$Fe(II)Hb + O_2 \longleftrightarrow Fe(II)O_2$$

where $Fe(II)O_2$ is oxyhemoglobin.

The structure of hemoglobin consists of four polypeptide chains with one heme group per chain located in crevices at the exterior of the molecule. Hemoglobin is an approximately spherical protein with dimensions of 64 x 55 x 50 angstroms. The iron atoms of the four heme groups form an irregular tetrahedron where the iron atoms are separated by 25 to 30 angstroms.

EXPERIMENTAL

Horse hemoglobin from Sigma Chemical Company was purified by gel filtration on a column of Sephadex G-100-120 (Sigma Chemical Company) using published procedures (28). Tris(hydroxymethyl aminomethane (99.9%) and cacodylic acid (hydroxydimethylarsine oxide) (99.5%) were used as received from Fisher Scientific Company. All solutions were prepared in purified water from a Milli-Q water system that exhibited a resistivity of 15 Mohm/cm. All experiments were carried out in nonbinding tris/cacodylic acid buffer at a pH of 7.0 and an ionic strength of 0.2M (29). Nitrogen was passed over hot copper turnings to scavenge oxygen and then through a bubble tower to saturate with water.

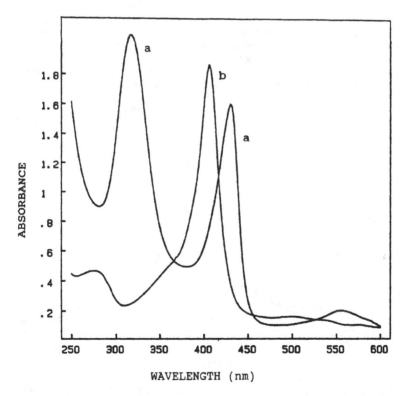

Figure 1. Hemoglobin spectra (a) ferrohemoglobin (b) ferrihemoglobin.

Derivative cyclic voltabsorptometry (DCVA) and potential step chronoabsorptometry (SPS/CA) experiments were carried out using a Princeton Applied Research Model 174 potentiostat. These data were acquired using an IBM PC AT computer. Spectra of the optically transparent thin layer electrode (OTTLE) were taken using a Perkin-Elmer Lambda 3B UV/Vis Spectrophotometer interfaced with an Epson Equity I+ computer. Spectropotentiometric measurements were controlled with a BAS CV-1 Cyclic Voltammograph. Electrochemical cells were of a conventional three electrode design. The cell used for DCVA and SPS/CA measurements has been described by King (13) and the OTTLE was a modification of the design by Heineman (30). The working electrodes, tin doped indium oxide film deposited on glass, were from Donnelly Corporation, Holland , MI. The reference electrode used for all measurements was an Ag/AgCl (1.0 M KCl) electrode. All reported potentials have been corrected to the normal hydrogen electrode. All measurements were made at room temperature, $22 \pm 1^{\circ}C$.

Prior to each experiment the working electrode was pretreated by published procedures (31). The working electrode was preconditioned for 20 minutes in electrolyte buffer by votammetric cycling over the potential range to be used in the hemoglobin measurement. All measurements were carried out under a blanket of nitrogen.

RESULTS AND DISCUSSION

Initial electrochemical investigations of hemoglobin in this study produced a faradaic response that was small compared to the background current. The use of background subtraction techniques have resulted in data that is not satisfactory for the extraction of quantitative results. Therefore it was decided to monitor the reaction spectrophotometrically. This requires that the oxidized and reduced forms of the hemoglobin have different absorption spectra. Figure 1 shows an absorption maximum at 409 nm for ferrihemoglobin and maxima at 431 nm and 320 nm for

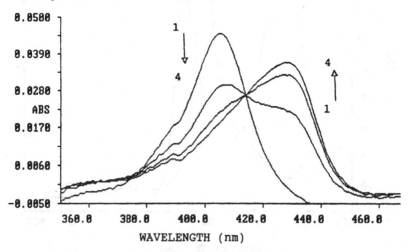

Figure 2. Hemoglobin spectra during reduction at an OTTLE at -.32 V vs NHE. Specta taken over 45 minutetime interval.

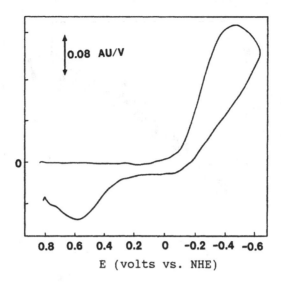

Figure 3. Derrivative cyclic voltabsorptammogram of 74 μM hemoglobin. Scan rate of 2 mV/sec.

ferrohemoglobin. Figure 2 shows the reduction of hemoglobin at a thin layer optically transparent indium oxide electrode. During the 45 minutes required for complete reduction the potential is poised at -.32 volts. The 409 nm ferrihemoglobin band decreases while the 431 nm ferrohemoglobin band increases during reduction. Of particular interest is the isobestic point that is maintained at 416 nm indicating a clean system with no side reactions and no denaturation of the protein. This shows that the heterogeneous electron transfer of hemoglobin can be driven electrochemically and monitored spectrophotometrically.

A second technique that has been investigated for probing the electron transfer reaction of hemoglobin is DCVA (32). Figure 3 shows a DCVA of hemoglobin exhibiting a reduction peak at -420 mV and an oxidation peak at +520 mV. The estimation of the formal potential based on peak separation yields a value of + 90 mV. This compares with a literature value of 132 mV. The difference in values is due to the peak separation not being symmetrical about the formal potential. Other methods will be used to more accurately establish the formal potential. The peak separation indicates that the kinetics are irreversible for these experimental conditions.

The third technique that has been investigated for use in this study is potential step chronoabsorptometry (SPS/CA). Figure 4 shows a plot of raw data of absorbance verses time for a 50 second potential step experiment where the absorbance of the reduced specie is monitored at 431 nm. Although it is clear from the increase in absorbance that the reduction is

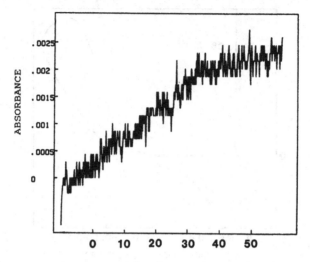

Figure 4. Single potential step chronoabsorptometry for 130 μM hemoglobin. Potential step from 220 mV to -380 mV vs. NHE

taking place, the long step time and noise level prevent the extraction of quantitative data from this experiment. The use of signal averaging to reduce the noise and a more sensitive spectrophotometer are expected to improve these signals.

The results of these initial investigations of hemoglobin show that hemoglobin can be oxidized and reduced at an unmodified indium oxide surface. The reaction can be driven electrochemically and monitored spectrophotometrically. The isobestic point in Figure 2 demonstrates that the reaction is clean with no side reactions and no denaturation during the time scale of the experiment. Usable techniques for the on going study are spectropotentiometry, derivative cyclic voltabsorptometry, and potential step chronoabsorptometry.

ACKNOWLEDGEMENTS

The authors thank Dr. Fred Hawdridge of Virginia Commonwealth University for the kind gift of indium oxide and the use of instrumentation. The financial assistance of the North Carolina Academy of Science and the Burroughs Wellcome Corporation are also gratefully acknowledged.

REFERENCES

1. E. F. Bowden, F. M. Hawkridge, H. N. Blount, in: Comprehensive Treatise of Electrochemistry, Vol. 10 (S. Srinivasan, Yr. A. Chizmadzhev, J. O'M. Bockris, B. E. Conway, E. Yeager, eds.) Plenum, New York, 1985, pp. 297-346.
2. F. A. Armstrong, H. A. O. Hill, N. J. Walton, Quart. Rev. Biophysics, 18, 1986, 261.

3. D. C. Johnson, M. D. Ryan, G. S. Willson, Anal. Chem., 60, 1988, 147R.
4. W. R. Heineman, R. M. Hawkridge, H. N. Blount, Electroanalytical Chemistry, A. J. Bard (ed.) Vol. 13, Marcel Dekker, New York , 1983, pp. 1-113.
5. R. Szentrimay, P. Yeh, T. Kuwana, Electrochemical Studies of Biological Systems, D. T. Sawyer (ed.), Vol. 38, ACS Symposium Series, American Chemical Society, Washington, D. C. 1977, pp. 143-169.
6. E. F. Bowden, F. M. Hawkridge, H. N. Blount, Bioelectrochem. Bioenerg., 7, 1980, 447.
7. J. Haladjian, R. Pilard, P. Bianco, P. Serre, Bioelectrochem. Bioenerg., 9, 1982, 91.
8. W. J. Albery, M. J. Eddowes, H. A. O. Hill, A. R. Hillman, J. Am. Chem. Soc., 103, 1981, 3904.
9. M. J. Eddowes, H. A. O. Hill, J. Am. Chem. Soc., 101, 1979, 4461.
10. S. Song, S. Dong, Bioelecrochem. Bioenerg., 19, 1988, 337.
11. P. Yeh, T. Kuwana, Chem. Lett., 1977, 1145.
12. K. B. Koller, F. M. Hawkridge, G. Fague, J. LeGall, Biochem. Biophys. Res. Commun., 145, 1987. 619.
13. B. C. King, F. M. Hawkridge, J. Electroanal. Chem., 237, 1987, 81.
14 D. J. Cohen, F. M. Hawkridge, H. N. Blount, C. R. Hartzell, Proceedings of the International Symposium on Bioelectrochemistry and Bioenergetics, Nottingham, England, 1983, pp. 19-31.
15. K. B. Koller, F. M. Hawkridge, J. Electroanal. Chem., 239, 1988, 291.
16. K. B. Koller, F. M. Hawkrikge, J. Am. Chem. Soc., 107, 1985, 7412.
17. D. E. Reed, F. M. Hawkridge, Anal. Chem., 59, 1987, 2334.
18. S. Sun, D. E. Reed, J. K. Cullison, L. H. Rickard, F. M. Hawkridge, Mikrochimica Acta, III, 1988, 97.
19. F. Scheller, M. Janchen, H-J. Prumke, Biopolymers, 14, 1975, 1533.
20. F. Scheller, M. Janchen, studia biophysica, 46, 1974, 153.
21. B. A. Kuznetsov, G. P. Shumakovich, N. M. Mestechkina, Bioelectrochem. Bioenerg., 4, 1977, 512.
22. S. R. Betso, R. E. Cover, J. C. S. Chem. Comm., 1972, 621.
23. M. F. Perutz, M. G. Rossmann, A. F. Cullis, H. Muirhead, F. Will, A. T. C. North, Nature, 185, 1960, 416.
24. G. Fermi, M. F. Perutz, B. Shaanan, R. Fourme, J. Mol. Biol., 175, 1984, 159.
25. P. Bianco, J. Haladjian, Bioelectrochem. Bioenerg., 8, 1981, 239.
26. P. Bianco, J. Haladjian, Electrochimica ACTA, 26, 1981, 1001.
27. P. Bianco, G. Fauque, J. Haladjian, Bioelectrochem. Bioenerg., 6, 1979, 385.
28. J. Reiland, Methods in Enzymology, S. P. Colowick, and N. O. Kaplan (eds.), Vol 22, Academic Press, New York, 1971, pp. 287-325.
29. E. F. Bowden, F. M. Hawkridge, H. N. Blount in: Electrochemical and Spectrochemical Studies of Biological Redox Components, (K. M. Kadish ed.), American Chemical Society, Washington, D. C., 1982, pp. 159-171.

30. W. R. Heineman, B. J. Norris, J. F. Goelz, Anal. Chem., 47, 1975, 79.
31. E. F. Bowden, F. M. Hawkridge, H. N. Blount, J. of Electroanal. Chem., 161, 1984, 355.
32. E. F. Bancroft, H. N. Blount, F. M Hawkridge, in: Electrochemical and Spectrochemical Studies of Biological Redox Components, (K. M. Kadish ed.), American Chemical Society, Washington, D. C., 1982, pp. 23-49.

GLUTAMATE RECEPTOR DESENSITIZATION

Peter N.R. Usherwood and Robert L. Ramsey

Zoology Department, The University
Nottingham, NG7 2RD
United Kingdom

INTRODUCTION

Receptors for neurotransmitters found in the surface membranes of excitable cells, such as neurones, muscle fibres and endocrine cells exhibit desensitization or tachyphylaxis when exposed to high concentrations of agonists. This phenomenon of agonist-induced refractoriness has been best studied for the nicotinic and muscarinic acetylcholine receptors of vertebrate animals, although it has not yet been fully evaluated for any transmitter receptor.

The rate of desensitization onset of nicotinic acetylcholine receptors (nAChR) depends on dose of agonist (Katz and Thesleff, 1957; Feltz and Trautmann, 1982), is enhanced by divalent cations and membrane hyperpolarisation (Magazanic and Vyskocil, 1970) and is raised by increasing temperature (Magazanic and Vyskocil, 1975), although different species of nAChR may vary in these respects. Initial studies on desensitization of nAChR led to the following descriptive model (1):

$$ACh + R \longleftrightarrow AChR \longleftrightarrow AChR' \quad (1)$$

where R is the receptor molecule. It was suggested that desensitization results from the gradual transformation of the receptor-ligand complex (which can open the receptor channel) into a refractory form AChR' (which cannot open the receptor channel), which reverts slowly to R after removal of acetylcholine (ACh). An alternative scheme proposed by Katz and Thesleff (1957) involves two simultaneous reactions, i.e. a depolarising step which reaches equilibrium rapidly and a slowly-developing desensitization step, but with the two reactions proceeding in parallel rather than in series (2):

$$ACh + R \underset{\searrow AChR'}{\overset{\nearrow AChR}{\diagup}} \quad (2)$$

These linear schemes were rejected as being unsuitable models for frog muscle nAChR desensitization by Katz and Thesleff (1957) in favour of two cyclic models. One of these is reversible (3); the other is irreversible and requires metabolic energy to drive it (4):

$$
\begin{array}{ccc}
\text{ACh} + \text{R} \longleftrightarrow \text{AChR} & \quad & \text{ACh} + \text{R} \longleftrightarrow \text{AChR} \\
\updownarrow \qquad\qquad \updownarrow \quad (3) & & \uparrow \qquad\qquad \downarrow \quad (4) \\
\text{ACh} + \text{R}' \longleftrightarrow \text{AChR}' & & \text{ACh} + \text{R}' \longleftrightarrow \text{AChR}'
\end{array}
$$

According to these cyclic models nAChR again exists in two forms, effective (R) and refractory (R'). The reversible model (3) is now gener- ally preferred. According to Katz and Thesleff (1957) data on desensit- ization of frog muscle nAChR can be fitted moderately well if the affin- ity of ACh to R' is higher than that to R. The proposal that the refrac- tory state of the nAChR is coupled to an increased binding affinity for ACh has received general confirmation in subsequent biochemical and pharm- acological studies of desensitization, where transitions between low and high affinity states of nAChR have been recorded (reviewed by Anholt et al., 1985). The desensitized or refractory nAChR is thought to be charac- terised by an approximately 300-fold increase of affinity for ACh. According to models (3) and (4) a fraction of nAChR of frog muscle end- plate and other cholinergic systems will exist in the desensitized state R', although for some nAChR this may involve only about 0.01% of the receptor population. However, the postjunctional nAChR of Torpedo elec- troplax has a 10% to 20% probability of being refractory in the absence of ACh (Boyd and Cohen, 1980). Apparently the nAChR of Torpedo are like those of vertebrate striated muscle since they share with the latter the property of slow recovery from desensitization (Feltz and Trautmann, 1982). Whereas ACh activation of its receptor requires at least 2 mole- cules of transmitter it seems that a single molecule of bound agonist is sufficient to cause desensitization of nAChR (Sine and Taylor, 1979). This infers that desensitization can proceed from a closed channel state of nAChR.

Single channels gated by nAChR of extrajunctional membrane of dener- vated frog muscle exhibit burst kinetics in the presence of desensitizing concentrations of ACh (Sakmann et al., 1980). At 20 μM, nAChR channel open- ings appeared in clusters of bursts, suggesting that desensitisation in this system comprises at least two reactions. The sequence of bursts and burst intervals reflects the kinetics of a fast component of desensitiz- ation with apparent rate constants of c. $2s^{-1}$ and c. $5s^{-1}$ for conversion to and from the desensitized state. The mean duration of the interval between clusters of bursts and the mean duration of the clusters relates to slow desensitization with onset of c. $0.2s^{-1}$ and recovery of c. $0.03s^{-1}$. These two desensitized kinetic states may relate to the two exponential components identified in the desensitization onset macro- studies of Feltz and Trautmann (1982).

Desensitization of other types of ligand gated receptors found on excitable cells has been investigated using mainly electrophysiological techniques. The postjunctional Quis-GluR found at excitatory synapses on locust leg muscle exhibits some of the properties which characterise nAChR (Usherwood, 1978; Sansom and Usherwood, 1989). During long-term, bath application of L-glutamate or agonist to a locust nerve-muscle preparation, the changes in membrane potential and membrane conductance caused by these compounds decline with time (Usherwood and Machili, 1968; Daoud and Usherwood, 1978) due to Quis-GLuR desensitization. However, unless high concentrations of agonist are applied, desensitization is incomplete.

The rate of onset of Quis-GluR desensitization in this system has been examined quantitatively using ionophoretic techniques for brief application of L-glutamate and agonists to the postjunctional Quis-GluR (Fig.1), whilst recording resultant excitatory currents under voltage clamp. Anis et al. (1981) showed that, in general, desensitization onset for Quis-GluR depends on agonist concentration and also on the potency of the agonist - the onset rate being higher for more potent agonists. The relative rates of desensitization onset for equipotent doses of L-glutamate and agonists are L-quisqualate > L-4 methylene glutamate > L-2amino-3-sulphino-propionate = L-glutamate.

In electrophysiological studies of nAChR desensitization, rapid application of agonist has shown that desensitisation onset is a multi-exponential process involving at least two reactions (Feltz and Trautmann, 1982; Slater et al., 1984), one of which may depend upon intracellular Ca concentration (Chesnut, 1982; 1983). Biochemical studies of reconstituted electroplax nAChR have demonstrated similar findings (McNamee and Ochoa, 1982; Hess et al., 1982). Anis et al. (1981) found that desensitization onset at locust glutamatergic nerve-muscle junctions was usually biphasic, although at some sites the decline in responsiveness of the postjunctional Quis-GluR population during repeated ionophoretic application of agonist can be fitted by a single exponential.

The rate of recovery from desensitization of locust muscle Quis-GluR populations is independent of the level of desensitization and in this respect it is similar to nAChR. However, unlike nAChR the rate of recovery of Quis-GluR from desensitization is agonist-dependent (Anis et al., 1981). The relative rates of desensitisation recovery for four agonists are; L-glutamate = L-2-amino-sulphinopropionate > L-4-methylene glutamate > L-quisqualate. This discovery led Anis et al. (1981) to conclude that the cyclic models for desensitization of Katz and Thesleff (1957) do not fit the Quis-GluR data, since they do not predict that desensitization recovery should be independent of agonist.

In ionophoretic studies desensitization recovery of Quis-GluR is monophasic for all agonists, although following prolonged ionophoretic application of glutamate (or agonist) recovery from desensitization is sometimes incomplete. Incomplete recovery from desensitization has also been observed during ionophoretic application of ACh to frog muscle end-plate (Magazanik and Vyskocil, 1970). Perhaps prolonged exposure of Quis-GluR and nAChR to agonist converts these receptors to a second desensitized state from which recovery is slow.

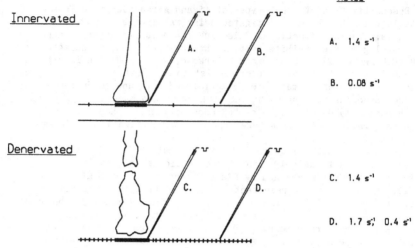

Innervated

A.   B.

A. 1.4 s⁻¹ → $1.4\ s^{-1}$

B. 0.08 s⁻¹

Denervated

C.   D.

C. $1.4\ s^{-1}$

D. $1.7\ s^{-1}$; $0.4\ s^{-1}$

Fig. 1.   Diagrammatic representation of the distribution of Quis-GluR on locust muscle and
heir rates of recovery from desensitization Top: L-glutamate is applied by iono-
phoresis from a micropipette, using brief
pulses of -ve current, to either postjunctional Quis-GluR located at a synapse
between a glutamatergic motor axon and a
muscle fibre (A) or to extra-junctional
Quis-GluR (B). Bottom: About 12 days
(at c. 29°C) after section of the motor
axon the postjunctional Quis-GluR remain as
before (C) but the population density of
the extrajunctional receptors has increased (D).

Innervated locust muscle fibres are endowed with extrajunctional
Quis-GluR which are present at much lower population densities than the
postjunctional Quis-GluR (junctional Quis-GluR, c. $4000m^{-2}$; extrajunctional Quis-GluR, c. $100m^{-2}$ (Usherwood, 1981; Dudel et al., 1988)). The
desensitization kinetics of the extrajunctional Quis-GluR are quantitatively different from those found postjunctionally at excitatory synapses
on these fibres. Desensitization onset is more rapid in the case of the
extrajunctional receptors and recovery rates slower (Fig.1), although
at most sites recovery from desensitization is usually still best described by a single exponential (Clark et al., 1979). Complete densensit-
ization of extrajunctional Quis-GluR is readily achieved during prolonged
application of agonist. Denervation of locust skeletal muscle leads to a
significant increase in the population density of extrajunctional Quis-
GluR (Usherwood, 1969; Clark et al., 1979). For a given concentration of
agonist the rate of onset of desensitization is less for Quis-GluR of
denervated muscle than for Quis-GluR of innervated muscle. Furthermore,
the recovery kinetics for denervated muscle are best described by two
exponential components, possibly corresponding to fast and slow phases of
desensitizion recovery. Desensitization onset rates are low and recovery
rates high for junctional receptors, whereas the opposite is true for
extrajunctional Quis-GluR of innervated muscle. The kinetics of desensitization onset and recovery for extrajunctional Quis-GluR of denervated

muscle fall between these two extremes. It follows, therefore, that Quis-GluR desensitization kinetics are related to the population density of this receptor. This may result from the influence of packing density on the conformational freedom of this receptor, although it has yet to be established that there is a causal relationship between receptor population density and desensitization kinetics.

Replacement of extracellular Na by equimolar concentrations of Li, NH_4 and guanidine, all of which permeate the cation-selective channel gated by locust muscle Quis-GluR (Anwyl and Usherwood, 1974; Kitts and Usherwood, 1988) accelerates onset and reduces the rate of recovery from desensitization (Clark et al., 1982). However, desensitization onset rate is unaffected by changing either extracellular Ca concentration or muscle membrane potential. The lack of action of Ca reported by Clark et al. (1982), contrasts with the results of Daoud and Usherwood (1978) who found that desensitization onset rate was reduced in Ca-free saline. However, quite different techniques for applying agonist were employed in these two studies.

Glutamate receptor - single channel studies

When extrajunctional Quis-GluR of locust leg muscle are exposed to agonist concentrations greater than c. 10^{-4} M they rapidly desensitize and show little evidence of recovery in the continuing presence of agonist (Anis et al., 1981). Using a mega-ohm seal recording technique (Patlak et al., 1979), Gration et al. (1980; 1981) showed that unless desensitization is inhibited by the lectin concanavlin A (Mathers and Usherwood, 1976) the single channel currents generated by these extrajunctional receptors are rarely seen even during long-term (60min) patch clamp recordings from single Quis-GluR. Similar results have been obtained subsequently using a giga-ohm seal recording technique where L-glutamate and agonist were applied rapidly to outside-out patches, excised from extrajunctional membrane of locust leg muscle fibres, using the liquid filament switch technique developed by Dudel and colleagues (Dudel et al., 1988). Macrosystem studies have shown that desensitization of extrajunctional Quis-GluR has a rapid, agonist concentration-dependent onset and a slow recovery (Mathers and Usherwood, 1976; Anis et al., 1981). Fig. 2 shows that when pulses of L-glutamate or agonist are applied to outside-out patches of locust muscle membrane, openings of the Quis-GluR channels are triggered (Fig. 2). The conductance of the open Quis-GluR channel is c. 150 pS (Patlak et al., 1979) and the gating kinetics of the channel have been extensively studied (e.g. Kerry et al., 1987, 1988). By averaging the responses to a large number (c. 100) of consecutive pulses a current waveform (averaged current) is obtained from which it is possible to estimate the channel opening rate from the rising phase of the current and desensitization onset kinetics from its falling phase. In preliminary studies by Dudel, Franke, Hatt, Ramsey and Usherwood, (in preparation) the averaged current obtained during pulsed application of 10^{-3} M L-glutamate had a rise time (10-90%) of 0.6ms, which increased to 2.7ms with 10^{-4} M L-glutamate. A time constant of 25ms was obtained for the exponential decay of the current for 10^{-4} M L-glutamate. For 10^{-3} M L-glutamate the decay was best fitted by 2 exponentials with time constants of 7.7ms and 43ms repectively.
The activation or channel opening rates derived from rise time measurements (assuming that the rise in concentration of the agonist at the level of the patch is more or less instantaneous) are c. $1.7 \times 10^{-6} M^{-1} s^{-1}$ for 10^{-3} M and c. $4 \times 10^{-6} M^{-1} s^{-1}$ for 10^{-4} M L-glutamate. The multi-component decay of the averaged currents for 10^{-3} M L-glutamate possibly reflects the multiple states of Quis-GluR desensitization, although given the complex gating kinetics of the Quis-GluR of locust muscle (Kerry et al., 1987; 1988) this may be too simple an explanation. Nevertheless, the single

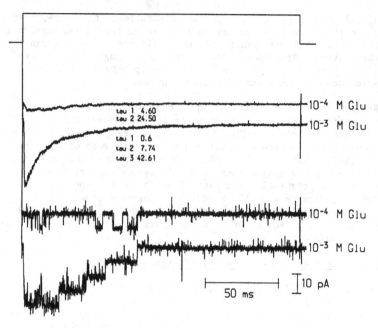

Fig. 2. Responses to Quis-GluR in outside-out patches excised from
locust extrajunctional muscle membrane to pulses of L-
glutamate (Glu) applied using a liquid filament switch.
Top trace; duration of Glu application. Bottom traces;
representative single channel currents of Quis-GluR
obtained with individual pulses of Glu (10^{-4} M and
10^{-3} M). Middle traces; averaged currents for
c. 100 pulses of either 10^{-4} M or 10^{-3} M Glu.
Tau values for multi-exponential fits to rise and
decays of averaged currents are presented below
each middle trace. (From Dudel, Franke, Hatt, Ramsey
and Usherwood in preparation).

channel data suggests that desensitization onset is very rapid. In fact,
a high rate of desensitization onset appears to be a general property of
Quis-GluR. Trussell et al. (1988) obtained desensitization onset half-
times of 3-15ms for L-glutamate-activated whole cell currents of chick
hippocampal neurones. Quis-GluR channel activity was inactivated in 3-8ms
during application of agonist to rat hippocampal neurones (Tang et al.,
1989). Mayer and Vyklicky (1989) used a fast perfusion system to apply
agonists to embryonic hippocampal neurones grown in culture and voltage
clamped in the whole-cell recording configuration. They found that with
100μM agonist desensitization onset had a time constant of c. 30ms for
DL-amino-3-hydroxy-5-methyl-4-isoxazolepropionic acid, L-glutamate and L-
quisqualate. Hatt et al. (1989) used the liquid filament switch (Franke
et al., 1987) to obtain a desensitization onset time constant of 5ms for
Quis-GluR channels in outside-out patches isolated from membrane of chick-
en spinal α-motoneurones.

Equilibrium patch clamp studies of locust muscle extrajunctional Quis-GluR suggest that recovery from desensitization is either very slow or that the receptor can enter its desensitized state(s) from its closed channel conformation(s). The fact that during repeated pulsed application of high concentrations of agonist to some outside-out membrane patches containing Quis-GluR single channel currents were not evoked with all pulses, suggests that desensitization can proceed from one or more closed channel states of Quis-GluR. Further support for this conclusion comes from experiments in which the responses of membrane patches to 10^{-2}M L- glutamate were compared before, during and after bath application of 10^{-4}M agonist. The bath application of glutamate failed to generate single channel currents but it completely abolished the responses to the pulsed agonist.

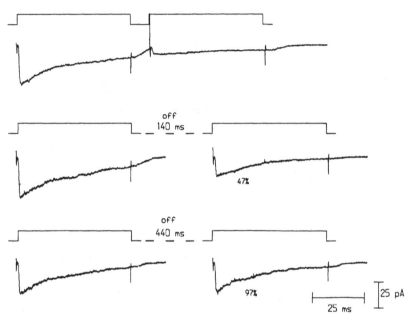

Fig. 3.   Averaged currents (lower traces) obtained with double-pulse application of 10^{-3}M L-glutamate (indicated in upper traces) to an outside-out patched excised from locust extrajunctional muscle membrane, with the interval between the two pulses in a pair varied. The averaged current during a pulse quickly peaks before decaying due to desensitization of Quis-GluR. With a brief inter-pulse interval the averaged current to the second pulse shows no recovery (top traces), but with an interval of 440ms (bottom traces) 97% recovery from desensitization was observed. (From Dudel, Franke, Hatt, Ramsey and Usherwood, in preparation).

In double pulse experiments where the interval between pulses in a pair was varied to study the time course of recovery from desensitization a time constant for this process of c. 200ms was obtained. This compares with a value of c. 16s obtained by Anis et al. (1981) in their ionophoretic studies of locust muscle Quis-GluR. Perhaps the application of agonist was so brief in the single channel experiments that the Quis-GluR failed to enter the desensitized state responsible for the slow recovery seen in the ionophoretic experiments of Anis et al. (1981).

In conclusion, it seems that the locust muscle Quis-GluR can enter a desensitized state(s) from its closed channel conformation(s). It is also likely that the receptor has at least two desensitized states from which it exits at different rates. These conclusions when set alongside the agonist-sensitivity of recovery from desensitization lead to the following model, which is an extension of that described by Kerry et al. (1987, 1988):

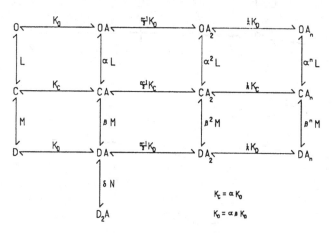

K_0, K_c, K_D, are microscopic binding constants for association of agonist (A) with open (O) and closed (C) channel and first desensitisation (D) state respectively

L,M,N, are equilibrium constants

α = relative affinity of agonist for open vs closed channel

β = relative affinity of agonist for first desensitisation state vs closed channel

δ = relative affinity of agonist for second desensitised receptor state (D)

Assume : $\delta > \beta > \alpha$

Inhibition of Quis-GluR desensitization

A major difference between nAChR and Quis-GluR resides in their relative responses to concanavalin A (ConA). ConA is one of a group of proteins known as lectins which possess well-defined sugar binding specificities. Treatment of locust muscle at pH 7 with 10^{-6} M ConA inhibits desensitization of Quis-GluR (Mathers and Usherwood, 1976). At lower pH, at which ConA dimerises this lectin no longer blocks desensitization of Quis-GluR (Evans and Usherwood, 1985). Succinyl-ConA, which is a dimer at pH 7, fails to block desensitization and this is true also for lectins which do not bind to mannose and glucose moieties of cell surface proteins. Significantly, when ConA at pH 7 is applied to desensitized Quis-GluR it does not reverse desensitization, and following removal of agonist it is no longer possible to inhibit desensitization of Quis-GluR with a further application of lectin. This suggests that ConA binds to the same site as the effective and refractory forms of the receptor, and that the lectin remains bound to the receptor during recovery from desensitization following removal of agonist. Presumably, this is sufficient to preclude appropriate interactions between Quis-GluR and ConA which are necessary to inhibit desensitization during future applications of agonist. Con A also selectiviely reduced desensitization of mammalian neuronal Quis-GluR (Mayer and Vyklicky, 1989).

REFERENCES

Anholt, R., Lindstrom, J., and Montal, M., 1985, The molecular basis of neurotransmission: Structure and function of the nicotinic acetylcholine receptor, In "Enzymes of Biological Membranes", A.N. Martonosi, Ed., Plenum, New York.

Anis, N.A., Clark, R.B., Gration, K.A.F., and Usherwood., P.N.R., 1981, Influence of agonists on desensitization of glutamate receptors on locust muscle, J. Physiol. Lond., 312:345.

Anwyl, R., and Usherwood, P.N.R., 1974, Voltage clamp studies of glutamate synapse, Nature, Lond., 252:591.

Boyd, N.D., and Cohen, J.B., 1980, Kinetics of binding of $[^3H]$-acetylcholine and $[^3H]$-carbamylcholine to Torpedo postsynaptic membranes: Slow conformational transitions of the cholinergic receptor, J. Biochem., 19:5344.

Chesnut, T.J., 1982, Components of desensitization at the frog neuromuscular junction and the effect of metabolic inhibtors, Cell Mol. Biol., 2:59.

Chesnut, T.J., 1983, Two-component desensitization at the neuromuscular junction of the frog, J. Physiol. Lond., 336:339.

Clark, R.B., Gration, K.A.F., and Usherwood, P.N.R., 1979, Desensitization of glutamate receptors on innervated and denervated locust muscle fibres, J. Physiol. Lond., 290:551.

Clark, R.B., Gration, K.A.F., and Usherwood, P.N.R., 1982, Influence of sodium and calcium ions and membrane potential on glutamate receptor desensitization, Comp. Biochem. Physiol., 72C:1.

Daoud, M.A.R., and Usherwood, P.N.R., 1978, Desensitization and potentiation during glutamate application to locust skeletal muscle, Comp. Biochem. Physiol., 59C:105.

Dudel, J., Franke, Ch., Hatt, H., Ramsey, R.L., and Usherwood, P.N.R., 1988, Rapid activation and desensitization by glutamate of excitatory, cation-selective channels in locust muscle, Neurosci. Lett., 88:33.

Evans, M., and Usherwood, P.N.R., 1985, The effects of lectins on desensitization of locust muscle glutamate receptors, Brain Res., 358:34.

Feltz, A., and Trautmann, A., 1982, Desensitization at the frog neuromuscular junction: a biphasic process, J. Physiol. Lond., 322;257.

Franke, C., Hatt, H., and Dudel, J., 1987, Liquid filament switch for ultrafast exchanges of solutions at excised patches of synaptic membrane of crayfish muscle, Neurosci. Lett., 77:199.

Gration, K.A.F., Lambert, J.J., and Usherwood, P.N.R., 1980, A comparison of glutamate single-channel activity at desensitizing and non-desensitizing sites, J. Physiol. Lond., 310:49P.

Gration, K.A.F., Lambert J.J., and Usherwood, P.N.R., 1981, Glutamate-activated channels in locust muscles, Adv. Physiol. Sci., 20:377.

Hatt, H., Zuffall, F., Smith, D.O., Rosenheimer, J.L., and Frank, Ch., 1989, Glutamate and quisqualate rapidly activate and desensitize channels in vertebrate central neurons, Proc. 15th Gottingen Neurobiol. Conf. (In Press).

Hess, G.P., Pasquale, E.B., Walker, J.W., and McNamee, M.G., 1982, Comparison of acetylcholine receptor-controlled cation flux in membrane vesicles from Torpedo californica and Electrophorus electricus: Chemical kinetic measurements in the millisecond region, Proc. Natl. Acad, Sci. USA, 79:963.

Katz, B., and Thesleff, S., 1957, A study of desensitization produced by acetylcholine at the motor end-plate, J. Physiol. Lond., 138:63.

Kerry, C.J., Kitts, K.S., Ramsey, R.L., Sansom, M.S.P., and Usherwood, P.N.R., 1987, Single channel kinetics of a glutamate receptor, Biophys. J., 51:137.

Kerry, C.J., Ramsey, R.L., Sansom, M.S.P. and Usherwood, P.N.R., 1988, Glutamate receptor channel kinetics: The effect of glutamate concentration. Biophys. J., 53:39.

Kitts, K.S., and Usherwood, P.N.R., 1988, Ion-selectivity of single glutamate-gated channels in locust skeletal muscle, J. exp. Biol., 138:499.

Magazanic, L.G., and Vyskocil, F., 1970, Dependence of acetylcholine desensitization on the membrane potential of frog muscle fibre and the ionic changes in the medium, J. Physiol. Lond., 210:507.

Magazanic, L.G., and Vyskocil, F., 1975, The effect of temperature on desenitization kinetics at the post-synaptic membrane of the frog muscle fibre, J. Physiol. Lond., 249:285.

Mathers, D.A., and Usherwood, P.N.R., 1976, Concanavalin A blocks desensitization of glutamate receptors of locust muscle. Nature, Lond., 259:409.

Mayer, M.L.,and Vyklicky, L., 1989, Concanavalin A selectivity reduces desensitization of mammalian neuronal quisqualate receptors, Proc. Natl. Acad. Sci. USA, 86:1411

McNamee, M.G., and Ochoa, E.L.M., 1982, Reconstitution of acetylcholine receptor function in model membranes, Neuroscience, 7:2305.

Patlak, J., Gration, K.A.F., and Usherwood, P.N.R., Single glutamate-activated channels in locust muscle, Nature, Lond., 278:643.

Sakmann, B., Patlak, J., and Neher, E., 1980, Single acetylcholine-activated channels show burst-kinetics in presence of desensitizing concentrations of agonist, Nature, Lond., 286:71.

Sansom, M.S.P., and Usherwood, P.N.R., 1989, Single channel studies of glutamate receptors, Prog. in Neurobiol., (In Press).

Sine, S., and Taylor, P., 1979, Functional consequences of agonist-mediated state transitions in the cholinergic receptor, J. Biol. Chem., 254:3315.

Slater, N.T., Hall, A.F., and Carpenter, D.O., 1984, Kinetic properties of cholinergic desensitization in Aplysia neurons., Proc. Roy. Soc. Lond., B233:63.

Tang, Ch., Dichter, M., and Morad, M., 1989, Quisqualate activates a rapidly inactivating high conductance ionic channel in hippocampal neurons, Proc, Natl. Acad. Sci. USA, 85:2834.

Trussell, L.O., Thio, L.L., Zorumski, Ch. F., and Fischback, G.D., 1988, Rapid desensitization of glutamate receptors in vertebrate central neurons, Proc. Natl. Acad. Sci. USA, 85:2834.

Usherwood, P.N.R., 1969, Glutamate sensitivity of denervated insect muscle fibres. Nature, Lond., 223:411.

Usherwood, P.N.R., 1978, Amino acids as neurotransmitters, Adv. comp. Physiol. Biochem., 7:227.

Usherwood, P.N.R., 1981, Glutamate synapses and receptors on insect muscle, In "Glutamate as a Neurotransmitter", G. Di Chiara and G. L. Gessa, Eds., Raven, New York.

Usherwood, P.N.R., and Machili, P., 1968, Parmacological properties of excitatory neuromuscular synapses in the locust, J. exp. Biol., 49:341.

EFFECT OF ELECTROCHEMICAL PROCESSES AND ELECTROMAGNETIC
FIELDS ON BIOLOGICAL SYSTEMS

MICROWAVES AFFECT THE FUNCTION OF RECONSTITUTED

AND NATIVE RECEPTOR MEMBRANES OF THE BRAIN

Oleg V. Kolomytkin, Vladimir I. Kuznetsov, and Inal G. Akoev

Institute of Biological Physics
Academy of Sciences of the USSR
Puschino, Moscow Region, U.S.S.R.

INTRODUCTION

The concept that the cellular membrane is a critical structure to be affected by microwaves has been generally prevalent in recent years[1-3]. Therefore much attention has been given to study on natural and artificial membranes. Artificial bimolecular lipid membranes (BLM) modified with channel-forming substances have proved adequate models to determine the effect of microwaves on the functioning of ionic channels in the membrane. We have succeeded in estimating the microwaves action on the parameters of both a single ionic channel and a membrane with a lot of channels. A substantial effect of microwaves on the conductance of ionic channels formed by antibiotics: alamecithin, amphotericin B, gramicidin A has been found. It has been shown that the effect of microwaves depends on microwave intensity, exposure time, electrolyte concentration and ion type[4].

The major conclusion of these studies is: It is impossible to explain the microwave action by total heating of the solution in the chamber. The mean specific adsorbtion rate (SAR) in the chamber was too low to explain the observed results. All other things being equal, the effect decreases with increasing microwave frequency (within interval studied 200-900 MHz) regardless of the nature of the channel-forming agent and the kind of the charge-transferring ion[4].

In the above-mentioned studies channel-forming antibiotics, which are not natural physiological compounds, have been used. The obtained data were to be checked on an experimental model approximating to a physiological one. A BLM with incorporated synaptic membrane fragments binding glutamate has been used for that purpose[5]. The membrane fragments were isolated from giant synaptosomes of rat brain hippocampus. The obtained planar membrane was a model of the postsynaptic membrane of the glutamatergic synapse. On this model the main results were confirmed obtained on artificial membranes modified with channel-forming antibiotics. Besides, we were able to study the microwave effect on a single channel. It was shown that microwaves increased the conductance of the open state of a single channel and slightly decreased the propability for a channel to be in the open state. In both experiments all effects were much more pronounced than it might be expected proceeding from the measured thermal heating of the whole electrolyte in the chamber. This made us assume that the heating of the solution in the perimembrane layers and in the membrane itself was much higher than that in the total electrolyte solution.

The main limitation of the latter study was that the time parameters of the model postsynaptic membrane were too low compared to those of a naturally occurring membrane. This discrepancy might be due to shortcomings of the technique for preparing BLM. Besides, it remained unclear if microwaves affect other receptor systems of the brain in vitro and in vivo. Investigations described below were carred out to solve the above problems.

The tasks of the present study were: (1) to isolate from rat brain and incorporate into a planar lipid bilayer the most frequently occurring inhibitory synaptic receptor of gamma aminobutyric acid (GABA), (2) to study the mechanism of microwave action on the GABA receptor in a model membrane, (3) to investigate the microwave effect on the GABA receptor in vivo, (4) to investigate the microwave effect on acetylcholinesterase activity in vivo, (5) to compare the results obtained in vitro and in vivo.

METHODS

Purification of the GABA receptor has been discribed earlier[6]. In the purified preparation 1 mg protein contained 10 pmol of GABA specific binding sites. The dissociation constant of the preparation with GABA was 250 nM. Fig.1,a shows an electrophoregram of the synaptic membrane preparation enriched in GABA binding sites. The electrophoresis was performed in 1% sodium dodecylsulphate[6]. This preparation was used to be incorporated into BLM. Fig.1,b shows an electrophoregram of the GABA receptor isolated by means of affinity chromatography[7]. There are two maxima corresponding to α and β subunits of the GABA receptor. The same maxima are present in Fig.1,a.

Incorporation of GABA receptor into BLM. One of the main purpose of our study was to show the possibility to reconstruct the synaptic GABA receptor into BLM. The obtained experimental evidence made us conclude that almost all the available techniques to incorporate the GABA receptor into BLM are not satisfactory. For successful reconstruction of the GABA receptor

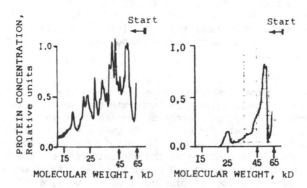

Fig.1. Electrophoregrams of the synaptic membrane preparations enriched in GABA binding sites: (a) preparation used to be incorporated into the BLM, (b) preparation of GABA receptor isolated by means of affinity chromatografy. The electrophoresis was performed in 1% sodium dodecylsulphate and 1% betta mercaptoethanol.

we modified the method proposed by Shindler for reconstruction of the ace-thylcholine receptor[6,8-10]. The synaptic membrane preparation enriched in GABA binding sites was implanted into a planar membrane in three stages: (1) purified synaptic membranes were built into liposomes by means of the deoxycholate dialysis method , (2) proteolipid monolayers were formed from the obtained proteoliposomes at the air-water interface, (3) the bilayer membrane was assembled of two obtained monolayers.

Irradiation of a model membrane by the high frequency electromagnetic field. The lavsan film with an artificial membrane was placed in the inside of the irradiator of condensor type (Fig.2). Two metal plates of the irra-diator were insulated from the solution with 100 μm thick polyethylene film. A high frequency voltage from generators GZ-20 and GZ-21 (0-1 W, 200-1800 MHz) was applied to the plates. The electrical component of the microwaves was directed perpendicular to the membrane plane.

Estimation of distribution of microwaves in the chamber. It is obvious that even in such a simple model system the field distribution is nonuniform. The simple geometry of the chamber used makes it possible to define qualita-tively the distribution of microwaves. The electrical component of the microwave field being perpendicular to the membrane plane, the field must be concentrated in the vicinity of the membrane. In oder to estimate the microwave intensity in proximity to the membrane we developed an original method to estimate the field distribution in the chamber.

Heating of the total electrolyte in the chamber under microwave action. The average specific adsorbtion rate (SAR) of microwaves in the electrolyte was estimated from the rate of the increase in the average temperature of

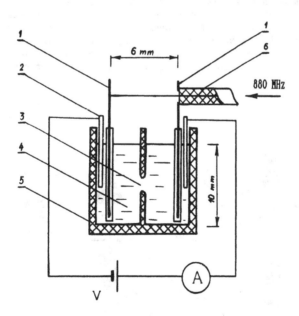

Fig.2. The scheme of the measuring cell. 1 - brass plates for
microwave irradiation of the membrane, the plates were
covered with polyethylene film; 2 - electrodes for
measuring electric current; 3 - aperture; 4 - electrolyte;
5 - teflon cell; 6 - H.F. cable.

the electrolyte in the chamber. The temperature was measured with a micro-thermistor MT-64. The average SAR was in the range of 0-200 W/kg.

Heating of the electrolyte in proximity to the hole on which the BLM was formed. In oder to make this estimation, we studied the effect of microwaves on the electroconductance of the electrolyte in the chamber in the absence of BLM. In this case the electroconductance was determined by the electroconductance of the electrolyte in proximity to the hole, since the area of the hole was much smaller than those of the cross-sections of electrolyte in other places of the cell. The dependence of the electrolyte conductance on temperature was known. Therefore it was possible to estimate the temperature rise in proximity to the hole under microwave action by measuring the conductance of the electrolyte.

Membrane heating. In oder to estimate more accurately the heating of the membrane itself under the action of microwaves we performed an experiment with BLM in the presence of dinitrophenol, the carrier of protons across the membrane. It is known that in the presence of this carrier the penetration of dinitrophenol anions through the lipid bilayer is a limiting stage of ion transport through an artificial membrane. Therefore changes in the current across the membrane would reflect the processes of ion transport in the membrane itself. The temperature dependence of the conductivity of BLM in the presence of dinitrophenol is known[11]. Suppose that the microwave action on the conductance of the membrane with the carrier dinitrophenol has a thermal mechanism. In this case we can estimate an equivalent rise in the temperature of the membrane under the microwave action.

The results obtained are presented in Fig.3. Under the action of microwaves the greater heating is registered in the parts of the solution adjacent to the membrane. The rate of the average temperature increase immediately after turning on microwaves was $0.025°$ K/s in the total solution, $0.3°$ K/s in vicinity of the membrane and $1.5°$ K/s in the membrane. Based on the obtained data we may conclude that when studying the mechanism of action of microwaves on ionic channels in BLM the contribution of the thermal effect of microwaves on the electrolyte surrounding the membrane should be taken into account.

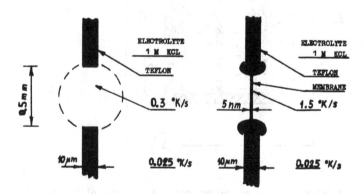

Fig.3. Scheme of distribution of the rate of temperature increase in the cell immediately after switching on microwaves. The diameter of the aperture as well as the widths of lavsan film and membrane are also presented.

Microwave irradiation of the animals. In oder to investigate the microwave action on the GABA receptor and acetylcholinesterase in vivo the whole animal was exposed to the field. Wistar male rats were used (150-200 g). The animals were kept under normal conditions, they were adopted to new conditions and manipulations during 7 to 10 days. Experiments were performed at the same time of the day. Three animals were irradiated in the chamber simultaneously. The microwave frequency was 880 MHz. Microwaves were modulated by rectangular pulses. The frequency of modulation was 3, 5, 7, 16 and 30 Hz (relative pulse duration 85%). Nonmodulated microwaves were used as well. The energy flux of microwaves in the exposure chamber was the same throughout the experiments and in the empty chamber was equal to 1.5 mW/cm^2. The exposure times in different experiments were 5, 15 or 60 minutes. In control experiments the animals were also placed in the chamber (without radiation) for the corresponding time intervals. From 6 to 9 rats were irradiated in each experiment. The number of control (nonirradiated) animals for each experiment was the same as the number of irradiated ones.

Preparation of the synaptic membrane fraction containing GABA receptors and acetylcholinesterase. The animals were decapitated immediately after the exposure. The neocortex was removed and homogenized at 4 $^{\circ}$C. In parallel, the control animals were decapitated. A fraction of synaptic membranes was obtained as described earlier[12].

Determination of muscimol specific binding by synaptic membranes. Radioactive (^{3}H)muscimol was used as an agonist of the GABA receptor. The obtained membranes were incubated with 0.12 μM of (^{3}H)muscimol. Then the total muscimol binding with the GABA receptor was determined by the filtration method[13]. The nonspecific binding of muscimol was determined in the presence of 10 mM GABA. The specific muscimol binding was obtained by subtraction of the nonspecific binding from the total binding. Each probe was determined three times.

Determination of acetylcholinesterase activity. The activity of the enzyme in the fraction of synaptic membranes was determined by the method of Ellman[14]. The method is based on the interaction of thiocholine, the product of acetylthiocholine hydrolysis, with the dithiobisnitrobenzoic acid resulting in production of 5-tio-2-nitrobenzoic acid. The rate of the reaction was measured at 25 $^{\circ}$C by using a spectrophotometer at a wave length of 412 nm.

RESULTS

Chemosensitivity of the model membrane with GABA receptors. The action of different substances on the model membrane was studied in order to prove the native state of the GABA receptor in the membrane. For that purpose substances were added into compartments of the cell. 10 μm acetylcholine when added to the cell did not influence the membrane conductance, neither 10 μm glutamate nor noradrenaline did. At the same time, 1 μm GABA added to the cell increased the ion conductance of the membrane (Fig.4). For different membranes the electroconductance increased 8-15 times compared to the initial level before the addition of GABA. Addition of another portion of 1 μm GABA resulted in an additional increase in membrane conductance. It has been known from electrophysiological studies that picrotoxin blocks the chlorine channels of the postsynaptic GABA receptor[20]. 10 μm picrotoxin introduced to the cell decreased the electroconductance of model membranes (Fig.4). We carried out more than 30 experiments to define the effect of the above substances on the model membrane. It should be noted that the data on the pharmacological properties of the membrane presented in Fig.4 are reliable and well reproducible.

In oder to record a single channel we formed a membrane from proteoliposomes with a protein-lipid ratio of 1:120. After adding 1 μm GABA to
the cell, jump-like changes in transmembrane current were observed at a
fixed membrane voltage (Fig.5). This result suggests that GABA molecules
open GABA-sensitive ion channels incorporated into BLM. Comparative analysis
of the model and natural membranes permits the conclusion that the considered electrical properties of model and natural membranes are quite
similar. Therefore, it can be assumed that the GABA receptors that were
built into the planar BLM were native.

Effect of high frequency electromagnetic field on a model membrane
with GABA receptors. The planar BLM with numerous incorporated GABA
receptors was considered as a model of the postsynaptic membrane of the
GABAergic synapse. Fig.6 shows the microwave effect on the ionic conductance
of the membrane in the presence of 2 μm GABA. It is seen that after the

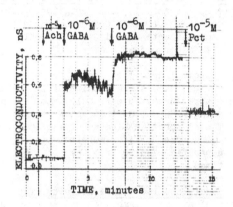

Fig.4. Effect of acetylcholine, GABA and picrotoxin on the integral
conductance of a model membrane. The experimental cell
contained 0.1 M KCl, 10 mM Tris-HCl, pH 7.4, membrane voltage
was 50 mV. The arrows show the moments at which the above
substances were added on one membrane side. At the moment of
addition of picrotoxin the recorder was disconnected from
the measuring system.

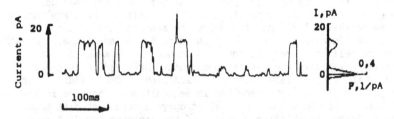

Fig.5. The record of a single channel of the GABA receptor.
The cell contained 1 μm GABA, 1 M KCl, 10 mM Tris-HCl,
pH 7.4. The membrane voltage was 100 mV. The right part
shows the probability density function of current amplitudes
during single channel events.

onset of irradiation the conductance of the membrane rapidly increases but
then the increase is sharply inhibited. After turning off microwaves the
conductance rapidly decreases down to the initial value and then remains
constant. Repeated switching on of the field resulted in repeating of the
effect. The effect was studied depending on the modulation frequency of
microwaves, the average SAR of the total solution in the chamber was the
same in all experiments (100 W/kg). The effects of the nonmodulated and
modulated (rectangular pulses of 5, 7, 16 and 30 Hz) microwaves were the
same (Fig.6).

To elucidate the mechanism of the microwave effect it was interesting
to measure the SAR dependence of the effect. Measurements were performed at
different average SAR (1 to 200 W/kg). The microwave effect on the model

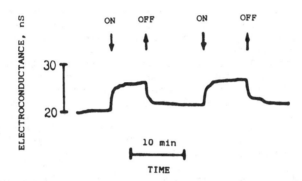

Fig.6. Time change in the electroconductance of the model membrane
with GABA receptors under the action of microwaves (880 MHz).
Instants of switching on and off microwaves are shown by
arrows. The mean SAR in the cell solution was $\approx$100 W/kg.
2μm GABA was present in the solution. The temperature before
switching on microwaves was 20° C.

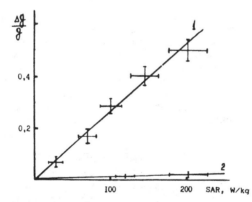

Fig.7. Dependence of a relative change in conductance of the model
membrane with GABA receptors on the average microwave SAR in
the total solution (Curve 1). Microwave frequency was 880 MHz.
For comparison, curve 2 shows a relative change in membrane
conductance due to uniform heating of the total solution during
20 s by means of an electric heater. 2μm GABA was present in
the solution.

155

membrane was reliably registered only at SAR values above ~100 W/kg. Fig.7 presents the experimental dependence of the relative change in membrane conductance on the average SAR absorbed by the solution in the chamber under irradiation. It is seen that the presented dependence is linear in the SAR interval studed.

The results presented in Fig.6 suggest that the synaptic GABA receptors are sensitive to microwaves. A question arises what is the nature of such effect. Whether this effect is due to heating and a high thermal sensitivity of the receptor or it may be explained by other reasons of the nonthermal nature. In oder to clarify this question, we heated the total solution in the chamber by an electric heater for 20 s. The dependence of the relative increase in membrane electroconductance on SAR of the heater was linear. However the effect was much lower compared to the case of microwave irradiation.

In the above two experiments the average SAR of the total solution was measured. Therefore if we assume that the microwave field in the chamber is uniform we can make an erroneous conclusion that the microwave effect has a nonthermal nature. However, as shown above the microwave field in the chamber is nonuniform. The microwave intensity in proximity to the membrane is higher than the average intensity. The obtained results was compared with the thermal dependence of model membrane conductance. This permited us to conclude that the observed effect of microwaves on BLM with the GABA receptor has a thermal nature. The high effect of microwaves is explained by the high SAR in a small region in proximity to the membrane compared to the average SAR. The reversibility of the effect is due to leveling off of the temperature over the whole solution after switching off microwaves. Therewith the temperature of the membrane falls practically to the initial value. Because of this the electroconductance of the membrane decreases to the initial value as well.

Hence, the assumption of the prevalent role of the heat mechanism in the effect of microwaves on synaptic GABA receptors was confirmed experimentally. It should be also noted that this conclusion is in accordance with the fact that the microwave effect on the model membrane does not depend on field modulation.

The obtained result is not trivial because at present a question about the nonthermal mechanism of action of relativly low microwaves ($\sim$1 mW/cm^2) on biological objects is being intensively discussed. Besides, it is known that the whole-body irradiation with low microwave fields may produce significant changes in some physiological parameters. Therefore we decided to study the effect of the whole-body microwave irradiation on the binding of muscimol (agonist of GABA) with the GABA receptor as well as on the acetylcholinesterase activity in rat neocortex. The choice was also determined by that GABA receptors are sensitive to various stress influences[15].

The effect of microwaves on the GABAergic system in rat brain. We studied muscimol binding by synaptic GABA receptors. Muscimol is an agonist of GABA receptors. The study of the concentration dependence of muscimol binding showed that the binding constant of the agonist is the same for irradiated and control animals. Only the concentrations of the agonist binding sites were different. Therefore the changes in muscimol binding are explained by changes in the concentration of GABA receptors.

Table 1 shows the data on the muscimol binding for various times of microwave exposure. The modulation frequency of microwaves was 16 Hz. The data presented show that even a relatively low intensity of microwave radiation results in statistically significant deviations from the control.

Table 1. (^{3}H)muscimol binding with GABA receptor depending on the time of microwave exposure (880 MHz, 1.5 mW/cm^2, modulation frequency of 16 Hz). B - the ratio of muscimol binding with GABA receptor for irradiated rats to muscimol binding with the receptor for the control (nonradiated) rats.

Exposure time, minutes	B	ΔB, standard deviation
5	0.70	0.05
15	0.74	0.04
60	0.82	0.08

The decrease of muscimol binding by GABA receptors was observed in all three experimental groups of animals (5, 15, 60 minutes). It suggests a decrease in the activity of the GABAergic system, which is the inhibiting synaptic transmission. It is seen that the decrease in muscimol binding compared to the control is most significant 5 minutes after the onset of irradiation. The increase of the exposure time results in lesser irradiation effect on muscimol binding by the GABA receptor.

Table 2 represents the data on the muscimol binding for various modulation frequencies of microwaves. The frequencies of 3, 5, 7, 16 and 30 Hz were used. Time of irradiation was 5 minutes. It was shown that the effect

Table 2. (^{3}H)muscimol binding with GABA receptor depending on the modulation frequency of microwaves (880 MHz, 1.5 mW/cm^2, exposure time 5 minutes). B - the ratio of muscimol binding with GABA receptor for irradiated rats to muscimol binding with the receptor for the control (nonradiated) rats.

Modulation frequency, Hz	B	ΔB, standard deviation
3	1.07	0.31
5	0.96	0.17
7	0.83	0.25
16	0.70	0.05
30	1.35	0.30
nonmodulated	1.12	0.13

Table 3. Dependence of acetylcholinesterase activity on the modulation
frequency of microwaves (880 MHz, 1.5 mW/cm^2, exposure time 60
minutes). A - the ratio of acetylcholinesterase activity
for irradiated rats to acetylcholinesterase activity for the
control (nonradiated) rats.

Modulation frequency, Hz	A	ΔA, standard deviation
5	0.85	0.04
16	0.76	0.03
30	0.70	0.04

of modulation frequency was considerable. At modulation frequency 16 Hz all
rats showed a decreased muscimol binding compared to the control. At the
same time in the absence of modulation as well as at modulation frequencies
of 3, 5, 7, 30 Hz no statistically significant difference between irradiated and control animals was observed.

The effect of microwaves on the activity of acetylcholinesterase in the
rat brain. First of all the activity of acetylcholinesterase of the synaptic
membranes from rat brain was studied depending on the time of exposure
(modulation frequency was 16 Hz). Microwave irradiation did not influence
the activity of the enzyme at exposure times of 5 and 15 minutes. At
exposure times of 5 and 15 minutes no statistically significant difference
in the activity of acetylcholinesterase between irradiated and control
animals was observed. Only 60-minute exposure resulted in considerable
changes of the enzyme activity. The ensyme activity decreased by 24 $\pm$ 3%.

Table 3 shows the changes in acetylcholinesterase activity after a
60-minute microwave exposure at different modulation frequencies of microwaves. At all frequencies used, 5, 16 and 30 Hz, the ensyme activity tended
to decrease. At higher frequencies the decrease of ensyme activity was more
significant. The decrease of acetylcholinesterase activity can be considered as an increase in the activity of the acetylcholinergic system.

DISCUSSION

Now let us consider the question of how the modulation frequency influences the biological efficiency of microwaves on the whole-body irradiation.
Earlier Adey et al. have shown that modulation at 16 Hz is especially efficient when microwaves affect the calcium ions efflux from brain tissues [3].
In our experiments at a modulation frequency of 16 Hz all rats showed a
decrease in the concentration of GABA receptors. With no modulation, as
well as at modulation frequencies of 3, 5, 7 and 30 Hz, no statistically
significant effect of microwaves on the concentration of GABA receptors was
revealed. Thus we obtained evidence that on the whole-body irradiation with
a relatively low microwave field (1.5 mW/cm^2) the effect of microwaves on
the binding characteristics of the GABA receptor and on the activity of rat
brain acetylcholinesterase depends on modulation frequency. This dependence

is hardly explainable by the microwave thermal effect only.

At the same time, the experiments with model membranes showed that the microwave effect on the GABA receptor incorporated into BLM is mainly due to heating (with regard to spatial nonuniformity of the field). For the effect to be pronounced a rather high field intensity (average SAR ~100 W/kg) is required. The effect does not depend on modulation frequency.

Based on the above results it can be concluded that in the organism there are systems that are capable of changing the GABAergic receptors even on irradiation with low-intensity microwave fields. The search for such systems will be the task of our future studies.

By the present time a great deal of data have been collected about the effect of microwaves on various physiological and biochemical parameters of the brain structures, single organs and whole organisms. Adequate theoretical explanation of the data obtained presents a complex problem. Nevertheless we believe that to understand the mechanism of action of microwave irradiation it would be reasonable to consider the microwave irradiation primarily as a stress effect.

We came to this conclusion when comparing our results on microwave effect with the literature data on the effect of stress influences (electroshock, pain stress, immobilization) on the number of GABA-receptor-binding sites[15,17]. Experiments with the whole-body irradiation of animals with modulated (16 Hz) microwaves showed that within first 5 min of treatment the most strong decrease of GABA receptor concentration took place. Further, 15 and 60 minutes after irradiation the effect became less pronounced. Of interest is that such a temporal dependence of the GABA receptor concentration in the cortex is observed during an immobilizing stress influence: a drastic decrease in the GABA receptor concentration in the first minutes, then a gradual return to the norm, and in 24 hours even an increase compared to the control concentration (Volkova, Kuznetsov and Kondrashova, personal communication). I.e. different phases in the change of GABA receptor concentration on microwave irradiation are analogous to the stages of the adaptation syndrome.

Besides, the data on the effect of microwave irradiation on the level of GABA in the brain[18] are in line with the results of many works about the effect of stress factors on this parameter[19].

Hence, we can suppose that microwave irradiation switches on a complex of stress reactions in the organism. Further, the organism reactions will probably follow the scenario of General Adaptation Syndrome. The hypothesis about the stress mechanism of microwave effect on animals requires further experimental support.

REFERENCES

1. E. H. Albert, Current status of microwave effects on the blood-brain barrier, J. Microwave Power, 14:281 (1979).
2. A. H. Frey, S. D. Field and B. Frey, Neural function and behaviour defining the relationship, Ann. N.Y. Acad. Sci., 247:433 (1975).
3. W. R. Adey, Tissue interactions with nonionizing electromagnetic fields, Physiol. Rev., 61:435 (1981).
4. I. G. Akoev, V. V. Tyazhelov, O. V. Kolomytkin, S. I. Alexeyev and P. A. Grigoriyev, Study of mechanism of microwaves influence on the model membraneous systems, Izvestia Acad. Sci. USSR, 1:41 (1985).
5. O. V. Kolomytkin, V. I. Kuznetsov and I. G. Akoev, Microwave effect on

ionic channels formed in a lipid bylaer by synaptic membrane
fragments binding glutamate, in: "Intersystem interactions upon
radiation damage", Acad. Sci. USSR, Puschino (1978).

6. O. V. Kolomytkin and V. I. Kuznetsov, Incorporation of gamma-amino-
   buturic acid receptor into planar bilayer, Studia Biophysica,
   115:157 (1986).

7. M. V. Karanova, V.I. Kuznetsov and A.K. Tonkikh, Isolation of GABA
   receptor from rat brain by affinity chromatography using fenazepam,
   Neurochemistry, 4:177 (1985).

8. H. Schindler and U. Quast, Functional acetylcholine receptor from
   Torpedo marmorata in planar membranes, Proc. Natl. Acad. Sci. USA,
   77:3052 (1980).

9. H. Schindler, Exchange and interactions between lipid layers at the
   surface of a liposome solution, Biochim. Biophys. Acta, 555:316
   (1978).

10. O. V. Kolomytkin, Structure of planar membrane formed from liposomes,
    Biochim. Biophys. Acta, 900:145 (1987).

11. A. F. Kozhakaru and V. A. Nenashev, Electrical characteristics of BLM
    in sticky solutions, deposite materials, VINITI N 5422-73, Moscow
    (1973).

12. C. Braestrup and R. F. Squires, Specific benzodiazepine receptors
    in rat brain characterized by high-affinity ^{3}H-diazepam binding,
    Proc. Nat. Acad. Sci. USA, 74: 3805 (1977).

13. J. F. Lever, Quantitative Assay of Binding of Small Molecules to
    Protein: Comparison of dialysis and Mermbrane Filter Assays,
    Analyt. Biochem., 50:73 (1972).

14. G. L. Ellman, K. D. Courthey, V. J. Andress and R. M. Featherstone,
    A new and rapid colorimetric determination of acetylcholinesterase
    activity, Biochem. Pharmacol., 7:88 (1961).

15. G. Biggio, M. G. Corda, G. Demontes, A. Concas and G. L. Gessa,
    Sudden decrease in cerebellar GABA binding induced by stress,
    Pharmacol. Res. Commun., 12:489 (1980).

16. S. M. Bawin and W. R. Adey, Sensitivity of calcium binding cerebral
    tissue to weak environmental electric fields oscillating at low
    frequency, Proc. Natl. Acad. Sci. USA, 73:1999 (1976).

17. W. P. Paul and G. B. Glavin, Restrain stress in biomedical research:
    A review, Neurosc. Behavioral review, 10:339 (1986).

18. K. M. Knieriem, M. A. Medina and W. B. Stavinoha, The levels of GABA
    in mouse brain following tissue inactivation by microwave
    irradiation, J. Neurochem., 28:885 (1977).

19. V. V. Andreev, Y. D. Ignatov, Z. C. Nikitina and I. A. Sytinsky,
    Antistress role of the brain GABAergic system, J. Vyshchei Nervnoi
    Deiatelnosti, 32:511 (1982).

20. G. A. K. Johnston, Neuropharmacology of amino acid inhibitory transmit-
    ters, Ann. Rev. Pharmacol. Toxicol, 18:269 (1978).

THE ACTIVITY OF RESPIRATORY ENZYMES IN MITOCHONDRIA OF HEPATOCYTES OF RATS EXPOSED TO CONSTANT MAGNETIC FIELDS

Eleonora Gorczynska*

Department of Biophysics
Pomeranian Medical Academy
70-111 Szczecin, Poland

INTRODUCTION

Despite the continuing controversy over the influence that magnetic fields might have on physiological processes and the structure of cells, recent experimental evidence suggests that humans exposed to static magnetic fields generated by medical and other technical eqipment may face serious health hazards (Feinendegen and Muhlensiepen 1988, Kavaliers and Ossenkopp 1988, Rudolph et al.1988). There is also evidence that magnetic fields modify biochemical processes of cells and, thereby, have a beneficial effect eg. on bone repair processes (Papatheofanis 1984, Blumlein et al. 1984). It has already been shown that enzyme control mechanisms in proliferating cells are modified by strong static magnetic fields (Feinendegen and Muhlesiepen, 1988) and that respiration processes are disturbed by changes in mitochondrial enzymes following magnetic field exposure (Gorczynska et al. 1986). Perhaps a magnetic field should be considered as a stress generating factor because peripheral blood level of stress indicators such as catecholamines and cortisol increase following exposure to a magnetic field (Barnothy and Sumegi 1969, Dixey and Rein 1982, Gorczynska and Wegrzynowicz 1989). It might be anticipated, therefore, that the changes in cellular respiration could be a consequence of a stress like reaction of an organism exposed to a

---
*Present address: Department of Biophysics, University of New South Wales, P.O.Box 1, Kensington 2033, Sydney, Australia

magnetic field. An increase in cortisol level in rats has been noted after two and three days of exposure to magnetic fields of 0.01 T and 0.001 T (Gorczynska and Wegrzynowicz 1989) but there is published no evidence for respiratory changes within the first week of constant exposure to a magnetic field.

In this paper I report the primary results of experiments which show that mitochondrial respiration of liver cells of rats exposed to constant magnetic fields of 0.01T and 0.001T is significantly altered and that these changes occur within the first days of exposure.

MATERIAL AND METHOD

Wistar male adult rats, aged 4 months and weighing 220-250g were used in the experiments. The first group was exposed to a magnetic field of induction 0.001T whilst the second group was exposed to a field of 0.01T. The third group was treated as controls. Each group contained 100 rats. The experimental animals were exposed to a magnetic field for one hour each day, for a period of ten days. The source of the magnetic field and the procedure of exposure have been described previously (Gorczynska and Wegrzynowicz 1986). Control rats were treated in identical fashion to the experimental animals except that they were not exposed to a magnetic field.

The biochemical studies were undertaken directly after decapitation and removal of the livers of both tested and controls rats. The methods for extracting mitochondria from the liver cells as well as the procedures for polarographic analysis of enzymes activities of the I,II and IV complex of the respiratory chain, i e. oxidoreductase NADH: ubiquinone, succinate dehydrogenase and cytochrome oxidase have been described previously (Gorczynska et al. 1986).

Results are presented below as mean values + standard error of mean for 10 animals in each group. Tests for significance of any differences were done using the Student t-test (P = 0.05 - confidence region).

RESULTS

During the first four days of exposure to magnetic fields there were no differences in enzyme activities between control and experimental animals (Table 1). However, after five days of exposure changes in respiratory enzyme activities were observed in both experimental groups. Rats exposed to 0.001T showed an increase in NADH dehydrogenase activity and succinate dehydrogenase activity. The activity of cytochrome oxidase at this time was not significantly different from controls.In the group of rats exposed to 0.01T the activity of NADH dehydrogenase was significantly increased but that of succinate dehydrogenase and cytochrome oxidase were not different from controls. A maximal increase in all enzymes activities of both experimental groups was observed at either the seventh and eight day of magnetic exposure. This peak occured after seven days in the group exposed to 0.01T when all three enzymes exhibited activities well above controls at this time. In rats exposed to 0.001T maximal activities were observed after eight days and again all three enzymes exhibited activities significantly greater than controls. The activity of respiratory enzymes then decreased, so that by the tenth day in both experimental groups the activities of NADH dehydrogenase and succinate dehydrogenase were much lower, although still higher than in controls. The activity of cytochrome oxidase, however, had returned to normal by this time.

Table 1. The activity of respiratory enzymes in mitochondria of liver cells of rats exposed to magnetic fields (expressed by respiration rates in uMO_2 x min^{-1} x mg^{-1}protein); mean+SEM n=10. Significantly different from control

Enzyme	Magnetic induction (T)	Time of magnetic fields action (days)			
		1-st	4-th	7-th	10-th
				**	**
O_2-uptake NADH dehydro-genase	0.001	1.62+0.1	1.85+0.1	3.52+0.2	2.63+0.2
				**	**
	0.01	1.59+0.1	1.96+0.1	3.76+0.2	2.43+0.1
	control	1.66+0.1	1.58+0.1	1.65+0.1	1.57+0.1
				**	**
O_2-uptake succinate dehydro-genase	0.001	3.17+0.1	3.52+0.1	5.26+0.2	4.04+0.1
				**	*
	0.01	3.08+0.1	3.67+0.2	5.63+0.1	3.80+0.2
	control	3.13+0.1	3.10+0.1	3.22+0.2	3.15+0.1
				*	
O_2-uptake cytochrome oxidase	0.001	6.28+0.4	6.42+0.5	7.61+0.4	6.85+0.5
				*	
	0.01	6.25+0.4	6.43+0.5	7.59+0.5	5.64+0.4
	control	6.33+0.4	5.90+0.4	6.12+0.5	6.22+0.4

* P 0.05   ** P 0.001

DISCUSSION

The absence of changes in the activity of mitochondrial enzymes of the I, II and IV complex of the respiratory chain during the first four days of exposure to a magnetic field suggests that neither functional disturbances nor structural in mitochondria have occured. However, sustained longer term exposure causes major structural changes in mitochondria of hepatocytes, ranging from swelling and disintegration of the matrix to reduction of the cristae (Gorczynska and Wegrzynowicz 1989). The present results show that between the fifth and tenth days of exposure there is also an increase in the activities of respiratory enzymes which reach a maximum after seven and eight days. It has been shown that the level of blood cortisol is maximal after two and three of exposure of rats to a magnetic field. This high level of blood cortisol is maintained for the duration of the exposure (Gorczynska and Wegrzynowicz 1989). Mitochondria seem to be highly sensitive indicators of stress. It is possible that disturbances of mitochondrial structure and function observed after five days of exposure to a magnetic field might be a result of stress. Usually stress reactions involve a mobilising phase when the activities of respiratory enzymes of mitochondria increase. However, studies performed previously over a seven week exposure period showed cyclic changes in the activities of respiratory enzymes in rats exposed to 0.008T (Gorczynska et al.1986).

In a medical or technical environment in which magnetic resonance imaging, NMR, EPR and thermonuclear devices are located the range of intensities of magnetic fields is much higher than those employed in the present studies. Numerous investigations, however, indicate that lower intensities of the magnetic fields seem to result in more obvious changes within cells, tissues and organis (Cremer-Bartels et al. 1984, Kavaliers et al. 1984, Welker et al. 1983). It was found that the application of a magnetic field of 1T did not cause a stronger respiratiry effect than the application of 0.008T (Cook et al. 1969). Thus to modify biochemical responses of cell both duration and intensity of applied magnetic field seem to be important.

The molecular mechanisms underlying the influence of magnetic fields on functioning of mitochondria have not yet been identified. The molecular structure of succinate dehydrogenase and NADH dehydrogenase (system Fe-S-Fe) is different from that of cytochrome oxidase (system Fe-Fe). Different values of the magnetic moments of paramagnetic particles might account for these differences. It has been suggested that the dynamic structure of phospholipides might be disturbed by magnetic forces (Speyer et al. 1987). The changes in the orientation of phospholipides in an internal mitochondrial membrane could influence the protein systems transmitting protons and electrons.

To understand the physical principles and to establish the molecular mechanisms of magnetic field action upon cell, further investigations are clearly needed. The data presented herein are an extention of previously performed studies concerning the respiratory processes in mitochondria of rat liver cells and show that even low intensities of magnetic fields applied over a period of only a few days might modify fundamental cellular processes.

REFERENCES

Barnothy, M.F. and Sumegi, I., 1969, Abnormalities in organs of mice induced by a magnetic field, Nature 225: 270.

Blumlein, H., Schneider U., Rahn, B.A., and Perven, S.M., 1984, Magnetfeld und Wechselstrom zur Behandlung von experimentellen Pseudarthrosen, Orthopade, 13:102.

Cook, E.S., Fardon I.C., and Nuttini L.G., 1969, Effects of magnetic fields on cellular respiration, Biological effects of magnetic fields. Plenum Press, N.York-London 2: 67.

Cremer-Bartels, G., Krause, K., Mitoskas, G., and Brodersen, D., 1984, Magnetic field of the earth as additional zeitgeber for endogenous rhythm, Naturwissenschaften 11:567.

Dixey, R., and Rein, G., 1982, 3-Noradrenaline release potentiated in a clonal nerve cell line by low intensity pulsed magnetic field, Nature, 296:253.

Feinendegen, L.E., and Muhlesiepen, H., 1988, Effect of static magnetic field on cellular metabolism in the living mouse Endeavour, 12:119.

Gorczynska, E., Galka, G., Wegrzynowicz, R., and Mikosza, H., 1986, Effect of magnetic field on the process of cell respiration in mitochondria of rats, Phys. chem. Phys. Med. NMR, 18:61.

Gorczynska, E., and Wegrzynowicz, R., 1986, Magnetic field provokes the increase of prostacyclin in aorta of rats, Naturwissenschaften, 73:675.

Gorczynska, E., and Wegrzynowicz, R., 1989, Structural and functional changes in organelles of liver cells in rats exposed to magnetic fields. Environ. Res., 1989, in review.

Kavaliers, M., Ossenkopp, K.P., and Hirst, M., 1984, Magnetic field abolish the enhanced nocturnal analgesic response to morphine in mice, Physiol. Behav., 32:261.

Kavaliers, M., and Ossenkopp, K.P., 1988, Day-night rhythms of opioid stress- induced analgesia: differential inhibitory effects of exposure to magnetic fields, Pain, 32:223.

Papatheofanis, F.J., 1984, A review on the interaction of biological systems with magnetic fields. Phys. Chem. Phys. Med. NMR, 16:251.

Rudolph, K., Wirtz-Justice, A., Krauchi, K., and Feer, H., 1988, Static magnetic fields decrease nocturnal pineal cAMP in the rat. Brain Res., 446:159

Speyer, J.B., Sripada, P.K., Gupta, S.K., and Shipley, G.G., 1987, Magnetic orientation of sphingomyelin-lecitin bilayers, Biophys. J., 51:687.

Welker, H.A., Semm, P., Willing, R.P., Commentz, I.C., Wiltschko, W.,and Vollrath, L. 1983, Effects of an artificial magnetic field on serotonin N-acetyltransferase activity and melatonin content of the rat pineal gland, Exp. Brain Res., 50:426.

CHARGE-FIELD INTERACTIONS IN CELL MEMBRANES AND ELECTROCONFORMATIONAL

COUPLING: TRANSDUCTION OF ELECTRIC ENERGY BY MEMBRANE ATPases

Tian Yow Tsong

Department of Biochemistry
University of Minnesota College of Biological Sciences
St. Paul, Minnesota 55108

R. Dean Astumian

National Institute of Standards And Technology
Gaithersburg, Maryland 20899

I. Introduction

Electron and ion transports, reductions and oxidations are common chemical reactions in cellular energy transductions. These reactions involve charges and are, thus, susceptible to influence by an electric field. Many laboratories have recently discovered that electric fields in certain ranges and frequencies can stimulate DNA, RNA, and protein biosyntheses (Blank & Findl, 1987; Cleary, 1987). Enhancement and suppression of enzyme activities have also been reported (Blank & Findl, 1987). This paper will discuss our experiments in which electric fields of defined frequency and amplitude were used to activate mitochondrial ATPases of rat liver and beef heart and the (Na,K)-ATPase of human erythrocytes (see Tsong et al., 1989 for references). Electro-conformational Coupling which we proposed earlier (Tsong & Astumian, 1986, 1987, 1988; Tsong et al., 1989a; Tsong, 1989; Astumian et al., 1989) will be used to explain these results. The following two papers (Astumian et al. & Robertson et al. in this volume) will elaborate the concept in greater details and analyze some simple reaction mechanisms

167

which are relevant to the present discussion. We will also show that non-linear effects arising from the coulombic interaction of membrane proteins and an electric field are responsible for the many observed effects.

II. Electric Field Induced ATP Synthesis by Mitochondrial ATPases

Submitochondrial particles (SMP) from rat liver and beef heart were used in these experiments. Cyanide and rotenone were added to the suspending medium to block the electron transport reaction. Any ATP synthesis from such a system would have to derive free energy from the applied electric field. In most of our experiments, electric pulses of 10 to 35 kV/cm and of duration 5 to 100 μsec were discharged into the SMP suspension through a capacitor. ATP formed in the suspension was detected by the luciferin/luciferase phosphorescence assay or by using ^{32}P-radioactive labeling of ATP. In the latter case, the newly formed ATP was rapidly converted to glucose-6-phosphate by hexokinase and the amount of glucose-6-phosphate was determined by extraction and radioactive counting. Both methods gave consistent result. Fig. 1 gives the result of one such experiment (Teissie et al., 1981) and Table 1 shows the effects of various inhibitors and ionophores on this activity. Several laboratories have obtained similar results in other ATP synthetic membranes and in purified and reconstituted systems (see Tsong & Astumian, 1986 for a review). The electric fields used in these experiments can generate transmembrane potential in the order of 200 mV, or an effective field of 400 kV/cm across the lipid bilayer. Later, low intensity alternating electric fields were also used (Chauvin et al., 1989). The results of these studies are summarized below.

1). When high intensity (30 kV/cm, an induced $\Delta\psi$ of 400 mV), exponentially decaying electric pulses (decay constant of 100 μsec) were used, the ATP yield was less than one per pulse per ATPase in the absence of DTT (dithiothreitol). In the presence of DTT the yield reaches 5-8 ATP per pulse per ATPase (beef heart, Fig. 2).

2). When an a.c. field of 60 V/cm was used the ATP yield was approximately 10^{-3} ATP per cycle per ATPase. The low yield per cycle was expected, as the field intensity was low and the interaction energy between the ATPase and the applied field would be small (see below). The yield decreased with increasing frequency and diminished around 10 kHz. The a.c. method permitted a prolong exposure of a sample to the electric field (lasting a few hours) and several ATPs per enzyme have been produced.

168

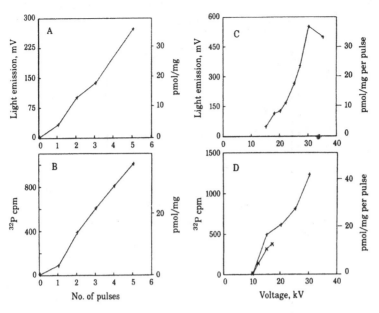

Fig. 1 ATP synthesis induced by pulsed electric field in rat SMP.
See Teissie et al. (1981) for experimental details.

Table 1

Effects of inhibitors of oxidative phosphorylation and of
ionophores on the electric pulse-driven ATP synthesis by SMP

Detection conditions	Luminescence		Radiolabeling	
	Signal, mV	ATP synthesized, pmol/mg protein per pulse	Signal, cpm	ATP synthesized, pmol/mg protein per pulse
Complete system	95 ± 10	10 ± 1	600 ± 100	17 ± 3
+Oligomycin				
1 μg	± 10	<1	ND	ND
2 μg	ND	ND	<50	<1.5
10 μg	± 10	<1	ND	ND
+Venturicidin				
0.6 μg	ND	ND	<50	<1.5
10 μg	± 10	<1	ND	ND
+Aurovertin				
5 μg	ND	ND	<50	<1.5
10 μg	± 10	<1	ND	ND
+DNP 10 μg	ND	ND	<50	<1.5
+DCCD 1 μg	ND	ND	<50	<1.5
+CCCP 1 μM	ND	ND	<50	<1.5
+FCCP 1 μg	ND	ND	<50	<1.5
+A23187 1 pM	50 ± 10	5 ± 1	ND	ND
+Valinomycin 1 μg	25 ± 10	2.5 ± 1	ND	ND
+Ethanol 1%	95 ± 10	10 ± 1	ND	ND

ND, not determined; DNP, 2,4-dinitrophenol; CCCP, carbonyl cyanide m-chlorophenylhydrazone; FCCP, carbonyl cyanide p-trifluoro-
methoxyphenylhydrazone. ± indicates SD. Counting uncertainty of
radiolabeling was ±50 cpm. Experimental conditions for both assays
were as described for Table 1. Inhibitors and ionophores were added
in ethanolic solution, except venturicidin, which was in methanol. See
text for details.

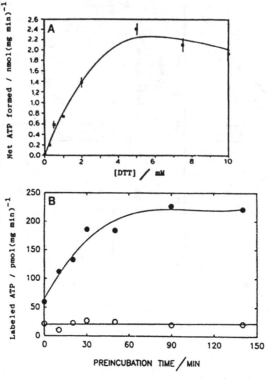

Fig. 2 ATP synthesis with pulsed electric fields in beef heart SMP.

A.   ATP yield of samples exposed to five electric pulses of 25 kV/cm, 100 μs decay time. Effect of DTT is shown. At the plateau, ATP yield was approximately 5-8 per pulse per enzyme.

B.   ATP formation with an a.c. field of 60 V/cm, 30 Hz. Open circles, ATP yield from samples in which the natural inhibitor peptide was not removed. Closed circles, ATP yield in samples where the natural inhibitor peptide was removed by an incubation of SMP in a high ionic strength buffer. X axis denotes the time of incubation.

3).   NEM (N-Ethylmaleimide), a sulfhydryl modifying reagent, inhibited both the d.c. and the a.c. inducible ATP synthetic activity. This fact and the fact that DTT was required suggest that certain -SH groups were involved in the transduction of electric field energy and the oxidation/reduction cycle of them enabled the enzyme to turnover.

4).   Inhibitors of F_0F_1ATPase and membrane ionophores suppressed electric field induced ATP synthesis.

5).   Purified and reconstituted mitochondrial ATPase could also synthesize ATP when exposed to high intensity electric pulses. However, in these cases, ATP yield was low, approximately 0.3 ATP per pulse (30 kV/cm) per enzyme.

III. A.C. Induced Cation Pumping by (Na,K)-ATPase

Human red blood cells were used for the experiment. Movements of sodium and rubidium ions into and out of the cytoplasm were monitored by using radioactive tracers, $^{22}Na^+$ and $^{86}Rb^+$. Erythrocytes in an isotonic suspension were exposed to an alternating electric field of up to 50 V/cm and frequency of up to 10 MHz. Na^+ and Rb^+ uptake and efflux were monitored and those due to the pumping of (Na,K)-ATPase were determined by comparing the activity of ouabain inhibited samples. We have only detected ouabain inhibitable Rb^+ uptake and Na^+ efflux. Rb^+ efflux and Na^+ uptake were not stimulated by the a.c. field. The conditions were chosen to demonstrate that the stimulated activities were active pumping of the cations by the (Na,K)-ATPase. Passive permeations of Na^+ and Rb^+, which were not blocked by 0.2 mM of ouabain in either direction, were insensitive to electric stimulation. Fig. 3A shows the frequency dependent Na^+ and Rb^+ pumping activity stimulated by an a.c. field of 20 V/cm. Other observations are summarized (Serpersu et al., 1983, 1984; Liu et al., 1989):

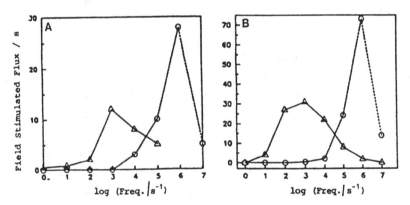

Fig. 3   Frequency dependence of a.c. induced cation pumping.

A.   Experimental observation using a 20 V/cm a.c. field. Triangles are data for Rb^+ uptake and open circles for Na^+ efflux.

B.   Simulation of the electroconformational coupling model based on the four state kinetics of Eq. (7). See Tsong et al. (1989b) for details.

1).   The electric field stimulated only the active transport modes of the enzyme. Ion exchange modes were insensitive to the applied field. And the stimulated activity was completely inhibited by ouabain, a specific inhibitor of (Na,K)-ATPase.

2). There was an optimum field strength of 20 V/cm (peak-to-peak) for activating both the Na^+ and the Rb^+ pumps. Lower and higher field strengths led to diminished activities. We interpret this phenomenon to be due to the effect of induced dipole moments which counter the effect of permanent dipoles (see below).

3). Under the optimum field strength of 20 V/cm, there was an optimum frequency of 1 kHz for the Rb^+ pump and 1 MHz for the Na^+ pump.

4). The maximum field induced activity was 15-30 ions/enzyme-sec for the Na^+ pumping and 10 to 20 ions/enzyme-sec for the K^+ pump. The 3 to 2 ratio of Na^+/Rb^+ transport was not strictly adhered to in each experiment, but the average of many experiments reflected this ratio.

5). ATP depleted red cells were used to demonstrate that the a.c. stimulated activity did not require ATP hydrolysis. When the ATP concentration was reduced to less than 10 μM, the a.c. stimulated activity was almost completely retained.

6). The K_m of internal Na^+ for the Rb^+ uptake and Na^+ efflux was 8 mM and for the external Rb^+ was 1.5 mM, with a Hill coefficient of approximately 1.5, consistent with ATP dependent pumping activity of the enzyme. This result suggests that the a.c. stimulated activity occurred by a mechanism similar to that of activation using the hydrolytic energy of ATP.

7). Most experiments were done at 4°C where the ATP hydrolysis activity of the enzyme was negligible. Experiments done at 26°C showed that although the enzyme could further be stimulated by the a.c. field, the extent of stimulation was greatly reduced. At 37°C, where the basal activity of the enzyme was at its maximum, the enzyme could no longer be stimulated by the a.c. field.

These results and the results of ATP synthesis by an applied electric field, in the absence of other energy sources, lead us to conclude that it was the free energy captured from the externally applied field that was doing the chemical work. Since the transmembrane potentials generated by these applied fields would be comparable to the in vivo membrane potentials of the membranes, it is assumed that the observed phenomenon is relevant to the energy transduction mechanisms, in vivo. Thus, electric activation is a valid approach to the study of the actions of these ATPases.

IV. Electroconformational Coupling

To understand how an enzyme can extract free energy from an electric field we have proposed that a membrane protein with charges or which is rich in helices can undergo a conformational transition when exposed to an electric field (Tsong, 1983; Tsong & Astumian, 1986; Tsong et al., 1989a). Effects that due to the interaction of the permanent electric moment will shift the equilibrium constant of the reaction,

$$P_1 \quad \text{<===>} \quad P_2, \tag{1}$$

according to the relationship,

$$K_E = K \exp{-(a\ E_e)}, \tag{2}$$

where K is the equilibrium constant under zero field and K_E is under an electric field of E_e. E_e is the effective field across the cell membrane in which the enzyme is embedded, and $a = \Delta M/RT$, where ΔM is the difference of the permanent electric moments of P_2 and P_1. E_e is related to the applied field E by the Maxwell relationship,

$$E_e = 1.5\ R\ E\ \cos\theta \tag{3}$$

where θ is the angle between the field direction and the normal of the vector describing the position of the enzyme on the membrane surface. This equation predicts that the maximum E_e is 1.5R/d times greater than the applied field, where d is the thickness of the membrane's hydrophobic

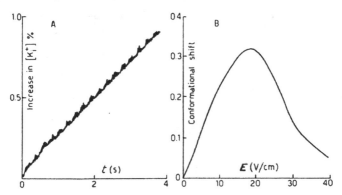

Fig. 4   Simulation of active transport and amplitude of conformational shift induced by an a.c. field.

A.   The four state model of Eq. (7) was shown to pump a substrate against its concentration gradient at a well tuned frequency.

B.   An optimum field strength for inducing conformational shift is demonstrated. This property of the model explains the results of Serpersu & Tsong (1983). For details see Tsong & Astumian (1986).

layer (roughly 5 nm). For a cell the size of human erythrocytes this factor is approximately 1000. Effects due to the interaction of an induced dipole moment and the electric field, on the other hand will shift the equilibrium according to,

$$K_e = K \exp{-(b\ E_e^2)} \tag{5}$$

where b is the difference in the polarizability of the two states (Tsong & Astumian, 1986). The combined effects will have the form,

$$K_e = K \exp[-(a \, E_e + b \, E_e^2)] \qquad (6)$$

From Eq. (6) we note that the effects of an induced dipole depends on the square of E_e and will always have a positive sign. At a high value of E_e the squared term dominates and the protein of Eq. (1) will be locked into either P_1 or P_2 state, thus, inhibiting the conformational oscillation which is critical for the enzyme to extract free energy from the electric field. Eq. (6) predicts that an electric field induced enzyme activation would exhibit a maximal activity at the field strength where the effect due to induced dipoles begins to cancel the effect that is due to the permanent dipoles. This property of the electroconformational coupling model can explain the optimum field strength for the a.c. activation of (Na,K)-ATPase (Fig. 4).

V. A Four State Kinetic Scheme Based on ECC Model

The most interesting results of the ECC model came from our analysis of the four state membrane transport model shown below (Tsong & Astumian, 1986; Westerhoff et al., 1986; Astumian et al., 1987; Astumian et al., 1989; Chen, 1987).

$$\qquad (7)$$

In this scheme a membrane embedded protein P is assume to have two conformational forms, P_1 and P_2. The protein is a pump which can hydrolyze ATP to drive the substrate S from the right hand side to the left hand side of a cell membrane. If P_2 has a higher molar electric moment than P_1 by ΔM and similarly for P_2S_R vs. P_1S_L, then a transmembrane electric field of E_e will shift the equilibrium of the system towards P_2 and P_2S_R and of $-E_e$ will shift the equilibrium towards P_1 and P_1S_L according to Eq. (6). Analysis of this scheme indicates that there are several conditions in which such a field enforced equilibrium shift can lead to a net flow of substrate, S, either to the left or to the right side of the membrane. The direction of the field induced flux would depend on the affinity of the substrate for the P_1 and for the P_2 states. Stronger affinity for P_2 would lead to a clockwise flux and for

P_1 a counter-clockwise flux. The directionality is independent of the polarity of E_e. This latter property suggests that an alternating electric field (a.c.) not only does not cancel out effects due to the positive phase and negative phase of the field but the two phases can combine to drive the kinetic wheel. Scheme (8) summarizes our calculation using a system where the affinity of S to P_2 is 500 times greater than that of S to P_1, and the binding and dissociation of S from the protein are faster than the electroconformational change steps. As shown, the positive a.c. phase induces mainly the flux from P_1 to P_2 and the negative phase induces the flux from P_2S_R to P_1S_L.

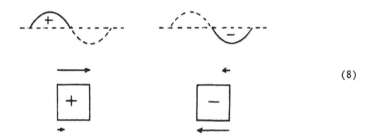

$$(8)$$

Our analysis of Scheme (7) also shows that when the kinetic constants of the scheme are assigned, the efficiency of the system to couple the a.c. field depends on the frequency of the field and that there is an optimal frequency where the coupling is most efficient. An example to simulate the results for (Na,K)-ATPase is shown in Fig. 3B. A more detailed discussion is given in the accompanying papers (Astumian et al. & Robertson et al., this volume).

VI. Enforced Conformational Oscillation of A Michaelis-Menten Enzyme

The four state scheme of Eq. (7) is a typical Michaelis-Menten enzyme embedded in a cell membrane (Tsong et al., 1989b). The catalytic cycle of a Michaelis-Menten Enzyme is,

$$E + S = ES = EP = E + P \qquad (9)$$

Since the enzyme, E, is recovered in each catalysis, the process is cyclic, and Eq. (9) may be re-written as,

E

S   P

ES   EP

$$(10)$$

If the enzyme state which favors binding of S is distinguishable from the state which favors binding of P, Eq. (10) is expanded to Eq. (7).

The membrane compartmentalizes the catalytic reactions and reactions occurring on each side of the membrane are often different. For example, in (Na,K)-ATPase, ATP approaches the enzyme from the side opposite the ouabain binding side but K^+ approaches from the same side. Na^+ binding and ATP hydrolysis occur in the same side. Reactions of this type are generally referred to as vectorial reactions and the transport of ions or substrates against their concentration gradients require energy input. The needed energy is derived from an exergonic reaction, e.g. the hydrolysis of ATP. This study shows that an oscillating electric field can be an energy source for fueling such reactions. Analysis indicates that any oscillating potential which the enzyme can interact with in a productive way will have a similar effect.

The theory of electroconformational coupling provides a logical understanding of cellular energy and signal transductions through the fundamental principle of thermodynamics. Where energy conversion and signal processing of a cell are performed by ordinary enzymes embedded in a cell membrane or in a supramolecular structures.

Acknowledgement

[The work of TYT is supported by a grant from Office of Naval Research.]

References

Astumian, R.D., Chock, P.B., Tsong, T.Y., Chen, Y.-D. & Westerhoff, H.V. (1987). Proc. Natl. Acad. Sci. USA, 84, 434-438.

Astumian, R.D., Chock, P.B., Tsong, T.Y. & Westerhoff, H.V. (1989). Phys. Rev. A, 39, June Issue.

Blank, M. & Findl, E., Editors (1987). "Mechanistic Approaches to Interactions of Electric and Electromagnetic Fields with Living Systems", Plenum Press, New York.

Chauvin, F., Astumian, R.D. & Tsong, T.Y. (1989). Submitted.

Chen, Y.-D. (1987). Proc. Natl. Acad. Sci. USA, 84, 729-733.

Cleary, S.F. (1987). IEEE Eng. Med. Biol., March Issue, 26-30.

Liu, D.-S., Astumian, R.D. & Tsong, T.Y. (1989). Submitted.

Serpersu, E.H. & Tsong, T.Y. (1983). J. Membr. Biol., 74, 191-201.

Serpersu, E.H. & Tsong, T.Y. (1984). J. Biol. Chem., 259, 7155-7162.

Teissie, J., Knox, B.E., Tsong, T.Y. & Wehrle, J. (1981). Proc. Natl. Acad. Sci. USA, 78, 7473-7477.

Tsong, T.Y. (1983). Biosci. Reports, 3, 487-505.

Tsong, T.Y. (1989). TIBS, 14, 89-92.

Tsong, T.Y. & Astumian, R.D. (1986). Bioelectrochem. Bioenerg., 15, 457-476.

Tsong, T.Y. & Astumian, R.D. (1987). Prog. Biophys. Mole. Biol., 50, 1-45.

Tsong, T.Y. & Astumian, R.D. (1988). Ann. Rev. Physiol., 50, 273-290.

Tsong, T.Y., Liu, D.-S., Chauvin, F. & Astumian, R.D. (1989a). Biosci. Reports, 9, 13-26.

Tsong, T.Y., Liu, D.-S., Chauvin, F., Gaigalas, A. & Astumian, R.D. (1989b). Bioelectrochem. Bioenerg., In press.

Westerhoff, H.V., Tsong, T.Y., Chock, P.B., Chen, Y.-D. & Astumian, R.D. (1986). Proc. Natl. Acad. Sci. USA, 83, 4734-4738.

CHARGE - FIELD INTERACTIONS IN CELL MEMBRANES AND ELECTROCONFORMATIONAL
COUPLING: THEORY FOR THE INTERACTIONS BETWEEN DYNAMIC ELECTRIC FIELDS AND
MEMBRANE ENZYMES

R. Dean Astumian,[*] Baldwin Robertson,[*] and Tian Yow Tsong[+]

* Bioprocess Metrology Group
Chemical Process Metrology Division
Center for Chemical Technology
National Institute of Standards and Technology
Gaithersburg, Maryland 20899, USA
and
+ Department of Biochemistry
University of Minnesota
St. Paul, Minnesota 55108

ABSTRACT

Free energy can be transduced from nonstationary electric fields to drive
the synthesis of ATP or the formation and maintenance of concentration
gradients across membranes. Even electrically silent reactions can be driven
against equilibrium by the nonlinear interaction between a dynamic electric
field and a membrane enzyme. There are two conditions necessary for an enzyme
to transduce free energy in an external field to its output reaction. First,
some conformational changes within the catalytic cycle of the enzyme must
involve the intramolecular movement of charge or a change in the dipole moment
of the enzyme. Second, the affinity of the enzyme for substrate must be much
different than the affinity for product. Based on these concepts, we will
compare the kinetic requirements which optimize a protein for transduction of
free energy with those which optimize a protein for rapid catalysis. This
model serves to bridge the classical chemiosmotic mechanism for free energy
transduction, with models based on energy dependent enzyme conformational
changes. Our electroconformational coupling model explains free energy
transduction in terms of the molecular mechanisms of enzyme catalysis.

INTRODUCTION

The conformational transitions of many membrane proteins involve the
intramolecular movement of charge· Among these proteins are voltage gated
channels, electron transport proteins, membrane ATPases, as well as many other
signal and free-energy transducing enzymes[1]. This movement of charge provides

a direct mechanism for the interaction between membrane proteins and external electric fields and endogenous potentials existing across the membranes of all living cells[2].

In another paper in this volume, Tsong and Astumian[3] discuss experimental results suggesting that the Na,K ATPase of erythrocytes can transduce free energy from an oscillating electric field. The field induced pumping of ions by the enzyme was strongly dependent on the frequency of the applied field. The K^+ pumping mode was stimulated most effectively at 1 kHz, while the Na^+ pumping was optimally stimulated at 1 MHz.

In this paper, we propose a four state model for a membrane transporter which is able to transduce free energy from an ac electric field. We show that the alternating, zero average, symmetric field modulates the free energy profile of the catalytic cycle in an asymmetric way, thereby inducing net cyclic flux. In some cases, the efficiency of transduction (power out/power in) is shown to be as high as 60%.

A major requirement for an enzyme to be able to transduce free energy from an ac field is that the affinity for binding of substrate must be much different than the affinity for binding of product. It is also necessary that at least one of the protein conformational transitions in the catalytic cycle involve the movement of charge or a change in dipole moment. It is not necessary that catalyzed substrate === product reaction itself be electrogenic or involve a change in dipole moment.

Free energy transduction occurs only within a band or "window" of ac frequencies for this model. The band of frequencies where flux against the gradient is induced is not related to the turnover time of the enzyme, but rather to the enzymes relaxation times and their spacing. The free energy transduction is due to chemical relaxation and there is no resonant interaction between the field and the model enzyme. It is required, however, that the interaction between the field and the protein conformational transitions be large enough that the intrinsic nonlinearity (equilibrium constants and hence rates and rate constants depend on the exponential of the interaction energy according to thermodynamics) be manifested.

THEORY AND CALCULATIONS

Four State Enzyme

Consider a four state cycle of a membrane enzyme or transporter which catalyzes the electrically silent reaction $S_1 = S_2$. A reasonable physical model of such a protein is a dipolar molecule imbedded in, and extending through a lipid bilayer of low dielectric constant[4]. A kinetic diagram describing the catalytic cycle may be written

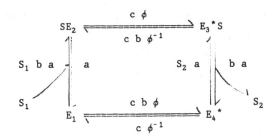

The numerical subscripts are introduced as a convenience in referring to the states. The rate coefficients are given in terms of three parameters, a, b, and c, which are characteristic of the enzyme and independent of the electric field. The a is a scaling factor which sets the time scale for the association - dissociation steps, the c sets the time scale for the conformational transition steps, and b is the zero field equilibrium constant between the states SE_2 and $E_3{}^*S$ and between the states E_1 and $E_4{}^*$. The field dependence of the conformational transitions is given by $\phi = \exp(x\,\psi)$. The x is the effective number of charges moved across the membrane in the conformational transition, and is known as the displacement charge. The ψ is the potential difference between the two surfaces of the membrane multiplied by $e/(2kT)$. Here, e is the electronic charge, k is the Boltzmann constant, and T is the Kelvin temperature. Notice that the field dependence is apportioned equally between the forward and reverse transitions as the simplest possibility. Also for simplicity, we have selected the parameters to display reflection symmetry when the field is zero. To most effectively highlight the role of the electric interaction between a field and an enzyme in free energy transduction, we have taken the reaction catalyzed to be non-electrogenic. Thus, while a dc membrane potential influences the kinetics of the catalysis, it plays no role in determining the driving force of the reaction.

The kinetic behavior of the enzyme in diagram 1 can be simulated by writing down the four differential equations, one for each enzyme state, and solving the system numerically.[5]

$$dE_1/dt = -(b\ a\ S_1 + c\ b\ \phi)\ E_1 + a\ SE_2 + c\ \phi^{-1}\ E_4{}^*, \tag{1}$$

$$dSE_2/dt = -(c\ \phi + a)\ SE_2 + c\ b\ \phi^{-1}\ E_3{}^*S + b\ a\ S_1\ E_1, \tag{2}$$

$$dE_3{}^*S/dt = -(b\ a + c\ b\ \phi^{-1})\ E_3{}^*S + c\ \phi\ SE_2 + a\ S_2\ E_4{}^*, \tag{3}$$

$$dE_4{}^*/dt = -(a\ S_2 + c\ \phi^{-1})\ E_4{}^* + c\ b\ \phi\ E_1 + b\ a\ E_3{}^*S. \tag{4}$$

Rather than solving non-linear differential equations and calculating the change in concentration of S_1 and S_2 as a function of time, as was done by Tsong and Astumian[2], it is typical to keep the concentrations S_1 and S_2 constant and evaluate the probability for the enzyme to be in each state as a function of time by solving the above equations (1)-(4).

These probabilities can then be used to calculate the instantaneous flux through each transition. At steady state where each of the time derivatives in eqs. (1) - (4) is zero and the magnitudes of the fluxes through each transition are equal, the flux around the cycle is clockwise when the concentration S_1 is greater than S_2, counterclockwise when $S_2 > S_1$ and zero when $S_1 = S_2$. The protein behaves as a typical Michaelis-Menton enzyme. This is true irrespective of the value of the dc membrane potential ψ_0.

When an ac field $\psi = \psi_0 + \psi_1 \cos(\omega t)$ is applied, the situation becomes quite different. When $S_2 = S_1$, the ac field causes the enzyme state probabilities to oscillate (see Fig. 1a), and also causes net clockwise flux to occur (see Fig. 1b). The reason for this can be partly understood in terms of the basic (state probability independent) free energy diagram shown in Fig. (2a-c). At zero field, the free energy diagram is characteristic of a very poor enzyme, with large activation barriers and very low energy intermediate states[6,7]. When a positive potential is applied the relative basic free energies of states SE_2 and $E_4{}^*$ are shifted such that $E_4{}^*$ becomes much more stable. Consequently the system must relax to the equilibrium specified according to its new "free energy surface", and the net effect is that SE_2 is converted to $E_4{}^*$ with the concomitant release of either S_1 or S_2. As seen in Fig. (2b), the free energy profile for the reaction 2 => 3 => 4 is smoothed by the field, and the equilibrium shift is accomplished by a lowering of the activation barrier for the net forward reaction. On the other hand, the shift in equilibrium for the process 2 => 1 => 4 is accomplished by raising the activation barrier for the reverse process. Consequently, most of the initial relaxation flux occurs by the former pathway. When the field is reversed, the equilibrium between 2 and 4 is shifted in favor of state 2, and the relaxation occurs on the free energy profile shown in Fig. c. Now the equilibrium shift is accomplished by raising the activation barrier for 4 => 3 => 2 and lowering that for 4 => 1 => 2 and so the bulk of the initial relaxation flux occurs via the latter pathway. This means that in a periodic or stochastic dynamic field, the flux 2 => 3 => 4 in positive field and flux 4 => 1 => 2 in negative field completes net turnover of the enzyme.

There is a frequency dependence of the stimulated cyclic flux as shown in Fig. 3, where it is seen that the frequency dependence of the ac field induced Na^+ and K^+ pumping by the Na,K ATPase can be approximately simulated within the context of the same model but with different values of the parameters.

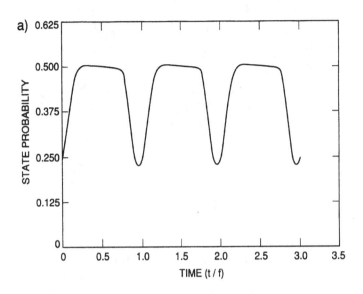

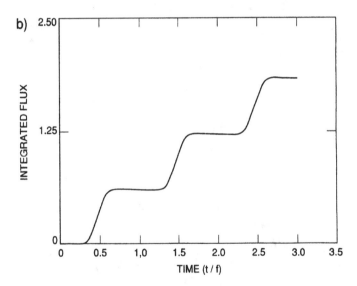

Fig. 1   Demonstration that an applied ac field induces a) enforced
          oscillation in the enzyme state probability SE_2, and b) net average
          clockwise flux even though the catalyzed reaction does not depend on
          the field and is at equilibrium ($S_1 = S_2 = 1$). The values of the
          parameters used for the calculation are a = 1, b = 500, c = 1, ϕ =
          exp [5cos(ωt)]

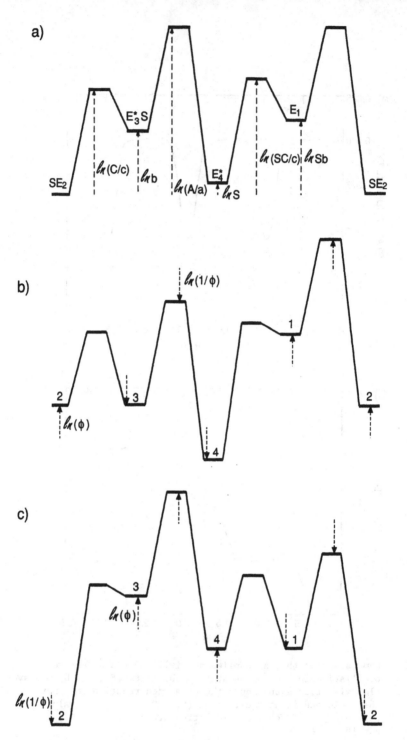

Fig. 2   Free energy profile for the four state enzyme mechanism of the text.
a) $\psi = 0$; b) ψ = positive; c) ψ = negative.

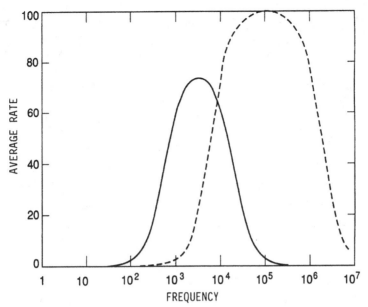

Fig. 3   Demonstration that the average field induced clockwise flux is
frequency dependent, and that the frequency optimum depends on the
kinetic parameters of the enzyme. The parameters used in the
calculation are $S_1 = S_2 = 1$, $b = 10$, $a = 10^4$, $\phi = \exp[1\cos(wt)]$,
and (_____) $c = 10^3$; and (------) $c = 10^5$.

dependence in greater detail. For the special case that the time scaling
factor for the conformational transition is the same as for the association-
dissociation steps, the frequency optimum is approximately $\underline{a} = \underline{c}$ as seen in
Fig. 4a, and the bandwidth is governed by $\underline{b}$ as seen in Fig. 4b.. The larger b
is, the larger the bandwidth, and in the limit that $b \to \infty$, the rate saturates
at high frequency and displays no optimum. If $\underline{a}$ ($\underline{c}$) is increased or decreased
keeping $\underline{c}$ ($\underline{a}$) constant, the optimal frequency goes up or down, respectively,
as seen in Fig. 4c (d) and the plateau becomes broader in both cases. As seen
in Fig. 5a, the field induced rate is also optimal under saturation conditions
for the substrate and product (i.e., when $S_1 = S_2 = 1$). This corresponds to
the concentration at which the free energy levels of SE_2 and $E_4{}^*$, and of $E_3{}^*S$
and E_1, are the same. At higher (oversaturated) or lower (unsaturated)
substrate levels, the field induced flux is less. Oversaturating the enzyme
significantly increases the optimal frequency for field induced cycling, but
decreasing the frequency does not lower the optimum frequency very much.
Interestingly, varying the ac amplitude influences the rate of cycling induced
by the field as expected, but does not significantly change the optimum
frequency, as seen in Fig. 5b.

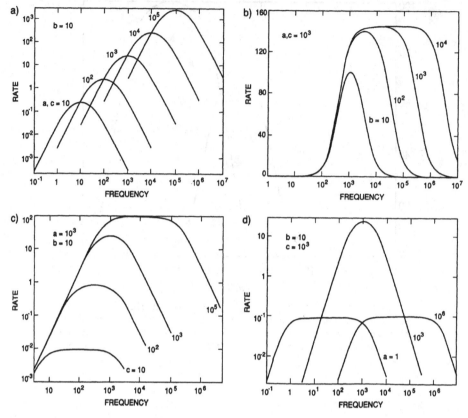

Fig. 4   Effect of varying the kinetic parameters on the frequency dependence
of the field induced rate of cycling. In all calculations, $\psi_1 = 1$;
$x = 2$; $S_1 = S_2 = 1$. a) Effect of varying a and c together when b =
10; b) effect of varying b when $a = c = 10^3$; c) effect of varying a
when b=10 and $c=10_3$; d) effect of varying c when b=10 and $a=10^3$

All of these results were obtained from calculations performed for the
case where the output reaction was at equilibrium ($S_1 = S_2$). Performing
transport experiments using radioisotopes under such conditions is the
equivalent of doing "equilibrium" isotope exchange in an ac field. The
results presented above suggest that these types of experiments may yield a
wealth of kinetic information about membrane transporters when done in
conjunction with simulations such as the examples shown here.

Results obtained for $S_1 = S_2$ do not demonstrate free energy transduction
since there is no load. As shown in Fig. (6), an ac field can drive the
enzyme to cycle in a clockwise direction even when $S_2 > S_1$. This represents
free-energy transduction from the ac field to do work on the electrically
silent reaction $S_1 = S_2$.

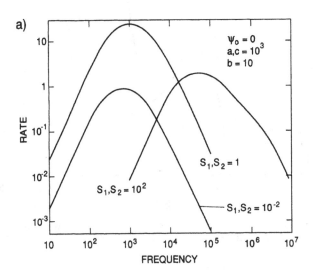

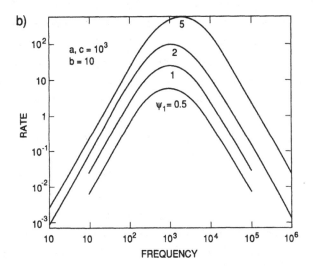

Fig. 5   Effect of changing a) saturation state of the enzyme; and b) ac
amplitude; on the frequency induced rate of cycling.   The parameters
used in calculation were b=10, and a=c=10³.

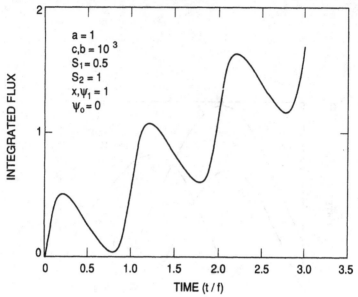

Fig. 6   Demonstration that the ac field can cause the enzyme to catalyze the production of S_2 from S_1 even when $S_2 > S_1$. In this case, $S_2 = 0.5$, and $S_1 = 1$.

The efficiency of free energy transduction can be calculated by taking the average power produced and dividing by the average power absorbed from the field by the protein. The average power produced is equal to the time average flux multiplied by the output free energy $\Delta G_o = \ln(S_2/S_1)$, and the average power absorbed is the time average of the protein displacement current times the potential, $\psi \, \Delta x \, d(E_1 + E_2 S)/dt$. When $a = 1$, $b = 10^7$, $c = 1000$, $x = 3$, $kT\psi_1/e = 125$ mV, $S_1 = 10^{-6}$, $S_2 = 1$, and the frequency, $f = 1$, the efficiency is 52%. When b is set equal to unity, the ability of the enzyme to transduce free energy (or to cause cycling when $S_1 = S_2$) from the ac field is lost. The same is true when a >> c. When b is less than one the enzyme regains its ability to transduce free energy, but the direction of ac field induced cycling is reversed. In general, making b much greater than one, and making c greater than a increases the efficiency. In the limit that b → ∞ , states 1 and 3 become steady state intermediates, and the four state diagram (1) can be reduced to a two state diagram. Effectively, states 1 and 3 have become the transition states for the top and bottom transition pathways, respectively. If we also let c >> a, the resulting diagram is

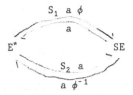

with the effective rate coefficients shown. As seen here, the result of increasing $\underline{b}$ and making $\underline{c}$ greater than $\underline{a}$ is that most of the field dependence is manifested in the forward rate coefficient for the top transition path which involves binding and release of S_1 and in the reverse rate coefficient for the bottom transition path which involves binding and release of S_2. Analytic results have been obtained for this model for square wave, sinusoidal, and stochastic variation in the membrane potential. In all cases, the dynamic perturbation induces a tendency to undergo net clockwise cycling, and free energy can be transduced from the ac electric field. Results for the two state model are discussed by Robertson et al.[8] elsewhere in this volume and previously by Astumian et al.[9] and Astumian and Robertson.[10]

Extension to Spherical Symmetry

The theory so far has been developed as though for describing the effects of ac membrane potentials on enzymes in planar bilayer membranes, while in actuality, many experiments are performed by applying ac electric fields to suspensions of spherical cells or vesicles which contain enzymes in their bilayer membranes. In this experimental case, the applied ac field of amplitude kT E_1/e induces a local membrane potential of amplitude $\psi_1 = 1.5$ R E_1 $\cos(\theta)$ which depends on θ, the angle formed between the normal to the membrane and the external field. The results obtained from the theory can be modified for spherical symmetry by integrating over the spherical surface. Quantities which are independent of the field are unchanged by this averaging, quantities which are linearly dependent, or depend on higher odd powers of the field average to zero, and the spherical average for quantities which depend on even powers m can be obtained by replacing $\psi_1{}^m$ with $(3$ R E_1 / 2$)^m$ / (m+1).

The net cyclic flux depends on the square and higher powers of the field, and consequently does not average to zero. The ac induced shift in the average state probabilities also is quadratic in the field and does not average to zero.

SUMMARY AND CONCLUSIONS

We have shown that an ac field will cause a typical enzyme which has electrically active conformational transitions to cycle even when the reaction it catalyzes is at equilibrium and electrically silent. We extensively analyzed the effect of the various kinetic parameters on the frequency dependence of the induced cycling. The results of this simulation suggest that a wealth of kinetic information can be obtained by experimentally determining the frequency dependence of an ac induced flux of an enzyme, and fitting the results to various possible mechanisms whose frequency dependence can be simulated[5,11]. Also, studying the effect of varying the ac amplitude can provide important information about the electrogenicity of the individual steps in the mechanism.

ACKNOWLEDGEMENTS

We thank Drs. Adolfas Gaigalas and Hans Westerhoff for many interesting and stimulating discussions.

REFERENCES

1.   T.Y. Tsong and R.D. Astumian, Prog. Biophys. Mol. Biol. **50**, 1 (1987).

2.   T. Y. Tsong and R. D. Astumian, Bioelectrochem. Bioenerg. **15**, 457 (1986).

3.   T. Y. Tsong and R. D. Astumian, this volume.

4.   S. J. Singer and G.L. Nicholson, Science **175**, 723 (1972).

5.   B. Robertson and R. D. Astumian, submitted, Biophys. Jour. (1989).

6.   L. M. Fisher, W. John Albery, and J. R. Knowles, Biochemistry **25**, 2529 (1986).

7.   W. P. Jencks, Methods in Enzymology **171**, 145 (1989).

8.   B. Robertson, R. D. Astumian, and T. Y. Tsong, this volume.

9.   R.D. Astumian, P.B. Chock, T.Y. Tsong, and H.V. Westerhoff, Phys. Rev. A **39**, 6416 (1989).

10.   R. D. Astumian and B. Robertson, J. Chem. Phys., in press (Oct. 1989).

11.   H.V. Westerhoff, T.Y. Tsong, P.B. Chock, Y.-D. Chen, and R.D. Astumian, Proc. Natl. Acad. Sci. U.S.A. **83**, 4734 (1986).

NONLINEAR EFFECTS OF PERIODIC ELECTRIC FIELDS ON MEMBRANE PROTEINS

Baldwin Robertson,[*] R. Dean Astumian,[*] and Tian Yow Tsong[+]

* Chemical Process Metrology Division
National Institute of Standards and Technology
Gaithersburg, MD 20899
and
+ Department of Biochemistry
University of Minnesota
Minneapolis, MN 55108

The nonlinear response of a two-state chemical reaction to an oscillating electric field is described. An interesting example is a conformational transition of a membrane protein in an applied ac electric field. Even a modest external field leads to a very large local field within the membrane and hence gives rise to nonlinear behavior. The applied ac field causes harmonics in the polarization and can cause a dc shift in the state occupancy, both of which can be observed and used to determine kinetic parameters. Fourier coefficients are given for the enzyme state probability in the ac field, exactly for infinite frequency, and in a series of powers of the field for finite frequency. The results are extended to the spherical symmetry relevant to suspensions of spherical cells or vesicles.

If the protein catalyzes a reaction, free energy is transduced from the electric field to the output reaction, even if that reaction is electrically silent. Many transport enzymes are ideal examples. The ac field can cause the enzyme to pump ions or molecules through the membrane against an (electro)chemical potential. The efficiency of this energy transduction can be as high as 30%.

INTRODUCTION

A reasonable physical model of a protein in the membrane of a biological cell is a dipolar molecule imbedded in a lipid bilayer of low dielectric constant. An integral membrane protein extends all the way through the bilayer and contacts both the outside and inside of the cell. Such proteins have very important roles in mediating signal transduction and in transporting material through the membrane. Membrane proteins also function as free energy converters. Some of these use ion electrochemical gradients to synthesize ATP (or other chemicals) against a free energy gradient. Others use the ATP thus accumulated to pump unwanted molecules out of the cell and concentrate needed molecules inside the cell. In carrying out these various functions, the protein must undergo a cycle of conformational transitions. Different conformational states may have very different dipole moments. The free energies of these conformations will thus be influenced differently by the membrane electrical potential.

Experiment shows that nonlinear electric fields dramatically affect the function of membrane proteins.[1] This is not surprising since a typical membrane potential of 60 mV represents a local field of over 10 MV/m within the bilayer. The theory for the effect of static electric fields is well established. Also much work has been done on linear effects of ac fields on membranes and membrane proteins. Only recently however has headway been made towards the development of a theory for nonlinear effects of nonstationary electric fields on membrane proteins.[2,3]

In the present paper we consider the effect of a single-frequency ac field on a two-state membrane protein. The conformational transitions involve the intramolecular movement of charge. We show that the ac field causes a dc shift in the average state occupancy and that protein oscillation currents occur at harmonics of the applied frequency. This can be used to determine the relaxation time of the system.[4,5]

These effects arise due to ac-field-induced shifts in the chemical equilibria rather than to rotation of the protein within the membrane. The whole protein cannot rotate significantly in response to the applied field since the activation energy would be prohibitive. Nonlinear effects can also arise due to induced dipoles. However, although this mechanism may sometimes be important, none of the effects discussed in this paper require the protein to have a nonzero polarizability.

The dc shift and harmonic generation occur for any electrically active chemical process in a strong electric field. The theory is particularly pertinent to interfacial processes since a small externally applied field causes a large local field near an interface.[6] Such reactions include the opening and closing of membrane channels, which regulate the permeability of biological membranes to various ions and molecules, and activation of membrane enzymes such as protein kinases.

In addition to the effects that may be expected for any protein in a membrane, there is another effect specific for a membrane enzyme.[6-9] We show that a membrane protein that acts as an enzyme can use energy from an ac electric field to drive a chemical or transport reaction away from equilibrium even if the reaction is electrically silent. When an enzyme catalyzes a chemical or transport process, a substrate molecule binds to the enzyme and is chemically converted to product, which then dissociates. In the chemical conversion step the enzyme usually undergoes some shape change, which may be very subtle. For transport enzymes, the shape change must be large enough that in one state the binding site for substrate is exposed to the inside, and in the other state to the outside of the cell. If charges move concomitantly with the conformational change the enzyme will interact with an electric field as well as with the chemical or transport reaction that it catalyzes. Many transport enzymes are known to involve significant intramolecular movement of charge during their catalytic cycle[10]. This provides a mechanism for a system to absorb energy from an ac electric field.

We show that for an enzyme in an ac field, the chemical interactions with substrate and product allow part of the free energy to do work on a chemical or transport reaction. Experimentally Na^+K^+ ATPase can use energy from an ac field to pump K^+ and Na^+ up their electrochemical gradients.[11-13] In this paper we show that it is not necessary for the output reaction to be electrogenic. An electrically active protein can transduce electrical free energy to do work on an electrically silent reaction. Further study of this effect should provide additional insight into the detailed mechanisms of free energy transduction by proteins.

In order to simplify exposition, we first consider an enzyme in a planar bilayer membrane. At the end of the paper we extend our results to the spherical symmetry relevant to suspensions of spherical cells or vesicles.

FOUR-STATE MODEL

Consider a large number of identical enzyme molecules in a planar bilayer membrane. Each is oriented in the same way in the membrane, and the average distance between the molecules is large enough that enzyme-enzyme interactions may be ignored. The enzyme catalyzes the reaction $S_1 \rightleftharpoons S_2$. It may exist in two electrically distinct conformational states,[9] E^* with the binding site for the substrate S_1 on one side of the membrane, and E with the binding site for the product on the other side. We assume S_1 and S_2 are constant concentrations and that the reaction $S_1 \rightleftharpoons S_2$ is non-electrogenic. The interaction with the electric field occurs only because of electroconformational coupling, i.e. the E^* forms have a different arrangement of charges than the E forms. Since the chemical or transport reaction $S_1 \rightleftharpoons S_2$ does not involve net movement of charge across the membrane, the free energy and hence the equilibrium constant of $S_1 \rightleftharpoons S_2$ are independent of the membrane potential. Nevertheless the electroconformational coupling is sufficient to drive the reaction. Our analysis can easily be generalized to electrogenic reactions.

We assume that all interconversions between states may be treated as unimolecular transitions, or as pseudo-first-order transitions. Then, a simple cycle for the enzyme can be described by the four-state kinetic model

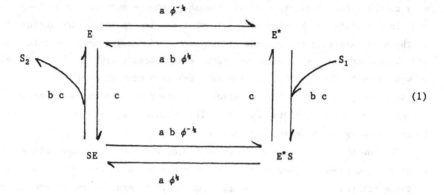

$$\phi = \exp (\Delta x \ \psi), \tag{2}$$

Δx is the effective number of charges that move across the membrane in a transition, and ψ is the potential difference between the left and right sides of the membrane multiplied by e/kT. We assume the field to be small enough that the induced dipole moment of the enzyme is negligible. For simplicity we have chosen the rate coefficients to have reflection symmetry in the absence of the applied potential. The parameter b is a bias factor that is a measure of the difference in affinity of the enzyme for substrate on the two sides of the membrane, a is a scaling factor for the conformational change steps, and c is a scaling factor for the association dissociation steps.

If c is much larger than a, there is no free-energy transduction from the field even at high frequency. We assume a is much larger than c. Then the association-dissociation reactions are slow compared to the conformational transitions, and the enzyme can use energy from an ac electric field to drive the reaction $S_1 \rightleftharpoons S_2$ away from equilibrium.

TWO-STATE MODEL

The four-state kinetic model can be reduced to a two-state model by the following rapid equilibrium and/or steady-state approximation. When the bias factor b is very large compared with one, the rate coefficients for transitions leaving states E^* and SE are much larger than the rate coefficients for transitions entering these states. As a result, the state probabilities E^* and SE will change only slowly in time. This can be used to eliminate E^* and SE in terms of E and E^*S, thus reducing the four-state model (1) to the two-state model

194

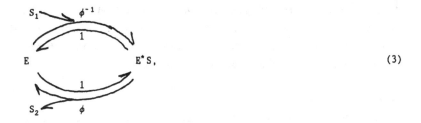

$$E \qquad E^*S, \qquad\qquad (3)$$

It follows that the equation for the rate of change of the probability for the enzyme to be in state E is

$$dE/dt + (\phi^{-1} S_1 + S_2 + 1 + \phi)\, E = 1 + \phi, \qquad\qquad (4)$$

where we have used conservation of probability, $E + E^*S = 1$.

The effective rate coefficients are combinations of elementary rate coefficients from the original four-state kinetic system[5]. For simplicity, we have selected c = 1 so that the four rate coefficients of the two-state model (3) are all equal to 1 when the potential ψ is zero. The reasonableness of this is apparent from the following. Without loss of generality, we can define time units such that one of the four rate coefficients is 1 and concentration units such that another is 1. Detailed balance requires that the two remaining rate coefficients be equal. We choose them to be 1 for simplicity. This is consistent with the ratio of the concentrations of a non-ionic substance across a membrane being 1 at equilibrium.

Free energy transduction can occur only if the electric potential introduces an asymmetry in the kinetics. If the potential-dependent expression ϕ were to appear symmetrically in the diagram (3), the field-induced cyclic flux would vanish. In order for a symmetric oscillating potential to produce an asymmetric response, the upper path must have a different response to the field from the lower path. For a simple two-state enzyme, this occurs when the amount of charge transported to the transition state is different for the two paths. For the upper path the transition state E^* looks electrically like E^*S, and for the lower path the transition state SE looks like E. The charge transported from E to the transition state E^* on the upper path is Δx, while that transported from E to the transition state SE on the lower path is zero. This asymmetry allows a symmetric ac potential to produce an asymmetric response.

Although we have considered only a membrane enzyme up to now, many of the results in this paper also hold for any two-state linear system such as for a simple membrane-channel transition C $\rightleftharpoons$ O, where C is the closed state and O is the open state. We can include this by setting $S_1 = S_2 = 1$ and omitting one

195

of the two paths in the diagram (3). We then will find that applying an oscillating electric field to a membrane containing channel proteins will change the dc conductivity. The frequency dependence of this change can be used to calculate the relaxation rate of the process.

OSCILLATING POTENTIAL

Applying an ac electric field normal to a planar membrane gives rise to an oscillating membrane potential. If the frequency of the field is small compared with the inverse relaxation time of the membrane double layer ($\approx 10^7$ Hz), the potential is given by

$$\psi = \psi_0 + \psi_1 \cos \omega t, \tag{5}$$

where ψ_0 is the dc potential and ψ_1 is the amplitude of the ac potential, which oscillates at the frequency $\omega/2\pi$. When the potential (5) is substituted into Eqs. (2) and (4), the coefficient of E and the right side of Eq. (4) will then be periodic. These expressions can each be expanded in a Fourier series so that Eq. (4) can be expressed neatly as

$$dE/dt + (P_0 + P_1 \cos \omega t + P_2 \cos 2\omega t + \cdots) E$$
$$= Q_0 + Q_1 \cos \omega t + Q_2 \cos 2\omega t + \cdots, \tag{6}$$

where the Fourier coefficients[5]

$$Q_n = \exp(\Delta x \ \psi_0) \ I_n(\Delta x \ \psi_1) + \delta_{n,0}, \tag{7a}$$

$$P_n = \exp(-\Delta x \ \psi_0) \ I_n(\Delta x \ \psi_1) \ S_1 + \delta_{n,0} \ S_2 + Q_0, \tag{7b}$$

$$n = 0, 1, 2, \cdots,$$

are independent of frequency $\omega/2\pi$. Here $I_n(z)$ is a modified Bessel function of the first kind, whose value can easily be looked up in a table[14] or computed using a widely available subroutine, and $\delta_{n,0}$ is 1 when $n = 0$ and is zero otherwise. The coefficient P_0 is the potential-dependent inverse relaxation time of the system.

SOLUTION

Equation (6) is a linear first-order differential equation with periodic coefficients. After any transients have decayed, the steady oscillation in the protein state probability E will be periodic and so can also be expanded in a Fourier series

$$E = A_0 + A_1 \cos \omega t + B_1 \sin \omega t + A_2 \cos 2\omega t + B_2 \sin 2\omega t + \cdots. \quad (8)$$

The Fourier coefficients A_k and B_k can be calculated exactly in the high-frequency limit and approximately at finite frequency.[5]

In the high frequency limit the exact solution to Eq. (6) is given by Eq. (8) with

$$A_0 \rightarrow Q_0 / P_0, \qquad\qquad A_k \rightarrow 0,$$

$$ (9) $$

$$\omega B_k \rightarrow 2 (Q_k - P_k A_0) / k, \qquad k = 1, 2, \cdots, \qquad \omega \rightarrow \infty.$$

This solution is accurate for any values of ψ_0 and ψ_1.

For finite frequency we expand A_k and B_k in a series of powers of ψ_1 and keep terms to second power in ψ_1. The results are

$$A_0 = Q_0 / P_0 - 2 P_1 (P_0 Q_1 - P_1 Q_0) / P_0 (P_0^2 + \omega^2). \quad (10)$$

$$A_1 = 2 (P_0 Q_1 - P_1 Q_0) / (P_0^2 + \omega^2), \quad (11)$$

$$B_1 = 2\omega (P_0 Q_1 - P_1 Q_0) / P_0 (P_0^2 + \omega^2), \quad (12)$$

$$A_2 = 2 \frac{(P_0 Q_2 - P_2 Q_0) P_0 (P_0^2 + \omega^2) - (P_0 Q_1 - P_1 Q_0) P_1 (P_0^2 - 2\omega^2)}{P_0 (P_0^2 + \omega^2) (P_0^2 + 4\omega^2)}, \quad (13)$$

$$B_2 = 2\omega \frac{2(P_0 Q_2 - P_2 Q_0) (P_0^2 + \omega^2) - 3(P_0 Q_1 - P_1 Q_0) P_0 P_1}{P_0 (P_0^2 + \omega^2) (P_0^2 + 4\omega^2)}, \quad (14)$$

A recursion relation for the leading term of the higher harmonics is given in Ref. 5. There we show that the leading term of the cosine amplitude A_k and the sine amplitude B_k of the kth harmonic satisfy Kramers-Kronig relations, and we introduce a response function for the kth harmonic.

Figure 1 gives the average state probability A_0, the first harmonics A_1 and B_1, and the second harmonics A_2 and B_2 of the state probability versus frequency $f = \omega/2\pi$ for substrate concentration $S_1 = 0.8$, product concentration $S_2 = 1$, dc interaction energy $\Delta x\ \psi_0 = 0$, and ac interaction energy $\Delta x\ \psi_1 = 1$. Graphs of the third harmonics A_3 and B_3 (not shown) have roughly the same

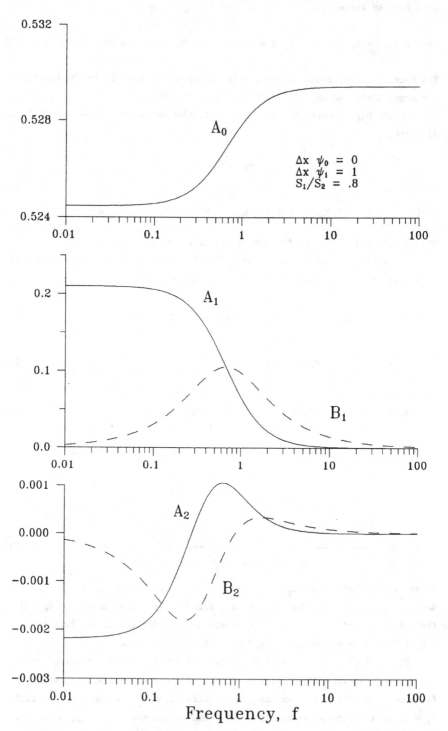

Fig. 1   Amplitudes of the zeroth, first, and second harmonics of
the protein state probability versus frequency.

shapes as those of the second harmonics A_2 and B_2. The qualitative shapes of the curves do not change when the kinetic rate coefficients are changed.

The frequency dependence of the first harmonic has the Lorentz or Debye shape of a single relaxation process. The in-phase amplitude A_1 decreases sigmoidally from a constant at low frequency to zero at high frequency. The frequency at which A_1 is half of its zero-frequency value is $P_0/2\pi$, which is the frequency at which B_1 is maximum. The second harmonic has a more complicated dependence, but B_2 is zero at low frequency and both A_2 and B_2 approach zero at high frequency.

The average state probability A_0 increases sigmoidally from one constant at zero frequency to a larger constant at high frequency. Its half-way point is also at frequency $P_0/2\pi$. If the concentration ratio is greater than unity, A_0 will decrease sigmoidally with increasing frequency. Figure 2 gives the dependence of the dispersion of the state probability amplitudes on the substrate levels S_1 and S_2. When $S_1 = S_2 = 1$, the average probability A_0 is constant, and the second harmonics A_2 and B_2 are zero at all frequencies. The dispersions of A_0 and A_2 are largest (most negative) when $\ln(S_1/S_2) = 2$, and the dispersion of A_1 is largest when $\ln(S_1/S_2) = 1/2$.

The dependence of the harmonic coefficients on substrate levels can be used to discriminate the signal due to the enzyme that catalyzes the reaction $S_1 \rightleftharpoons S_2$ from that due to other sources. Signal that does not have this dependence must be due to something that does not bind S_1 and S_2 and so can be subtracted out.

RATE

The net instantaneous flux through the upper path in the two-state model (3) is

$$J_1 = \phi^{-1} S_1 E - (1 - E). \tag{15}$$

If the applied potential oscillates, the time integral of this quantity increases with time in an oscillatory way as shown in Fig 3a, where $\Delta x \ \psi_0 = 0$, $\Delta x \ \psi_1 = 1$, frequency $\omega/2\pi = 1$, and $S_1/S_2 = 0.8$. The upward trend for the time integral of the flux J_1 indicates the accumulation of net clockwise flux, i.e. a net conversion of S_1 to S_2. During the positive phase of the potential cycle, there are net transitions from left to right, more by the upper path than by the lower one. During the negative phase of the cycle, there are net transitions from right to left, more by the lower path than by the upper one.

This means that the time average of the flux

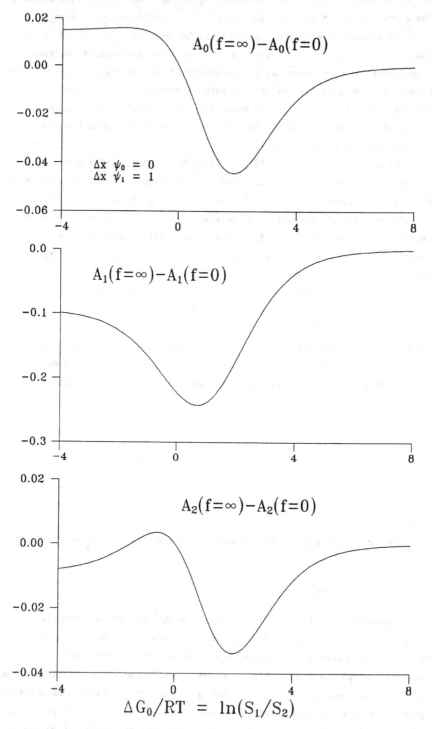

Fig. 2   Amplitudes of the dispersion of the harmonics vs the change in the Gibbs free energy for the reaction $S_1 \rightleftarrows S_2$.

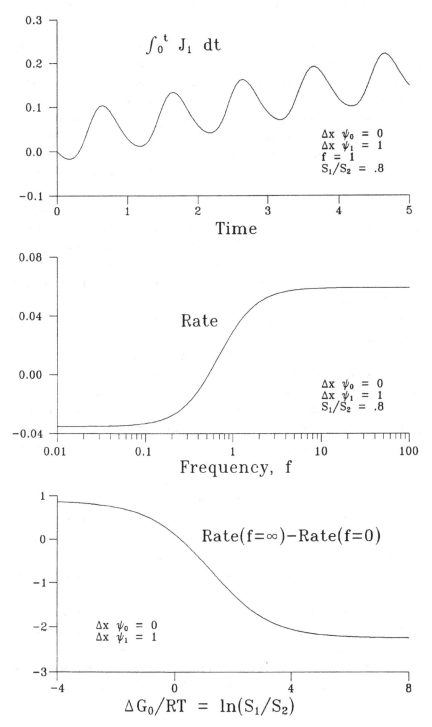

Fig. 3   a) Time integral of clockwise flux vs time.   b) Transport
rate vs frequency.   c) Dispersion of rate vs free energy.

$$\overline{J_1} = \exp(-\Delta x\ \psi_0)\ [I_0(\Delta x\ \psi_1)\ A_0 - S_1\ I_1(\Delta x\ \psi_1)\ A_1] - (1-A^0), \qquad (16)$$

is positive. This is the average rate, which is plotted vs frequency in Fig. 3b for ac interaction energy $\Delta x\ \psi_1 = 1$, and concentration ratio $S_1/S_2 = 0.8$. At low frequencies, the average rate is negative, which means that the net flux is from S_2 to S_1, the spontaneous direction based on the relative concentrations of S_1 and S_2. At high frequencies the rate becomes positive. This means that the oscillating potential causes the enzyme to convert S_1 into S_2 even though the concentration of S_2 is greater than that of S_1. This is free energy transduction from the ac field to the electrically silent reaction $S_1 \rightleftarrows S_2$.

The rate increases sigmoidally with frequency because we have assumed that the bias parameter b of the four-state model is infinite in order to reduce that model to the two-state model. Whether the rate increases or decreases with increasing frequency depends on the concentration ratio as shown in Fig. 3c. For $S_1 = 1$, the rate at high frequency vanishes when $S_2 = 1 + (\Delta x\ \psi_1)^2/2$. This is static head.

The effect of various values of ac and dc potentials is described in more detail in Fig. 4, in which rate is plotted vs the change in free energy $\Delta Go/RT = \ln(S1/S2)$. The upper graph shows that, with the ac potential equal to zero, our two-state model is a normal Michaelis-Menten enzyme. The rate is negative when S1 is less than S2, vanishes when S1 equals S2, and is positive when S1 is greater than S2. The slope is smaller when the dc potential is turned on in either direction. This means that the dc potential inhibits the transport. However, the dc potential does not shift the value of S1/S2 for which the rate is zero.

The lower graph in Fig. 4 shows that our model, with the ac potential amplitude equal to 1, is different from a Michaelis-Menten enzyme. We see that, for $\ln(S_1/S_2)$ greater than -0.47, i.e. for S_1/S_2 greater than 0.625, the ac potential causes the rate to be positive even though S1 is less than S2. An ac potential can cause an enzyme to pump molecules through the membrane against a concentration gradient. This is free energy transduction from the ac field by the enzyme, to maintain the concentration gradient.

The dc potential still inhibits the transport and still does not shift the value of S1/S2 for which the rate is zero. Note that all the curves cross at rate equal to zero, i.e. the static head does not depend on the dc potential. The concentration ratio at which the rate is zero depends on the ac potential amplitude and frequency but not on the dc potential.

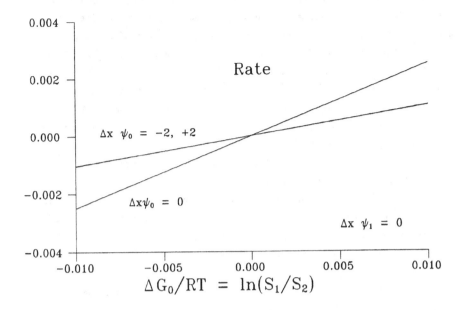

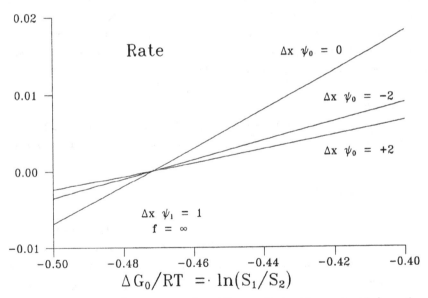

Fig. 4   Rate vs free energy for (a) no alternating potential, and
(b) unit alternating potential with a unit charge Δx.

EFFICIENCY

The efficiency is the power produced by the enzyme pumping against a concentration gradient divided by the power it absorbs from the alternating potential. The average power produced is the average rate of pumping substrate times the free energy increase per molecule pumped, $-\overline{J_1} \Delta G_o/RT = \overline{J_1} \ln(S_2/S_1)$. The average power input is the time average of the potential times the current, $\overline{\psi \Delta x\ dE/dt} = \Delta x\ \psi_1\ \omega\ B_1\ /\ 2$. Note that this is proportional to the out-of-phase amplitude of the first harmonic. The efficiency is plotted vs frequency in Fig. 5a, where it is seen to increase with increasing frequency and to be positive for frequency greater than 1. For the relatively small ac potential to which our finite-frequency solution is limited (i.e., for $\Delta x\ \psi_1 \approx 1$), the efficiency is small, typically below 10%. Our high-frequency solution on the other hand is valid for arbitrarily large alternating potential. In Figs. 5b and 5c the efficiency is plotted vs ac potential and free energy change at high frequency. This shows that the efficiency reaches a maximum of 25% at $\Delta x\ \psi_1 = 6.5$ and $S_1/S_2 = 0.02$, in zero dc potential.

An applied dc potential shifts the value of two of the rate coefficients in the kinetic model (3). The effect of the dc potential is shown in Fig. 6, where A_0, the rate, and the efficiency are plotted vs $\Delta x\ \psi_0$ at the values of ac potential, frequency, and concentrations that maximized efficiency in Fig. 5. This shows that values near $\Delta x\ \psi_0 = -2$ give the maximum effect, and the efficiency reaches a maximum of 30% there.

SPHERICAL SYMMETRY

When an ac field of amplitude E_1 is applied to a suspension of spherical cells or vesicles, the induced amplitude of the membrane potential oscillation is[10]

$$\psi_1 = 1.5\ R\ E_1\ \cos\ \theta. \tag{17}$$

Here R is the radius of the cell or vesicle, and θ is the angle between the field and the normal to the membrane. If we assume that the protein remains uniformly distributed in the membrane when the electric field is applied, the spherical averages $\langle A_n \rangle$ and $\langle B_n \rangle$ can be calculated by expanding Eq. (16) in powers of the ac potential ψ_1. Terms involving an odd power of ψ_1 average to zero, and terms involving an even power do not. For example, a term independent of ψ_1 is unchanged by the average, and quadratic terms are evaluated by replacing ψ_1^2 with $3(E_1R)^2/4$. The averages $\langle A_1 \rangle$ and $\langle B_1 \rangle$ are found to be zero, and the averages $\langle A_2 \rangle$ and $\langle B_2 \rangle$ and the potential- and frequency-dependent terms in $\langle A_0 \rangle$ are not.

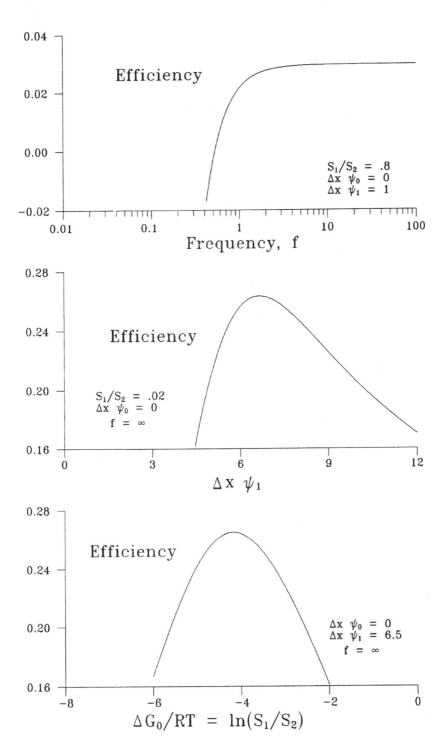

Fig 5. Power produced / power absorbed vs (a) frequency, (b) ac potential times charge, and (c) free energy.

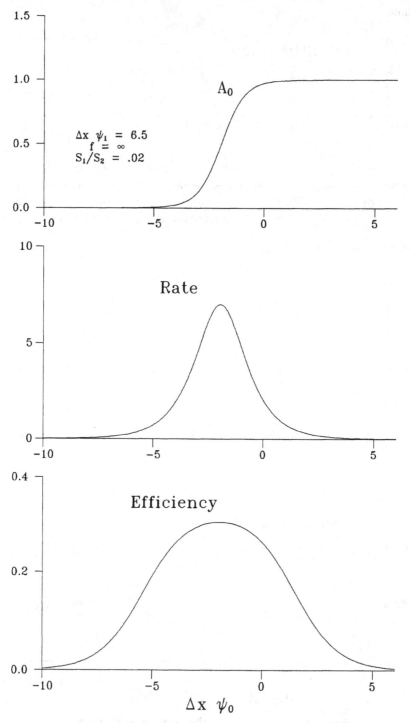

Fig. 6   (a) Average probability, (b) Rate of transport, and (c) Power produced / power absorbed vs dc potential.

This shows that the average value of any physical property, such as fluorescence, that is proportional to the enzyme state probability A_0 may change when an ac field is applied to a suspension of cells containing the enzyme. Measuring this parameter as a function of frequency allows calculation of the relaxation time of the system.

The rate of conversion of S_1 into S_2 averaged over the spherical surface is also not zero. High-frequency pumping against a gradient still occurs.

SUMMARY

We have given the solution to the kinetic equations for the time-dependent state probability of a two-state membrane protein in an ac electric field. The solution is exact at high frequency, and is a series in powers of the membrane potential at finite frequency. There are three phenomena not anticipated by linear response theory that our results reveal.

First, the ac field can cause a shift in the average state probability. Protein states often differ with respect to some physical property such as fluorescence. Monitoring the average value of the fluorescence, for example, as a function of the frequency of an applied field can be used to determine the kinetic relaxation time of the protein conformational transitions. Measuring the average fluorescence as a function of the amplitude of the applied field can determine electrical properties of the protein such as the amount of charge displaced across the membrane due to a conformational change. The shift in the state probability can also have important physiological ramifications. If the two protein states differ from one another with regard to some catalytic activity, an ac field may thereby elicit a significant biological response.

We also demonstrated that an ac field can cause a polarization current, due to the protein conformational change, at twice the frequency of the applied ac field. The amplitude of the second harmonic depends strongly on the amount of substrate and product (or ligand if the protein has no catalytic function). By varying the amount of substrate or product one can discriminate the signal due to the protein of interest from that due to other sources.

Both of the above are nonlinear effects, which can arise only at the bilayer membrane since the local field due to a modest externally applied potential is large only very near the membrane. Thus almost all of the nonlinear signal must come from processes occurring in the membrane itself or in the double layer around it.

We extended the theory to spherical geometry for suspensions of spherical cells or vesicles by averaging over all angles with the applied field. The first harmonic amplitudes of the protein state probability vanish. The dc shift, the second harmonic amplitudes, and the protein contribution to the power absorbed do not.

We have presented results for a polarizable protein with a dc field added to the applied ac field. These results also apply to many other chemical reactions such as adsorption-desorption processes at surfaces. Nonlinear relaxation spectroscopy provides an important method for investigating the kinetics of a wide variety of reactions occurring in nonhomogeneous media. The techniques exploit the fact that small externally applied fields cause large local fields in membranes and at interfaces.

In addition to the above effects, which may be expected for any protein in a membrane or indeed for any electrically active chemical process exposed to a strong ac electric field, there is an effect specific for a membrane enzyme. We showed how a four-state enzyme can under certain reasonable conditions be reduced to a two-state model. We showed for this model that movement of charge within a catalytic cycle provides a mechanism for the enzyme to absorb energy from an ac electric field and use that energy to do work on a chemical or transport reaction. The output reaction does not itself have to involve the displacement of charge. An electrically active protein can mediate free energy transduction between an ac electric field and an electrically silent reaction. We computed the turnover rate and the efficiency of the free energy transduction.

The ability of an enzyme to carry out free energy transduction is determined by the transition rate coefficients and by the electrical properties of the protein. Our electroconformational coupling model explains free energy transduction in terms of the mechanisms of enzyme catalysis.

ACKNOWLEDGEMENTS

We thank Drs. Adolfas Gaigalas, Hans Westerhoff, Douglas Kell, and Adrian Parsegian, for helpful and interesting discussions.

REFERENCES

1.   T. Y. Tsong and R. D. Astumian, Prog. Biophys. Mol. Biol. 50, 1 (1987).

2.   R. D. Astumian, P. B. Chock, T. Y. Tsong, and H. V. Westerhoff, Phys. Rev. A 39, (June, 1989).

3.   H. V. Westerhoff, R. D. Astumian, and D. B. Kell, Ferroelectrics, 86, 79 (1988).

4.   M. Eigen and L. DeMayer, Techqs. Chem. 6, 219 (1973).

5.   R. D. Astumian and B. Robertson, submitted to J. Chem. Phys.

6.   T. Y. Tsong and R. D. Astumian, Bioelectrochem. Bioenerg. 15, 457 (1986).

7.   H. V. Westerhoff, T. Y. Tsong, P. B. Chock, Y.-D. Chen, and R. D. Astumian, Proc. Natl. Acad. Sci. U.S.A. **83**, 4734 (1986).

8.   R. D. Astumian, P. B. Chock, T. Y. Tsong, Y.-D. Chen, and H. V. Westerhoff, Proc. Natl. Acad. Sci. U.S.A. **84**, 434 (1987).

9.   T. Y. Tsong and R.D. Astumian, Ann. Rev. Physiol. **50**, 273 (1988).

10.   P. Lauger and P. Jauch, J. Memb. Biol. **91**, 275 (1986).

11.   E. H. Serpersu and T. Y. Tsong, J. Biol. Chem. **259**, 7155 (1984).

12.   D. S. Liu, R. D. Astumian, and T. Y. Tsong, J. Biol. Chem., submitted (1989).

13.   T. Y. Tsong, D. S. Liu, F. Chauvin, and R. D. Astumian, Biosci. Rep., **9**, 13 (1988).

14.   M. Abramowitz and I. A. Stegun, <u>Handbook of Mathematical Functions</u>, National Bureau of Standards Applied Mathematics Series 55 (1964), Eqs. 9.6.34, 9.6.10, and 9.6.6.

FUNCTIONAL ALTERATION OF MAMMALIAN CELLS BY DIRECT HIGH FREQUENCY

ELECTROMAGNETIC FIELD INTERACTIONS

Stephen F. Cleary, Li-Ming Liu and Guanghui Cao

Bioelectromagnetics Laboratory
Department of Physiology
Virginia Commonwealth University
Richmond, VA 23298-551

INTRODUCTION

Numerous and diverse effects of high frequency electromagnetic radiation (EMR) on living systems have been reported during the past 40 years. Reviewing these reports is perplexing due to inconsistencies that in some instances lead to apparently contradictory conclusions. Early studies of EMR bioeffects were faulted as a consequence of inadequate dosimetry and densitometry. During the past decade, experimental and theoretical advances have more clearly defined the nature of problems that have beset this area of inquiry. These major problem areas fall into two broad interrelated categories: physical and biological interactions.

Investigations of dielectric properties have revealed living systems to be highly nonuniform, principally due to variation in the state and distribution of water. One significant consequence of this is highly nonuniform distribution of induced-EMR fields and energy absorption in tissue. Recent developments in EMR dosimetry and densitometry have well-documented and defined the nature of this problem with respect to EMR bioeffects research (1). Although there are practical limitations to our present ability to fully quantitate nonuniform EMR energy distributions in living tissue, anticipated technological advances should overcome such difficulties in the near future.

Problems related to biological-EMR interactions have proven less tractable than physical interactions, principally due to the highly complex, interactive nature of living systems. Attempts to assess the human health implications of exposure to EMR, based on data derived in vivo using laboratory animals, are rife with uncertainties (2). The most salient aspect of such uncertainties derives from a lack of understanding of the basic mechanisms of interaction of EMR with living systems. As a result of recent advances in EMR dosimetry and densitometry and exposure procedures, the question of interaction mechanisms has emerged as a central issue. It is now obvious that EMR can alter living systems via mechanisms other than indirectly as a consequence of heating (3). Thus other interaction mechanisms must be invoked to explain the growing body of data that are inconsistent with the thermal hypothesis. Although alternative mechanisms have proven elusive, recent data has begun to provide some insight. The purpose of this communication is to discuss such data and areas of research that may lead to a better understanding of nonthermal EMR interaction mechanisms and effects in living systems.

CELLULAR STUDIES IN VITRO

In view of the complexity of physical coupling and internal EMR field distributions, and the interactive nature of highly organized living systems such as mammals, in vivo studies have provided data of very limited usefullness in the development of mechanisms of interaction. At best, such studies in which comparisons of EMR versus non-EMR-induced heating were made, provide confirmatory evidence of mechanisms other than indirect thermal effects. Molecular-level studies of EMR-interactions, representing the opposite extreme in terms of organizational complexity, have yielded little data of direct usefullness to the understanding of EMR interaction mechanisms. This may be attributed to the nonquantized nature of EMR-interactions in this range of frequencies, and the essential involvement of nonequilibrium processes characteristic of more complex and metabolically active systems, such as cells. The majority of the data in evidence of direct interactions of EMR with living systems has been obtained in in vitro studies of cells or membranes (3). The cell provides a metabolically functional entity that can be exposed to EMR under well-controlled conditions, thus affording a means of studying direct effects of EMR on relevant physiological endpoints (1).

Erythrocytes

Evidence of direct nonthermal interactions with living systems was provided by investigations of EMR effects on cation permeability of mammalian erythrocytes. Liu, et al. (4) reported that exposure of rabbit, canine, and human erythrocytes to 2.45-, 3.0-, and 3.95-GHz continuous wave (CW) EMR at specific absorption rates (SARs) of from 0 to 200 W/kg resulted in dose rate-dependent increased Na^+ and K^+ transmembrane fluxes, hemoglobin release, and hemolysis. Sham-exposure at the same temperatures, however, produced the same effects. Subsequent studies revealed, however, that membrane cation permeability could be directly altered by EMR when erythrocytes were exposed at specific temperatures that were subsequently found to be associated with gel to liquid-crystal membrane phase changes (5). Olcerst, et al. (6), for example, reported alteration of passive Na^+ and rubidium (Rb^+) fluxes in rabbit erythrocytes exposed to 2.45-GHz EMR, but only at temperatures of 8 to 11°C, 22.5°C, and 36°C. Cleary, et al. (5) reported statistically significant specific K^+ release from rabbit red cells exposed to CW or pulse-modulated 8.3-GHz EMR at a temperature of 24.6°C, but not at other temperatures in the range 20- to 30°C. Liburdy and Penn (7) and Liburdy and Vanek (8) reported specific alteration of Na^+ passive permeability in rabbit erythrocytes exposed to 2.45-GHz EMR only in the temperature range 17.7 to 19.5°C. In this range there was a dose-rate dependent linear increase in permeability which plateaued at an extrinsic EMR electric field strength of approximately 600 V/m. In addition to altered cation fluxes these authors also reported membrane protein shedding, effects of antioxidants and membrane cholesterol, and a dependence on cell suspension oxygen tension (7,8). The results of Fisher, et al. (9) provided additional confirmation of direct interaction of EMR with both passive and active membrane cation transport associated with a phase change.

A direct effect of EMR on erythrocytes leading to hemolysis has also been reported. Cleary, et al. (10) exposed whole rabbit blood at 22.5°C for 2 h to RF EMR. Hemolysis occurred at an applied field strength of 400 V/m, or greater when cells were exposed to CW 100-MHz EMR, whereas a field strength of greater than 900 V/m was required for 10-MHz radiation. Comparison of these results with the results of studies conducted at higher frequencies indicated direct frequency dependent effects of EMR on erythrocyte membranes (10). Serpersu and Tsong (11) who investigated the effect of lower frequency EMR on active transport of Rb^+ by erythrocyte membranes, arrived at a similar conclusion. More specifically, Serpersu and Tsong (9) concluded that 1-kHz EMR directly interacted with the membrane active transport enzyme complex, Na^+/K^+ ATPase, resulting in a field-induced conformational change. A similar interaction mechanism was proposed by Allis and Sinha-Robinson (12) who reported direct inhibitory effects of 2.45-GHz EMR on the same erythrocyte membrane active transport enzyme complex.

Although detailed mechanisms for EMR effects on membrane cation transport are not available, studies of effects on membrane fluidity provide insight. Allis and Sinha (13) investigated the viscosity of human erythrocyte membranes during exposure to 1-GHz CW EMR at temperatures of 15 to 40°C. Activation energies of fluorescent membrane probes were unaffected by EMR. Fluorescent membrane probes were also used by Kim, et al. (14) to investigate membrane fluidity in erythrocytes and erythrocyte ghosts exposed to 340- or 900-MHz EMR. Exposure decreased membrane lipid viscosity, altered protein-lipid contact regions, and decreased lipid protein shielding. However, it was concluded that these effects were indirect thermal alterations (14). A similar conclusion was arrived at by Ortner et al (15), who investigated EMR effects on membrane protein and lipid structure.

Taken as a whole, studies of EMR on cation transport and membrane structure and fluidity indicate direct field interactions with specific membrane macromolecules, or macromolecular complexes, rather than more generalized interactions resulting in global changes in membrane properties.

Lymphocytes

The temperature dependence of lymphocyte transformation is well-documented (16,17). It is thus not surprising that in vivo investigation and epidemiological studies of effects of EMR exposure on lymphocytes have produced highly varied results due to induced tissue heating. In vitro studies in which EMR-induced heating was measured and/or controlled provide evidence of direct effects on lymphocyte transformation. Stodolnik-Baranska (18), for example, reported an increase in spontaneous blast transformation and mitotic index in unstimulated and stimulated human lymphocytes exposed to 2950-MHz at 37°C at incident intensities of 7 or 20 mW/cm^2. Exposure caused a maximum intrasample temperature elevation of 1°C. Increased spontaneous blast transformation was also attributed to exposure of lymphocytes to pulse-modulated 2.95- or 10-GHz EMR at incident power densities of 5- or 15 mW/cm^2 (10). Roberts (20) and Hamrick and Fox (21), on the other hand, detected no increase in the rate of lymphocyte transformation in vitro upon exposure to 2.45-GHz at SARs of 4 W/kg or less. Likewise there was no effect of 0.29- or 4 W/kg pulse modulated 2.45-GHz EMR on stimulated or unstimulated human lymphocytes (22).

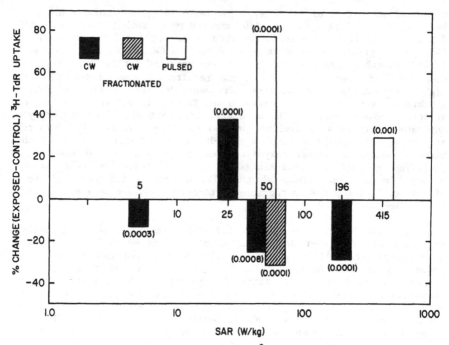

Fig. 1   Mean percentage change in ^{3}H-TdR incorporation rate
(6 h pulse label) in human peripheral lymphocytes 3 d
after exposure to 27-MHz EMR in vitro for 2 h at
$37\pm0.2°C$; shown as a function of time-averaged SAR.
Cells were cultured at a density of 2.5×10^5 cells/ml
in RPMI-1640 supplemented with 10% human AB serum and
20 μg/well PHA. Dose fractionation consisted of two
1-h exposures to 27-MHz CW EMR, 24 h apart. Pulse-
modulation consisted of exposure for 2 h to 27-MHz
EMR with a 2 s pulse each 5.3 s (duty cycle 0.38).
Numbers in parentheses are p-values for two-tailed F-
tests.

    We have investigated the effects of EMR on human peripheral
lymphocytes (23). Human whole blood was exposed or sham-exposed in vitro
for 2 h to CW or pulse-modulated 27-MHz RF radiation or to CW 2.45-GHz
microwave radiation under isothermal conditions (37 ± 0.2°C). Pulse-
modulation of 27-MHz EMR consisted of a 2 s RF-pulse every 5.3 s (duty
cycle 0.38). In the case of pulse-modulated EMR exposure, results are
given in terms of time-averaged SAR. The effect of CW EMR dose-
fractionation was investigated by exposing blood for 1 h, followed 24 h
later by another 1 h exposure to the same SAR. After EMR exposure
mononuclear cells were separated by Ficoll density gradient
centrifugation and cultured for 3 d at 37°C in the presence of various
concentrations of the mitogen phytohemagglutinin (PHA). EMR effects on
unstimulated lymphocytes were studied also by culture in the absence of
PHA. Lymphocyte proliferation was determined by 6h pulse labelling with
^{3}H-thymidine (^{3}H-TdR). Cells were exposed to 27-MHz EMR at SARs of from
5- to 415 W/kg. Exposure to 2.45-GHz occurred at SARs of 5- to 50 W/kg.
Each exposure condition was replicated at least twice. Exposure effects
were evaluated by unbalanced randomized block analysis of variance
(ANOVA), using type III sums of squares and two-tailed F-tests.

Exposure to either frequency EMR caused biphasic dose (or dose-rate) dependent alteration in the rate of ^{3}H-TdR incorporation. Results of studies of effects of CW, split-dose CW, and pulse-modulated 27-MHz EMR are summarized in Figure 1. The PHA concentration was 20-μg/culture well, the concentration that resulted in maximum mitogenic stimulation in EMR or sham-exposed cultures. There was a highly statistically significant decreased uptake of ^{3}H-TdR in cells exposed for 2 h to 50- or 196 W/kg (208-, 814 V/m respectively) CW EMR, relative to sham-exposed controls. Exposure to this same type of EMR at a SAR of 25 W/kg (104 V/m), on the other hand, caused a significant 38% increase in ^{3}H-TdR in exposed lymphocytes. Exposure at 5 W/kg resulted in 13% reduction in ^{3}H-TdR. Similar dose-responses occurred at lower PHA concentrations (viz. 1-, 10-μg/well) but there was greater data variability due to lower counting rates at suboptimum mitogen concentrations. In the absence of PHA exposure at 5 W/kg resulted in a statistically significant (p-value 0.004) 58% increase in ^{3}H-TdR uptake; at 25 W/kg there was a marginally statistically significant 65% increased uptake. Split-dose exposure at 50 W/kg caused a statistically significant decrease in ^{3}H-TdR incorporation similar in magnitude to single-dose exposure at this SAR. Exposure to pulse-modulated 27-MHz EMR at a time-averaged SAR of 415 W/kg led to a statistically significant 30% increased ^{3}H-TdR. Pulse-modulated exposure at 50 W/kg resulted in a highly statistically significant 78% increased incorporation of ^{3}H-TdR, in direct opposition to the effect of single- or split-dose CW exposure at this SAR, both of which suppressed ^{3}H-TdR incorporation.

Exposure of mononuclear cells to 2.45-GHz CW radiation led to similar biphasic effect on ^{3}H-TdR in PHA or unstimulated lymphocytes. Data for EMR exposure effects at a PHA concentration of 20 μg/culture well are summarized in Figure 2. At the largest SAR used in these experiments, 50 W/kg (145 V/m), ^{3}H-TdR uptake was significantly suppressed relative to sham controls. However, exposure at 25- or 39.5 W/kg (72.5- or 114.5 V/m, respectively) resulted in statistically significant increased ^{3}H-TdR incorporation. At 39.5 W/kg uptake was increased by 45% (p<0.001) at a PHA concentration of 20-μg/well. There were also significant increases at lower PHA concentrations. In the absence of PHA, uptake in EMR-exposed cells was increased 142% (p value 0.03) at this SAR. Similar increases in ^{3}H-TdR were observed in cells exposed to 25 W/kg and cultured in the presence of PHA concentrations of less than 20 μg/well. At a SAR of 5 W/kg (14.5 V/m) there was a 20% greater ^{3}H-TdR incorporation rate by lymphocytes cultured at 20 μg PHA/well and an 8% increase in unstimulated cells. Although these data were not statistically significant, logarithmic-transformed data were significant (p<0.05). Split-dose exposure at 50 W/kg resulted in a nonstatistically significant reduction in ^{3}H-TdR uptake. Pulse-modulated exposure was not undertaken at 2.45- GHz.

The results of these experiments indicate that: a) human peripheral lymphocytes may be activated directly by either 27-MHz or 2.45-GHz EMR under isothermal exposure conditions at 37 $\pm$ 0.2°C; b) activation is dose-dependent and biphasic; c) EMR affected lymphocyte proliferation in PHA stimulated and unstimulated cultures; d) dose-fractionation, as used in this study, produced generally similar effects on lymphocyte ^{3}H-TdR uptake; e) pulse-modulated 27-MHz EMR-induced qualitative and quantitative differences in ^{3}H-TdR uptake relative to exposure to CW radiation at the same SAR; f) generally similar effects occurred following exposure to 2.45-GHz or 27-MHz EMR, but lymphocytes exhibited somewhat greater sensitivity to the latter frequency EMR. Comparison of these data with results of studies of temperature-dependent lymphocyte activation indicates qualitative and quantitative differences in the effects of EMR and heat (23).

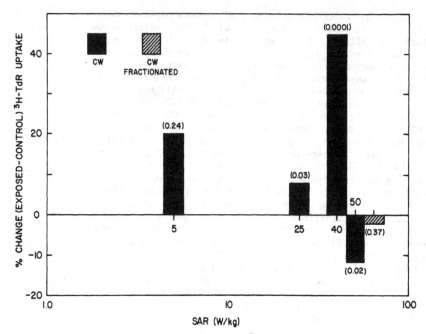

Fig. 2. Mean percentage change in ^{3}H-TdR incorporation rate
(6 h pulse label) in human peripheral lymphocytes 3 d
after exposure to 2.45-GHz EMR in vitro for 2 h at
37±0.1°C; shown as a function of time-averaged SAR.
Cells were exposed simultaneously with cells exposed
to 27-MHz EMR and cultured in an identical manner.
Numbers in parentheses are p-values for two-tailed F-
tests.

Glioma

Using identical in vitro exposure-methods a human glioma cell line
LN71 was exposed to EMR radiation (24). Five milliliter aliquots of
glioma (10^6cells/ml) suspended in Dulbecco's Modified Eagle Medium (DMEM)
were simultaneously exposed or sham-exposed to 27-MHz or 2.45-GHz CW EMR
at 37°C for 2 h. Following exposure, cell concentration was determined
by cell counting and viability assayed by Trypan dye exclusion. Glioma
suspensions were transferred to culture plates, cultured in DMEM at 37°C
in the presence of 5% heat-inactivated FCS for 1-, 3-, or 5 days. Cells
were pulse-labelled for 6 h with ^{3}H-TdR or tritiated uridine (^{3}H-UdR) and
precursor incorporation determined by liquid scintillation counting.
Sample radioactivity (dpm) was normalized to equal cell concentrations
and relative percentage precursor uptake (viz exposed-control) was
determined. Each exposure condition (i.e. SAR and EMR frequency) was
replicated at least twice. Exposure effects are summarized graphically
in Figures 3-, 4- and 5. Percentage uptakes were weighted by sample
size. Normalized data, in the form of radioactive disintegration rates
(dpm), and logarithm (dpm) of ^{3}H-TdR and ^{3}H-UdR, were statistically
analyzed by unbalanced randomized block analysis of variance (ANOVA, type
III sum of squares), as in the case of lymphocyte data discussed above.
Cells were fixed for histological examination by light or
electronmicroscopy. Precursor uptake was determined while cells were in
log-phase, prior to confluent growth limitation. There were no
detectable effects of EMR exposure on cell viability or morphology.

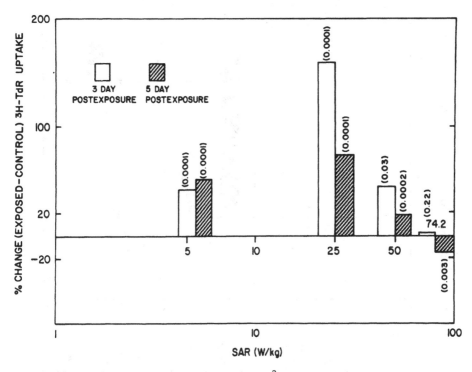

Fig. 3.   Mean percentage change in ³H-TdR incorporation rate
          (6 h pulse label) in human glioma (LN71) 3- or 5 d
          after exposure to 2.45-GHz CW EMR _in_ _vitro_ for 2 h at
          37±0.1°C; shown as a function of time-averaged SAR.
          Cells were cultured at a density of 2.5×10^5 cells/ml
          in DMEM supplemented with 5% heat-inactivated FCS.

Figure 3 summarizes the effects of 2.45 GHz EMR on 3- and 5 day
postexposure percentage uptake of ³H-TdR in glioma. Exposure at SARs of
5-, 25- or 50 W/kg caused statistically significant ($p < 0.05$) increased
uptake of ³H-TdR both 3- and 5 d after exposure. Exposure at 74.2 W/kg
caused a slight, statistically insignificant, increased uptake 3 d
postexposure.   Five days after exposure there was a statistically
significant 15% reduction in ³H-TdR in EMR exposed cells. Uptake of ³H-
UdR followed the same pattern as ³H-TdR.

Figure 4 summarizes the effects of 27-MHz EMR on 3- and 5 d
postexposure incorporation of ³H-TdR in glioma as a function of SAR.
Exposure to this frequency EMR resulted in statistically significant
increased ³H-TdR incorporation at SARs of 5- or 25 W/kg, 3- or 5 days
postexposure. The 96% increased mean uptake 3 d after exposure at 25
W/kg was not statistically significant.   However, uptake for log-
transformed data for this SAR at this time postexposure was highly
statistically significant ($p = 0.0001$).   At 50 W/kg or greater, 27-MHz EMR
exposure caused statistically significant reductions in ³H-TdR
incorporation 3 d postexposure. Five days after exposure this trend was
reversed, with cells exposed to 50 W/kg or 200 W/kg having statistically
significant increased ³H-TdR incorporation rates compared to sham-exposed
cells.   Incorporation of ³H-UdR followed similar trends to ³H-TdR
incorporation.

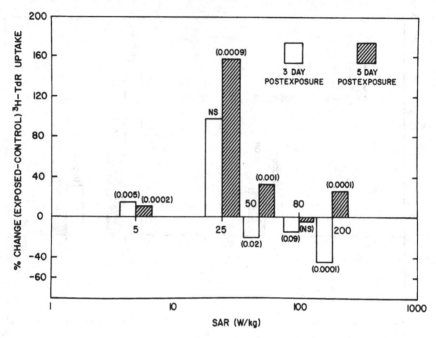

Fig. 4.   Mean percentage change in ^{3}H-TdR incorporation rate
(6 h pulse label) in human glioma (LN71) 3- or 5 d
after exposure to 27-MHz CW EMR in vitro for 2 h at
37±0.2°C; shown as a function of time-averaged SAR.
Cells were exposed simultaneously with cells exposed
to 2.45-GHz EMR and cultured in an identical manner.
Numbers in parentheses are p-values for two-tailed F-
tests.

Postexposure time-dependent alterations in ^{3}H-TdR incorporation
rates in glioma exposed to 25 W/kg 27-MHz or 2.45-GHz CW EMR are
summarized in Figure 5.   At this SAR, ^{3}H-TdR uptake was greater in
exposed glioma at 1-, 3-, or 5 d after exposure.   Whereas the ^{3}H-TdR
incorporation rate was maximum at 3 d after exposure to 2.45 GHz EMR,
incorporation monotonically increased in cells exposed to 27-GHz over
this postexposure interval.   Incorporation at later times was not
determined.

Over comparable SAR ranges, comparison of 27-MHz and 2.45-GHz EMR
effects on 3- and 5 d postexposure ^{3}H-TdR or ^{3}H-UdR incorporation
indicates differences in cellular kinetic responses.   For this 2 d
postexposure period, glioma exposed to 27-MHz EMR at SARs of 50 W/kg or
greater exhibited a 45% increase in relative average ^{3}H-TdR incorporation
rate, and a 54% overall average increase in ^{3}H-UdR uptake rate.   Exposure
to 2.45-GHz at 50 W/kg or greater, on the other hand, caused 13- and 19%
reductions in incorporation rates of ^{3}H-TdR and ^{3}H-UdR, respectively,
over this same postexposure interval.   Compared to the reductions for ^{3}H-
TdR and ^{3}H-UdR incorporation respectively for glioma exposed to 2.45-GHz
EMR, the 45- and 46% increase in incorporation rates of these precursors
from day 3 to day 5 after exposure to 27-MHz indicates a different
latency for these frequencies for this cellular endpoint.

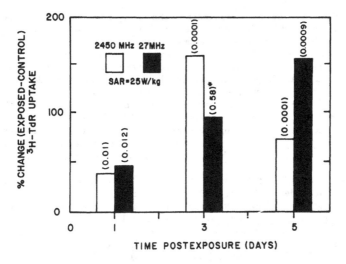

Fig. 5.   Mean percentage change in ^{3}H-TdR incorporation rate
(6 h pulse label) in human glioma (LN71) 1-, 3-, or 5
d after exposure to 27-MHz or 2.45 GHz CW EMR for 2 h
at 37°C at a SAR of 25 W/kg.   Cells were exposed
simultaneously to either EMR frequency and cultured
in an identical manner.

In summary, a 2 h exposure of glioma to 27-MHz or 2.45-GHz EMR,
under isothermal conditions, at SARs of 50 W/kg or less resulted in a
greater than 100% increase in incorporation of ^{3}H-TdR or ^{3}H-UdR, 3- or 5
d after exposure.   SARs of 80 W/kg decreased uptake 3 d after exposure by
40% or less.   This effect, however, was dependent upon time after
exposure.

These results support the hypothesis that EMR, under the conditions
employed in this study, induced shifts in glioma cell cycle.   The greater
stimulatory effect of EMR at SARs of less than 50 W/kg on ^{3}H-TdR and [3]H-
UdR uptake, contrasted with the suppressive effect of higher SARs could
be due to a difference in the fraction of cells that are sensitive to
mitogenic modulation by this type of EMR.   The decreased mitogenic
responsiveness of glioma to high SARs indicates SAR-dependent effects on
cells in specific phases of the cell cycle.   Indeed, we have detected
SAR-dependent effects of EMR on synchronized Chinese hamster ovary (CHO)
cells exposed for 2-h isothermally at 37°C to 27-MHz or 2.45-GHz CW
radiation.   Mitotic cycles were altered in cells exposed in G_0/G_1-, but
not in S- or G_2/M-phase cultures (25).

DISCUSSION

In vitro studies of the effects of EMR exposure of mammalian cell
under conditions of controlled temperature indicate direct mechanisms of
interaction   resulting   in   protracted   functional   alterations   not
attributable to indirect thermal effects.   Studies of effects of high
frequency EMR exposure of erythrocytes document the cell membrane as a

site of interaction. Considering the frequency-dependent dielectric properties of cell membranes, EMR-induced transmembrane potentials, at these frequencies are of such a small magnitude relative to the resting membrane potential that it is difficult to understand how they could affect membrane function. Studies of effects on membrane molecular complexes, such as Na^+/K^+ ATPase, on the other hand, provide evidence of localized direct molecular-level EMR interactions. Again, considering the known frequency dependence of biomacromolecular-EMR field interactions it appears necessary to consider alternate modes of high frequency EMR energy coupling to membrane biomolecules. The most likely mode of coupling is via EMR absorption in water associated with biomolecules in the form of bound water or water of hydration.

Effects of high frequency EMR on lymphocyte and glioma mitogenesis in vitro provide evidence of direct dose or dose-rate dependent alteration of a highly physiologically significant cellular endpoint. Although in vitro data cannot be extrapolated directly to in vivo responses the experimental conditions of these in vitro studies suggest that qualitatively similar effects may be induced by in vivo exposure to EMR fields of these frequencies. In applying in vitro data obtained under isothermal exposure conditions to in vivo exposure effects it must be kept in mind that in the latter case attainment of isothermal exposure conditions is difficult if not impossible. Finally, it may be concluded that in at least a general sense, the results of in vitro cellular studies are consistent with reported effects of high frequency EMR on laboratory animals and human beings. The biphasic dose-dependence detected in in vitro studies provides an explanation for the troublesome and previously unexplained variability of in vivo data.

References

1. S.F. Cleary, Biological Effects of Nonionizing Radiation, in: "Encyclopedia of Medical Devices and Technology," Vol. 1, E. Webster, ed., John Wiley & Sons, New York, N.Y. (1988).
2. S.F. Cleary, Microwave radiation effects on humans, Bioscience 33:269-273 (1983).
3. S.F. Cleary, Cellular effects of electromagnetic radiation, Special issue: IEEE Engineering in Medicine and Biology, EMB306:26-30(1987).
4. L.M. Liu, F.G. Nickless, S.F. Cleary, Effects of microwave radiation on erythrocyte membranes, Radio Science 14:109-115 (1979).
5. S.F. Cleary, F. Garber and L.M. Liu, Effects of X-band microwave exposure on rabbit erythrocytes, Bioelectromagnetics 3:453-466 (1982).
6. R.B. Olcerst, S. Belman, M. Eisenbud, W.W. Mumford and J.R. Rabinowitz, The increased passive efflux of sodium and rubidium from rabbit erythrocytes by microwave radiation, Radiation Res. 82:244-256, (1980).
7. R.P. Liburdy and A. Penn, Microwave bioeffects in the erythrocyte are temperature and pO_2 dependent: Cation permeability and protein shedding occur at the membrane phase transition, Bioelectromagnetics 5:283-291 (1984).
8. R.P. Liburdy and P.F. Vanek, Microwaves and the cell membrane II. Temperature, plasma, and oxygen mediate microwave-induced membrane permeability in the erythrocyte, Radiation Res. 102:190-205 (1985).
9. P.D. Fisher, M.J. Poznarsky and W.A.G. Voss, Effect of microwave radiation (2450 MHz) on the active and passive components of $^{24}Na^+$ efflux from human erythrocytes, Radiation Res. 92:411-422 (1982).

10. S.F. Cleary, L.M. Liu, F. Garber, Erythrocyte hemolysis by radiofrequency fields, Bioelectromagnetics 6:313-322 (1985).

11. E.H. Serpersu and T.Y. Tsong, Stimulation of a ouabain-sensitive Rb$^+$ uptake in human erythrocytes with an external electric field, J. Membrane Biol. 74:191-201 (1983).

12. J.W. Allis and B.L. Sinha-Robinson, Temperature-specific inhibition of human red cell Na$^+$/K$^+$ ATPase by 2,450-MHz microwave radiation, Bioelectromagnetics 8:203-212 (1987).

13. J.A. Allis and B.L. Sinha, Fluorescence depolarization studies of red cell membrane fluidity. The effect of exposure to 1.0-GHz microwave radiation, Bioelectromagnetics 2:13-22 (1981).

14. Y.A. Kim, B.S. Fomenko, T.A. Agafonova and I.G. Akoev, Effects of microwave radiation (340 and 900 MHz) on different structural levels of erythrocyte membranes, Bioelectromagnetics 6:305-312 (1985).

15. M.J. Ortner, M.J. Galvin, D.I. McRee, C.F. Chignell, A novel method for the study of fluorescence probes in biological material during exposure to microwave radiation, J. Biochem Biophys Methods 5:157-167 (1981).

16. N.J. Roberts, Jr., R.T. Steigbigel, Hyperthermia and human leukocyte functions: Effects on response of lymphocytes to mitogen and antigen and bactericidial capacity of monocytes and neutrophils, Infect Immun 18:673-679 (1977).

17. J.B. Smith, R.P. Knowlton, S.S. Agarwal, Human lymphocyte responses are enhanced by culture at 40°C, J. Immunol 121:691-694 (1978).

18. W. Stodolnik-Baranska, The effects of microwaves on human lymphocyte cultures, in: "Biologic Effects and Health Hazards of Microwave Radiation", P. Czerski et al., ed., Polish Medical Publishers, Warsaw (1974).

19. P. Czerski, Microwave effects on the blood forming system with particular reference to the lymphocyte, Ann N.Y. Acad. Sci. 247:232-242 (1975).

20. N.J. Roberts, Radiofrequency and microwave effects on immunological and hematopoietic systems, in: "Biological Effects and Dosimetry of Nonionizing Radiation," M. Grandolfo, ed., New York, Plenum Press: 429-460 (1983).

21. P.E. Hamrick, S.S. Fox, Rat lymphocytes in cell culture exposed to 2450 MHz (CW) microwave radiation, J. Microwave Power 12(2):125-132 (1977).

22. N.J. Roberts, S. Michaelson, S.T. Lu, Exposure of human mononuclear leukocytes to microwave energy pulse modulated at 16 or 60 Hz, IEEE Trans MTT-32 8:803-807 (1984).

23. S.F. Cleary, L.M. Liu, R.E. Merchant, Lymphocyte proliferation modulated in vitro by isothermal radiofrequency radiation exposure, Submitted to Bioelectromatnetics (1989).

24. S.F. Cleary, L.M. Liu and R.E. Merchant, Glioma proliferation modulated in vitro by isothermal radiofrequency radiation exposure, Radiation Research (in review) (1989).

25. S.F. Cleary, L.M. Liu, G. Cao and R.E. Merchant, Modulation of mammalian cell proliferation by in vitro isothermal radiofrequency radiation exposure, Abstracts of Eleventh Annual Meeting of the Bioelectromagnetics Society, Tucson, Arizona, June 18-20 (1989).

ACKNOWLEDGEMENTS

This research was supported by the National Institute of Occupational Safety and Health (1R010H02148) and the Office of Naval Research (N00014-K-0539).

EFFECTS OF MICROWAVE RADIATION ON INDUCABLE ION

TRANSPORT OF RAT ERYTHROCYTES

Kim Yu. A., Kim Yu. V., Fomenko B. S.,
Holmuhamedov E. L., and Akoev I. G.

Institute of Biological Physics
Academy of Sciences of the USSR
Pushchino, Moscow Region, USSR

INTRODUCTION

The activity and viability of the animal cells strongly depend on the state of the cell membrane. One of the main functions of the latter is known to maintain non-equilibrium distribution of various inorganic ions, specifically, potassium, sodium and calcium concentration gradients. Such state of the cells is reached due to operation of several types of the membrane-located ion-transporting systems including passive ion channels and active ion pumps. At present, because of unique bioelectrochemical properties of cell membrane it's role as the primary acceptor of microwave radiation is widely discussed[1-3]. As a result of microwave radiation the changes in passive sodium and potassium fluxes across the erythrocyte's membrane were reported[4-6]. In most cases these changes were connected with the integral heating of the samples by the microwaves. However, in several reports the evidences have been presented indicating that the active transport through Na/K-ATPase significantly affected by 2450-MHz microwave radiation but only within the narrow region near the point of inflection on an Arrhenius plot[6-9] Some specific interactions between the selected membrane components and microwave radiation have been suggested[7,9].

In the present work the effects of 2450-MHz microwave radiation on chloride-dependent ion fluxes and calcium-activated potassium channels have been examined. The results obtained have demonstrated the possibility both of thermal and some specific effects of the microwaves on membrane components of the red cells.

MATERIALS AND METHODS

The protonophore carbonyl cyanide m-chlorophenylhydrazone (CCCP), the ionophore of divalent cations A23187 and saponin were obtained from Sigma Chem. Co. (USA). Choline chloride was purchased from Fluka (Switzerland) and tris(hydroxymethyl)aminomethane-HCl (Tris-HCl) was from Serva (West Germany). Other reagents were home-produced materials of the highest purity grade.

Red cells were obtained from the rats of the Vistar line weighting 180-200 g. After rat decapitation the blood was gathered in the medium containing 150 mM NaCl and 1 mM EDTA as anticoagulent. After centrifuga-

223

tion at 1400 g for 10 min at 4° C the erythrocytes were washed three times at the same conditions but without EDTA. The last pellet of cells was kept in the cold at 4° C and was used in expirements at the same day.

All experiments were performed in specially constructed thermostated measuring vessel (Fig. 1.). The vessel was made from perspex. The internal diameter was 20 mm and the height - 30 mm. The thickness of the bottom - 1 mm and of the wall - 3 mm. In the vessel pH electrode from Radiometer (Denmark) and K^+- and Cl^--selective electrodes from Radelkis (Hungary) were installed. Electrode responses were calibrated in each time after disruption of erythrocytes by 0.007 % saponin and following additions of standard solutions of KCl and HCl or NaOH.

To measure specific absorbtion rate (SAR) and for temperature control the calibrated semiconductor thermocouple was placed in the upper part of the vessel (3 in Fig. 1.). The incubation medium was constantly stirred by electrically driven perspex stick with rotational speed - 2800/min.

Chloride-dependent ion fluxes across the red cells membrane were registered in medium with low ionic strength containing in a total volume 10 ml : 300 mM sucrose, 0.1 mM KCl, 1 mM Tris-HCl, pH 7.4, at 20° C and a final cell density 10^9 per ml. In some experiments NaOH instead of Tris-HCl was used to bring the pH of the medium to 7.4 .

The oscillations of ion fluxes were induced in unbuffered medium containing 150 mM choline chloride, 1 mM KCl, 0.15 mM MgCl and 5 mM glucose, initial pH 7.4, at 37° C and final hematocrit - 5 % . Ion fluxes were induced after succesive additions of 1.2 μM CCCP, 3.3 μM A23187 and 1.3 mM $CaCl_2$. Other detailes are specified in the figure legends.

The red cells were exposed to 2450-MHz microwaves with the use of the standard generator " Luch-3 " produced in the USSR for medical purposes. Two regimes of exposition were used : (i) 20 min preliminary irradiation of the samples in the strip-line waveguide with SAR - 280 W/kg. In this case an initial temperature of the samples was 21° C and the final one 39° C ; and (ii) direct irradiation of cell suspension during ion transport with various SARs (3 - 180 W/kg). In the latter case, under the bottom of the

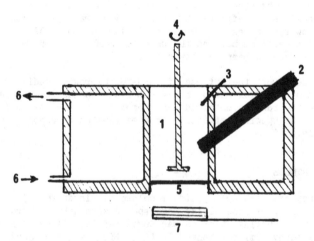

Fig. 1. The scheme of the measuring vessel. 1 - measuring chamber;
2 - ion-selective electrodes; 3 - semiconductor thermocouple;
4 - electrically driven perspex stirrer; 5 - bottom of the
chamber; 6 - recirculating thermostated water; 7 - microwave
radiator or water heater.

vessel the standard ceramic radiator (∅ 20 mm) of cylindric form from the generator accessories were clamped (7 in Fig. 1). In control experiments a heating system connected to water thermostat was placed instead of the radiator. SAR was measured by heating and cooling curves.

RESULTS AND DISCUSSION

Effect of microwave exposure on chloride-induced ion fluxes in red cells

On addition of red cells into the medium with low ionic strength a re-distribution of inorganic ions through the native channels and carriers have been observed[10-12]. The electrochemical gradient of chloride ions is usual-ly to be a motive force for this process. In Fig. 2. a typical example of such the changes in the medium containing unbuffered isoosmotic sucrose is shown. After addition of rat erythrocytes a rapid and slow phases of Cl^-, H^+ and K^+ transport across the cellular membrane could be distinguished. The rapid one continued for about 1-2 min and was characterized by a strong acidification of the medium and appearance of potassium ions. Then followed the slow phase that continued until the establishment of a steady state in the system. The main feature of this phase were ouabain-insensitive K^+ efflux from the cells and a slow increase of pH of the medium.

It was elucidated that at the first phase Cl^- and H^+ effluxes (or here indistinguishable OH^- influx) from the cells took place directly in medium as a result of $Cl^-/^-OH$ exchange through the known Jacobs - Stewart cycle[10]. At the same time the appearance on this phase of K^+ in the medium was due to potassium leakage during preincubation period. For quntitative analysis two parameters were selected : (i) the amount of Cl^- , H^+ and K^+ extruded at the first phase and the rate of transfer of these ions across the cell membrane at the slow phase.

Table 1 demonstrates the data obtained during registration of ion

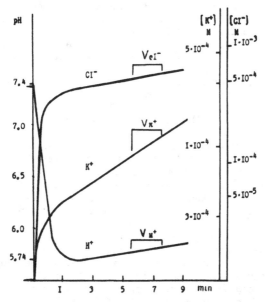

Fig. 2. Typical curves of the changes of K^+ , H^+ and Cl^- concentra-tions after addition of rat erythrocytes in the medium conta-ining unbuffered 300 mM sucrose. V_i -the rates of K^+ , H^+ and Cl^- fluxes measured at the indicated period of time.

Table 1. Ion fluxes in rat erythrocytes after 20 min exposure
at 2450 MHz, 280 W/kg

ION	Control c N	d V	Exposed N	V	39° C -treated N	V
K⁺	1.9+0.1	115.0+0.1	5.3+0.2 *	60.5+0.1 *	5.2+0.2 ᵃ	62.0+0.1 ᵃ
Cl⁻	54.5+0.2	59.5+0.2	53.9+0.3 ᵃ	54.5+0.3 ᵇ	56.7+0.6 ᵃ	66.0+3.0 ᵇ
H⁺(OH⁻)	107+1	25.5+0.1	119+2 **	21.0+0.1 **	115+1 ᵃ	24.0+0.1 ᵇ
$\dfrac{N(H^+)}{N(Cl^-)}$	2.0		2.2		2,2	
$\dfrac{V(K^+)}{V(Cl^-)}$	2.0		1.1		1.0	
$\dfrac{V(K^+)}{V(Cl^-)+V(H^+)}$	1.4		0.8		0.7	

N - amount of ions released at the rapid phase in 10^{-7} mole per 10^9 cells
V - rate of ion efflux at 6-th min in nmol per min per 10^9 cells
* - $p < .001$; ** - $p < .01$; ᵃ - n.s ; ᵇ - $p < .05$
Probability levels (p) were calculated according to Student's t-test
microwave-treated and 39° C-treated against control

fluxes in the medium after addition (i) of the cells kept at 3 - 4° C (cold
control); (ii) microwave-treated cells and (iii) the cells kept at 39° C for
20 min (temperature control) The most significant changes ($p < .001$)
have been observed in K⁺ fluxes. Microwave radiation caused about 3-fold
increase of K⁺ leakage from the cells during the exposure and a decreased
rate of K⁺ efflux at the slow stage. Some small changes in H⁺(OH⁻) fluxes
was also observed, while Cl⁻ fluxes remained practically unchanged.

The stoichiometries of ion fluxes across the red cell membrane are
presented in the last three lines of Table 1 . In cold control the ratios
of N (OH⁻) / N (Cl⁻) in the rapid phase and V (K⁺) / V (Cl⁻) or
V (K⁺) / V (Cl⁻) + V (H⁺) in the slow phase were significantly higher than
one. This meant that some leakage of another anions such as bicarbonate,
lactate or phosphate occurred to keep the electroneutrality of charge transfer across the membrane. In microwave-irradiated cells a small increase of
N (OH⁻) / N (Cl⁻) ratio and a strong decrease of V (K⁺) / V (Cl⁻) or
V (K⁺) / V (Cl⁻) + V (H⁺) ratios were registered. It appeares that these
changes can be explained by the increasing and cessation of the effluxes of
non-registered anions, respectively, at the rapid and slow phases of ion
transport.

Nearly the same changes as in the microwave-treated cells were monitored in the cells incubated at 39° C for 20 min, i.e. at the temperature established in cell suspension during irradiation (Table 1). It seems that
slightly different temperature regimes of the cell incubation (at constant
39° C in comparison with continuous heating from 21° C to 39° C during microwave radiation) were responsible for the observed small variations in
the values of ion fluxes (compare, for instance, V (Cl⁻) and V (H⁺) in
Table 1) between the two types of treatment.

Thus, the experiments demonstrate that preliminary microwave exposure of rat erythrocytes have caused the changes in Cl^--dependent K^+ and H^+ (OH^-) fluxes across the membrane basically of thermal nature.

Oscillations of ion fluxes in red cells after microwave exposure

Due to involvement of the amplifier cascade(s), feedback loops, etc. the oscillatory fenomena in cells seems to be one of the most suitable models for investigation of microwave effects, particularly, of a weak nature. Earlier it was found that in unbuffered medium an addition of protonophore CCCP, divalent cations ionophore A23187 and calcium the red cells had undergone a train of oscillation of ion fluxes across the membrane . Schematically, the oscillations arised from the increase of A23187-mediated Ca^{2+} influx, subsequent polarization of the cell membrane due to opening of Ca^{2+}-induced K^+ channels, followed by CCCP-mediated H^+ influx. Then activation of CaATPase caused a decrease of Ca^{2+} concentration in cytoplasm, closing of K^+ channels and subsequent H^+ efflux and membrane depolarization[13,14]. However, at-present, the detailed mechanism of the oscillations remaines unclear. A key role of CaATPase possessing a hysteretic characteristics have been suggested[15] .

A typical example of the oscillations of ion fluxes in rat erythrocytes is presented in Fig. 3. Normally, five sometimes six-seven cycles of oscillations of K^+ and H^+ fluxes have been registered (Fig. 3 and Fig. 4 A). At the same time, a preliminary 20-min microwave irradiation of the isolated red cells caused a decrease both of the number of cycles by 1-2 and of the amplitude of the oscillations by 25 - 30 % against the control (Fig. 4).

The same changes in oscillations of ion fluxes were simulated when the red cells were incubated at 39° C for 20 min. Therefore, continuous preliminary microwave irradiation of the red cells suspension has not definitely revealed the possibility of specific effects of the microwaves, that haven't to be linked with integral heating of the samples.

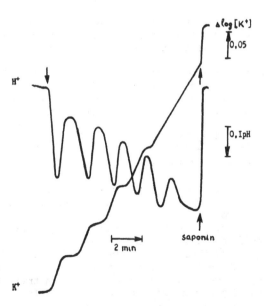

Fig. 3.   Oscillations of H^+ and K^+ fluxes in suspension of red cells in the presence of protonophore CCCP and Ca^{2+}-ionophore A23187 The oscillations were induced by addition of 1.2 mM $CaCl_2$.

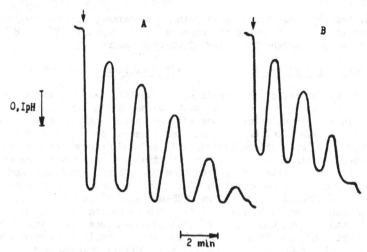

Fig. 4.   A decrease of the number of cycles and of the amplitude of oscillations in medium containing the cells treated by microwave
radiation (A). B - untreated cells.

Microwave-induced oscillation of ion fluxes

At the optimal proportion of Ca^{2+} and A23187 the oscillations of ion
fluxes proceeded until the exhaustion of potassium concentration within the
cells. Reducing the concentration of the added Ca^{2+} or A23187 it was
possible to induce only one-two cycles of the oscillations followed by an
establishment of a new steady state (Fig. 5A). In this state the addition
of the next portion of Ca^{2+} (or A23187) resumed the oscillations of H^+
and K^+ fluxes. The latter suggested that damping of the oscillations was
caused by a decrease of calcium influx possibly because of continuous binding of Ca^{2+} or A23187 to some cellular components.

To examine the possibility of the weak effects of microwave radiation
on ion fluxes a special system allowing to measure the process of ion transport directly during irradiation have been constructed (Fig. 1).

In conditions of damped oscillations switch on of the electromagnetic
field evoked the generation of one cycle of oscillations of ion fluxes
(Fig. 5 B). When the output power of the generator was 16 W, corresponding to estimated SAR - 200 W/kg, the opening of K^+ channels occurred after a
time delay of about 10 s. Decreasing the output power of the generator from
16 W to 1.5 W (corresponding SAR changed from 200 W/kg to 30 W/kg) resulted in increasing of the time delay from 10 s to 110 s. In semilogarithmic
plot between the two parameters a linear relationship was observed (Fig. 6).

After the switch on of the microwave radiation till the opening of
Ca^{2+} -induced K^+ channels, when the field was switch off, the temperature of
the medium increased from $37°$ C to 38.5 - $39°$C despite the intensive stirring and thermostat (Fig. 5 C). Therefore the thermal effects of microwave
field on the red cells could be expected. However, analogous heating of the
medium by the system described in " Materials and Methods " section (keeping the similar absolute values of the initial and final temperatures and
the rate of heating) haven't caused any effect on ion transport in erythrocytes after damping of the oscillations. The only result of such heating
was temperature-dependent slow drift of the electrode readings (data not
shown).

It should be noted here that microwave radiation was without of the

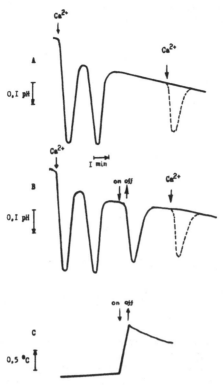

Fig. 5. Damped oscillations of H^+ fluxes after addition of nonoptimal
concentration of $CaCl_2$ (1.03 mM). A - control. B - induction of a new cycle of oscillation by microwave radiation
(the first two arrow - switching on and off of the generator).
In (A) and (B) the cycle of oscillation induced by addition of
another portion of Ca^{2+} (0.5 mM) is shown (break lines).
C - the curve of the temperature changes in the medium registered by thermoucouple during microwave exposure.

effect on ion fluxes in the absence of the added Ca^{2+} or A23187. Moreover,
when the oscillations of ion fluxes were inhibited by addition of 2 μM Co^{2+},
the switch on of the field again caused the opening of K^+channels and one
cycle of the oscillation (data not shown). In this case, an integral heating of the system was also without of the effect.

The data presented in this series of experiments show that transient
microwave radiation can induce the opening of Ca^{2+} -activated K^+channels
most likely by increasing the net Ca^{2+} influx into the cells. The latter
can be a result of two processes: (i) a decrease of Ca^{2+} efflux through
CaATPase or/and (ii)an increase of A23187-mediated Ca^{2+} influx. The former
seems less probably since in control experiments the parameters of the full-
scaled oscillations practically unchanged during continuous microwave exposure. At the same time, the increase of A23187-mediated Ca^{2+} influx can
result from release of Ca^{2+} from Ca^{2+}-binding sites of the membrane. Despite
different conditions of microwave radiation the possibility of microwave-
induced release of bound Ca^{2+} from the external side of neuronal membranes
has been demonstrated[16-18]. In addition, the unbindig of Ca^{2+} from the red
cell membrane resulting in protein shedding was also assumed by Liburdy and
Vanek [9].

Another possible explanation of microwave effects can be the presence

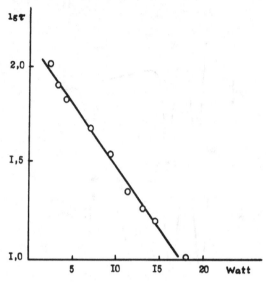

Fig. 6. Plot of the output power of the microwave generator against a
time delay between the switching on of the generator and the
opening of Ca^{2+} -induced K^+ channels.

of hot spots in the system. The experiments directed to elucidate such possibility are now in a progress. Preliminary data have indicated that hot
spots really exist in the system and in model experements using local micro-
heater an induction of one cycle of the oscillation of ion fluxes have been
observed . However, both in the latter case and in the case of Ca^{2+} release
from binding sites the linearity in semilogarithmic plot between the time
delay of the opening of K^+ channels and the output power of the generator
(or SAR) remaines unclear. Accepting the overheating of the cells in hot
spots then continuous K^+ loss could be expected instead of the observed
simulteneous opening of Ca^{2+}-activatedd K^+ channels. Further detailed analysis of Ca^{2+} distribution in the cells during microwave radiation as well
as model experiments seems help to resolve the problem.

In conclusion, after continuous microwave irradiation of red cells
basically thermal effects on ion fluxes across the membrane have been observed. However, direct registration of ion fluxes in red cells during irradiation revealed the effects that have not been explained by integral heating
of the samples. Some other mechanism(s) such as the formation of local temperature gradients in membrane-water interface or the existence of hot spots
in the system should be considered.

REFERENCES

1. W. R. Adey, Tissue interactions with nonionizing electromagnetic fields,
   Physiol. Rev., 66:435 (1981).
2. W. R. Adey, The sequence and energetics of cell membrane transductive
   coupling to intracellular enzyme systems, Bioelectrochem. Bioenerg. ,
   15: 447 (1986).
3. N. J. Roberts, jr., S. M. Michaelson and S. T. Lu, The boilogical effects
   of radiofrequency radiation: a critical review and recomendations,
   Int. J. Radiat. Biol., 50:379 (1986).
4. E. S. Ismailov, Mechanism of the effect of microwaves on the permeabi-
   lities of RBS's for K and Na , Biol. nauki , 3:58 (1971) (In
   Russian).

5. R. B. Olcerst,S. Belman ,M. Eisenbud ,W. W. Mumford and Rabinowitz, The increased passive efflux of sodium and rubidium from rabbit erythrocytes by microwave radiation, Radiat. Res., 82:244 (1980).

6. P. D. Fisher, M.J. Poznansky and W. A. G Voss, Effect of mocrowave radiation (2450 MHz) on the active and passive components of Na efflux from human erythrocytes, Radiat. Res., 92:411 (1982).

7. J. W. Allis and B. L. Sinha-Robinson, Temperature-specific inhibition of human red cell Na /K ATPase by 2450-MHz microwave radiation, Bioelecmagnetics, 8:203 (1987).

8. R. P. Liburdy and A. Penn, Microwave bioeffects in the erythrocytes are temperature and pO dependent: Cation permeability and protein shedding occur at the membrane phase transition, Bioelectromagnetics, 5: 283 (1987).

9. R .P. Liburdy and P.F. Vanek, jr. Microwaves and the cell membrane. iii. Protein shedding is oxygen and temperature dependent: evidence for cation bridge involvement, Radiat. Res., 109:382 (1987).

10. H. Passow, Passive ion permeability of the erythrocyte membrane, Progr. Biophys. and Molec. Biol., 19, part 2:425 (1969).

11. P. A. Knauf, G. F. Fuhrmann, S. Rothstein and A. Rothstein, The relationship between anion exchange and net anion flow across the human red blood cell membrane, J. Gen. Physiol., 69:363 (1977).

12. G. S. Jones and P. A. Knauf, Mechanism of the increase in cation permeability of human erythrocytes in low-chloride media, J. Gen. Physiol., 86:721 (1985).

13. B. Vestergaard-Bogind and Bennekou P., Calcium-induced oscillations in K conductance and membrane potential of human erythrocytes mediatede by the ionophore A23187, Biochem. Biophys. Acta, 688:37 (1982).

14. J. H. Sadykov, E. L. Holmuhamedov and Evtodienko Y. V., Effects of pH and proton buffer on oscillations of ion fluxes in rat erythrocytes, Eur. J. Biochem., 113:369 (1984).

15. O. Scharff, B. Foder and U. Skibsted, Hysteretic activation of the Ca^{2+} pump revealed by calcium transients in human red cells, Biochem. Biophys. Acta, 730:295 (1983).

16. S. M. Bowin and W. R. Adey, Sensitivity of calcium binding in cerebral tissue to weak enviromental electric fields oscillating at low frequency. Proc. Natl. Acad. Sci. USA, 73:1999 (1976).

17. S. K. Dutta, A. Subramoniam and R. Parshad, Microwave-radiation induced calcium efflux from human neyroblastoma cells in culture, Bioelectromagnetics, 5:71 (1984).

18. C. F. Blackman, S. G. Benane, D. E. House and W. T. Joines, Effects of ELF (1-120Hz) and modulated (50 Hz) RF fields on the efflux of calcium ions from brain tissue in vitro, Bioelectromagnetics, 6:1 (1985).

ELECTROPORATION OF LIPID VESICLES

BY INNER ELECTRIC FIELD

Yu.A.Chizmadzhev and V.F.Pastushenko

A.N.Frumkin Institute of Electrochemistry
USSR Academy of Sciences, Moscow, USSR

Electroporation of lipid membranes by outer electric field has been studied in detail in plane bilayers bordering on menisci (see, for instance, the review [1]). Vesicle systems require other methods of analysis, since they lack the tension-providing menisci. Of significance in this case is also the drop of the electric potential, caused by the increase in the pore size, especially in studies of the breakdown by the membrane potential. Our paper discusses these problems.

1. Free Energy of the Vesicular System as a Function of the Pore Radius

Free energy F of the vesicular system which includes electric energy F_φ, energy F_γ of the pore edge and elastic energy F_e equals

$$F = F_\varphi + F_\gamma + F_e \tag{1}$$

Let radius R of the initial vesicle be large as compared with pore radius a

$$R \gg a \tag{2}$$

Any outer electric field is assumed to be absent but a membrane potential is present which is due to the difference in ion concentration between the vesicle interior and the environment. For simplicity we can consider the membrane to be permeable only for ions of one type, the membrane potential then being reduced to the Nernst value. Under condition (2) the electric field at about the pore will be the same as in the vicinity of a similar pore in a plane membrane. Away from the pore the difference of potentials across the membrane is $2\varphi_0$. Electric energy will be calculated as the result of the action of the local force at the boundary of two dielectric media. Thus, let a membrane with permitivity ε_m has the pore of radius a. Conductance and permittivity of the electrolyte solution are equal to $\mathfrak{æ}$ and ε_s, respectively. The electric potential satisfies to the following conditions:

$$\varphi(r,z) \to \varphi_0, \qquad r^2 + z^2 \to \infty, \quad z > \delta/2$$

$$\varphi(r,z) \to -\varphi_0, \qquad r^2 + z^2 \to \infty, \quad z < -\delta/2 \tag{3}$$

where r and z are the coordinates of the cylinder system; δ is the thickness of the membrane. Along with conditions (3) the symmetry of the system requires also that the following condition

$$\varphi(r,0) = 0 \tag{4}$$

be satisfied. This enables us to confine the consideration of the electric field within the area $z > 0$. Let us assume the concentration of the electrolyte to be sufficiently high so that the thickness of the diffuse layer (the Debye length) is small as compared with the membrane thickness. Then in the area $z > \delta/2$ the potential distribution is described by the Laplace equation

$$\Delta \varphi = 0 \tag{5}$$

Passing over into the pore, the situation is complicated by a number of circumstances: the presence of image forces and their dependence on the radial coordinate; the presence of mouth regions in the pore which are affected by the dependence of image forces on coordinate z and, finally, the presence of a strong electric field leading to a nonlinear dependence of the pore conductivity on the difference of potentials across the mouth regions. Due to this reason, the problem of the electric field within a pore deserves special consideration.

This paper will use the following approach. Let us assume that the mouth regions of a pore are equipotential, i.e.

$$\varphi(r,\delta/2) = \varphi_a = \text{const}, \quad r \leqslant a \tag{6}$$

Let us also take the standard chemical potential of ions within the pore to be independent on coordinates r and z. Then in the constant field approximation the electrolyte within the pore should be regarded as an ohmic conductor of resistance

$$R_p = \frac{\delta}{\pi \varkappa a^2} \exp\left(\frac{\mu_0}{kT}\right) \tag{7}$$

where μ_0 is the standard chemical potential of ions within the pore; kT is the characteristic thermal energy of a particle. To estimate μ_0, let us use the result of work[2] :

$$\frac{\mu_0}{kT} = \frac{e^2}{\varepsilon_m kT} P\left(\frac{\varepsilon_m}{\varepsilon_s}\right) \cdot \frac{1}{a} \tag{8}$$

$P(\varepsilon_m/\varepsilon_s)$ here is the known function.

Finally, let us define the boundary conditions for eq.(5). In addition to conditions (3) and (6) we consider as a boundary condition the absence of the normal current component on the membrane surface:

$$\frac{\partial \varphi}{\partial z} = 0, \quad (r > a, \quad z = \delta/2) \qquad (9)$$

The distribution of the potential within the nonconducting material of the membrane can be found by solving eq.(5) at a given value of the potential at the boundary of the region. For this eq.(2) is used as well as the solution of the problem determined by eq.(5) with additional conditions (3), (6), (9).

Let us calculate the free energy change occurring as a result of applying on the medium of infinitesimal deformations described by vector $\vec{l}$:

$$dF_\varphi = \frac{1}{2} \iint \left\{ E^2 \, grad \, \mathcal{E} - grad \left(E^2 \rho \frac{\partial \mathcal{E}}{\partial \rho} \right) \right\} \vec{l} \, dN \qquad (10)$$

$$dV = 2\pi r \, dr \, dz$$

Here E is the electric field strength, ρ is the density of the medium; $\mathcal{E}$ is the permittivity of the medium as a function of space coordinates. If the compressibility of the dielectrics is neglected then, as can be shown, the second term in eq.(10) becomes zero. The field of deformations is chosen so that it corresponds to the radial displacement of the cylindrical wall of the pore by a value da. Then the required change in free energy, as suggested by eq.(9), has the form

$$dF_\varphi = -\pi a \delta (\mathcal{E}_s - \mathcal{E}_m) E_a^2 \, da \qquad (11)$$

where E_a is the electric field strength on the wall of the pore:

$$E_a = 2\varphi_a / \delta \qquad (12)$$

Potential φ_a occurring in this expression depends on current J through the pore:

$$\varphi_a = J R_p / 2 \qquad (13)$$

Solution of problem (5) with conditions (3), (6), (9) yields the value of the current:

$$\mathcal{I} = (\varphi_o - \varphi_a) \cdot 4 \varkappa a \qquad (14)$$

From eqs.(13), (14) it follows that

$$\varphi_a = \frac{\varphi_o}{1 + \lambda} \qquad (15)$$

By λ we designate here the ratio of input resistance $R_i = (2\varkappa a)^{-1}$ to pore resistance R_p:

$$\lambda = \frac{\pi a}{2\delta} \exp\left(-\frac{\mu_o}{kT}\right) \qquad (16)$$

Integrating eq.(11), we obtain the final expression for the free energy of the system which takes into account the flow of electric current:

$$F_\varphi = -\frac{4\pi}{\delta}(\varepsilon_s - \varepsilon_m)\varphi_o^2 \int\limits_0^a \frac{x\,dx}{[1 + \lambda(x)]^2} \qquad (17)$$

As seen from the derivation, the electric term in the free energy describes the work done by the electric field at the change in the pore radius. It appears from eq.(11) that this work can be interpreted as a result of a force radially applied to the cylindrical wall of the pore. As is known, a force at the interface between two dielectrics in the presence of an electric field is directed towards the dielectric with the smaller permittivity. Calculations show that the work of this force coincides with expression (11).

It is easy to see that at $\lambda = 0$ eq.(17) coincides with the trivial result in which the effect of the pore conductance is ignored. It is clear that at $\lambda > 0$ the free energy of the system is higher than at $\lambda = 0$. This means that the neglect of the pore conductance leads to an underestimated value of the energy barrier overcome by the pore in its change. Correspondingly, the lifetime of the membrane proves to be underestimated.

Using eq.(8) and the typical values of the parameters, we obtain

$$\lambda(\xi) = \frac{\pi\xi}{2} \exp(-1/\xi) \qquad (18)$$

where $\xi = a/\delta$.

Let us now consider the elastic energy of the vesicle. At the change of the pore radius the curvature of the vesicle

236

surface changes. Area S of the vesicle membrane is considered to be constant; the vesicle, to be a sphere. If R_c is the radius of the vesicle containing a pore, its elastic energy equals

$$F_e = SM/R_c^2 \qquad (19)$$

where M is the bending elasticity modulus of the membrane. From the condition S = const we obtain

$$4R^2 = 4R_c^2 - a^2 \qquad (20)$$

Calculating from here curvature radius R_c and subtracting from eq.(19) the constant term $4\pi M$, which corresponds to the subtraction of elastic energy from energy of a closed membrane, we have

$$F_e = -\pi Ma^2 / (1 + a^2/4R^2)R^2 \qquad (21)$$

Taking into account condition (2) we obtain the squared dependence of elastic energy on the pore radius:

$$F_e \approx -\pi Ma^2 / R^2 \qquad (22)$$

As for the energy of the pore edge, we shall assume the work of formation of the edge with the unit length $\gamma (J/m)$ to be independent on its radius. Then the energy of the edge is equal to

$$F_\gamma = 2\pi a\gamma \qquad (23)$$

Substituting eqs.(22), (23) and (17) in eq.(1) we obtain the free energy of a pore vesicle

$$F = 2\pi\gamma a - \frac{4\pi\varepsilon_s \varphi_o^2}{\delta} \int_0^a \frac{x\,dx}{[1+\lambda(x)]^2} - \frac{\pi M}{R^2}a^2, \quad (\varepsilon_m \ll \varepsilon_s) \qquad (24)$$

Formally, this expression is equivalent to that for a planar membrane with the difference that instead of surface tension δ we have here the value $\pi M /R^2$. At $M \sim 10^{-19}$ J, $R \sim 10^{-6}$ m the elastic energy is very small compared with the other terms and can be neglected ($M /R^2 \sim 10^{-7}$ J/m^2 whereas $\delta \sim 10^{-3}$ J/m^2). It is convenient to measure the energy in kT and introduce new variables $E = F/kT$, $\Gamma = 2\pi\gamma\delta/kT$, $\Phi = 4\pi\varepsilon_s\delta\varphi_o^2/kT$:

$$E = \Gamma\xi - \Phi \int_0^\xi \frac{x\,dx}{[1+\lambda(x)]^2} \qquad (25)$$

Function $E(\xi)$ determined from eq.(25) is shown in Fig.1 at $\Gamma = 19.1$ and different values Φ corresponding to different voltages. As seen from Fig.1, as voltage increases, an intermediate minimum corresponding to a metastable state of the pore occurs on curve $E(\xi)$. Any subsequent increase of Φ leads to

237

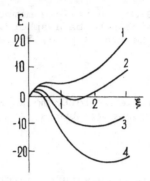

Fig.1. Dependence E (ξ) at different
values of parameter Φ: 1, 2.3;
2, 3; 3, 4; 4, 5.

these pores becoming stable and those of smaller dimensions
metastable so far as the intermediate minimum satisfies to
condition E < 0.

2. Breakdown Modes

At sufficiently high Φ the main role is taken over by
large pores which can decrease the initial membrane potential.
Referred to Fig.1, it implies the transfer from type 3 curves
to those of type 1 which have no intermediate minimum. To elucidate the dynamics of pore distribution in the space of radii,
let us write down the equation for voltage across the membrane
having a pore:

$$\left(\frac{C_o}{G_o}\right) \frac{\partial \varphi}{\partial t} = \varphi_o - (1 + \frac{1}{w})\varphi \tag{26}$$

Here $W = SG_o/g$ is a parameter characterizing the voltage generator power, C_o is the specific capacitance of the membrane,
G_o is the specific conductivity of the membrane with respect
to the ions, g is the pore conductivity. The stationary solution of this equation has the form

$$\varphi_{st} = (1 + \frac{1}{w})^{-1} \cdot \varphi_o \tag{27}$$

The characteristic time of voltage transfer from φ_o to φ_{st}
equals

$$\tau_\varphi = C_o (1 + 1/W)^{-1}/G_o \tag{28}$$

(a) Strong Generator (W >> 1)

238

In this case $\varphi_{st} \approx \varphi_o$, i.e. the voltage decrease due to the current leakage through the pore is small. The equilibrium distribution of pores in the space of radii will be established if characteristic time τ_p of pore diffusion is significantly smaller than relaxation time τ_c of the potential-determining ion concentration. In the stationary state the current coming through the pore and basically transferred by indifferent electrolyte ions is equal to the current of the potential--determining ion through the membrane generator. This current leads to the equilibration of the potential-determining ion concentrations, drop of φ_o, shift to the type 1 curves (Fig.1) and recuperation of the pores. If the membrane has no pumps, the prosess is complete at that. To estimate τ_c, use can be made of the relationship $\tau_c \approx 2R^3/aD$, where D is the diffusion coefficient of ions in the solution. Since $\tau_p \approx a^2/D_p$, where D_p is the pore diffusion coefficient in the space of radii, we have the following condition for the establishment of the pore equilibrium with respect to radii:

$$a^3 \ll D_p R^3/D \qquad (29)$$

Taking for the estimation $D_p \approx 10^{-8}$ cm^2/c, $D \approx 10^{-5}$ cm^2/c, we obtain a < 0.2 R. According to condition (2) it implies that in the case of a strong generator the equilibrium distribution of the pores is established in the energy space of the type 3 curve (Fig.1), which then slowly relaxes to the type 1 curve.

(b) Weak Generator ($W \ll 1$)

In this case the pore equilibrium in the space of radii is established on condition of $\tau_p \ll \tau_\varphi$, i.e. $R^2 \gg a^3/6D_p C_o \mathscr{æ}^{-1}$, where $\mathscr{æ}^{-1}$ is the specific resistance of the electrolyte. If the relaxation of the potential-determining ion concentration occurs slower than the electric relaxation ($\tau_c \ll \tau_\varphi$) the system undergoes aperiodic oscillations the character of which is as follows. As a result of thermal fluctuations the pore achieves the dimensions which correspond to the intermediate minimum (Fig.1) and the discharge of the membrane capacitance begins. As a result, curve E (ξ) changes over from type 3 to type 1, the intermediate minimum disappears and the pores pass over into the region of small dimensions. Further on, the membrane generator restores the voltage close to the initial value and, following a stochastic waiting time of a new thermal fluctuation, the cycle is reiterated. The numerical estimation shows that the process can be realized only in the case of sufficiently large vesicles with R > 0.01 mm.

REFERENCES

1. Yu. A. Chizmadzhev and V. F. Pastushenko, Electric Breakdown of Bilayer Lipid Membranes, in: "Thin Lipid Films", J. B. Ivanov, ed., Marcel Dekker, Inc. N.Y. (1988).
2. V. A. Parsegian, Energy of an Ion Crossing a Low Dielectric Membrane, Nature, 221:844 (1969).

EFFECT OF RADIATION ON BIOLOGICAL SYSTEMS

Marko S. Markov

Department of Biophysics
Sofia University
8 Tzankov Blvd., Sofia 1000, Bulgaria

Recent developments in modern technology have increased the number of devices creating electromagnetic fields which may affect biological systems. Every particular living system is exposed to increasing values of electromagnetic pollution. Thus, the problems of evaluating electromagnetic fields and their various physical parameters have become of great importance. Biotechnology, in the last decade, has made large advances in methods to analyze the effects of different physical parameters of electromagnetic fields on living systems.

It seems to us that the following problems are open for evaluation:
- the search for mechanisms of electromagnetic field effects on living systems
- the discovery of targets of electromagnetic radiation on cellular and subcellular levels
- the adaptations of living systems when exposed to applied magnetic fields
- the comparisons between the actions of ionizing and non-ionizing radiation
- the creation of standards for electromagnetic fields in occupational conditions and everyday life
- the evaluation of the window effects
- the long-lasting after effects of electromagnetic exposure.

Some of these problems have been studied for many years, but science is still far away from identifying and solving the problems of the mechanisms of action and of targets for electromagnetic radiation.

Investigations on the long-lasting effects of electromagnetic fields and the comparison of the actions of ionizing and non-ionizing radiation indicate a number of things. As it seems to us, the introduction of the term "non-ionizing" radiation could accelerate the change in strategy for investigation of electromagnetic field effects on living systems. The generalization of ionizing and non-ionizing radiation studies methodology would permit one to look for new aspects of electromagnetic field influence on living systems.

To further analyze these general problems, we could also stress the ability of different systems to exist on changed electromagnetic backgrounds without any significant changes in the course of basic biochemical and physiological processes.

In spite of the large number of studies on electromagnetic field action on biosystems, there is still absent a general approach to magnetobiological experiments. Two main reasons for this fact seem quite probable: (1) participation in such experiments of scientists with various university backgrounds of education -- physicists, biologists, medical doctors, engineers, etc., (2) electromagnetic fields include varying frequencies, intensities, field-shapes, and physical factors. For this reason, it seems quite probable that no general mechanism probably exists. A systematic study of the electromagnetic field action on any particular system has to consider and explain the following parameters of electromagnetic fields:
- type of field
- intensity or induction
- gradient
- vector
- frequency
- pulse shape
- component (electric or magnetic)
- localization
- time of exposure

As it was explained elsewhere[1], one should also take into account the amount and state of water in living systems in general as well as of the water that is in close vicinity of membranes and ions. It is quite probable that the modification of some physical-chemical properties of water solutions under electromagnetic field influences could play a role in the realization of the observed effects[2,3].

The search for possible targets of electromagnetic action is developing in different aspects. The first direction is to look for the existence of magnetic material at the level of the cell[4,5]. The second is to search for biogenic deposits of inorganic materials, such as ferric/ferrous oxide in birds, vertebrates, and humans. Bauer et al.[6] found that the bones forming the walls of the sphenoid and ethnoid sinuses showed a level of magnetic remanence significantly greater than that of the background. At the same time, a wide range of animals are known to be able to detect the geomagnetic field without any specific magnetic sense organs[7]. This gives credit to the idea that the cell membrane may function as a target for electromagnetic field actions.

WINDOW EFFECTS

The sensitivity of the biological tissues to small variations of the electromagnetic field may explain the window effect described elsewhere. The existence of "windows" for different fields would favour system transductive properties that are cooperative and thus nonequilibrious in character[8]. Membrane surface glycoproteins with a polyanionic structure may function as sensing sites.

In a number of experiments directed to study large intervals of inductions of constant magnetic fields, applied to different levels of organization biosystems (microorganisms, plants, animals)[9,10] a specific dependence of the extent of magnetic field effects on the value of the effects was found at induction B=45 mT. It has been proposed that there is an existence of "permitted" levels of magnetic induction which biosystems could attain under the electromagnetic influence of constant magnetic fields. When the field induction is equal to that necessary for the transition, a "stationary" state will be achieved. It can remain in this state for a certain period of time. Any other energy will bring the system to a state different

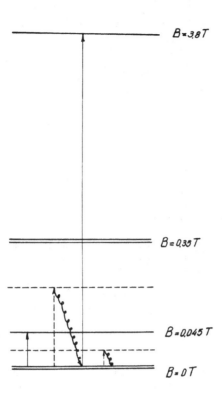

Fig. 1 Specific "resonance" levels
of magnetic induction which are of
importance in living systems

from that of the stationary one. Because the system intensively exchanges
energy with the environment, the observed effects are lower and quickly
disappear at these "non-permitted" states. In many respects, this idea is
similar to the idea for "window" action of electromagnetic fields.

It had been observed that maximum sensitivity to the microwave field
occurred with a 16 Hz modulation coinciding with the frequency-dependent
responses in brain tissue [11,12]. Calcium release from isolated cerebral
tissues was decreased in electromagnetic fields with corresponding electrical gradients of 5-100 V/m. The degree of inhibition was maximum at fields
with a 16 Hz frequency and a gradient of 10 V/m - this is a demonstration
of both "frequency and amplitude windows". A high differential frequency
sensitivity in the low frequency range has been modeled in terms of quasi-
particle behaviour in linear macromolecules such as transmembrane lipoproteins[13].

The finding of a transient sensitivity of C-AMP independent protein
kinase activity to continuing microwave exposure suggests a "window in
time". Disclosure of the "time window" suggests transient induction of a
persisting but reversible molecular state in the elements of the transductive system. There is evidence that these field-induced changes involve
the cell membrane[14].

It seems quite probable that different kinds of "windows" exist in biological systems. These particular "windows" react towards field strengths,
frequencies, or times of exposure.

Having in mind the idea of "biological windows", one could explain many
of the apparent inconsistencies in the literature.

ELECTROMAGNETIC FIELDS EFFECTS

In living organisms, electromagnetic fields of sufficiently high intensity and size to affect chemical processes are encountered only within cell
or membrane phases, but also near the surface of membranes and protein or
lipid organizations of aggregates.

During the last decade, it has been shown that chemical reactions
through radical pairs are influenced by magnetic fields. A laser-photolysis
study of magnetic field effects on photochemical reactions in micelles show
that the triplet-singlet conversion rate is reduced by the applied magnetic
field. The magnetic field effects in micelles are interpreted in terms of
relaxation mechanisms[15].

An externally applied electromagnetic field is considered as a
vectorial perturbation in the chemical or orientational distributions of
interacting molecules or molecular organization. Major structural-chemical
changes in electromagnetic fields require the presence of ions, ionized
groups, and permanent or induced dipolar charge configurations preferably
in macromolecular structures. In most considerations of electromagnetic
fields effects, the theoretical expressions are based on the "homogeneous-
field approximation" of the continuum relationship between the total
polarization and field strength.·

When an electromagnetic field is applied to a biological system, the
field experienced by the cell membrane is affected by two important time-
dependent mechanisms: dielectric changes of water molecules which occur
with a time constant of the order of 10^{-11} sec and time-dependent free
charge separation over the membrane.

One can assume that the electromagnetic fields affect the water bound
to outer membrane components of the erythrocyte ghosts and that this leads
to a disturbance of the hydrogen bond formation near the membrane surface.
When the intensity of the electromagnetic field is not high enough to get
energy for molecular rearrangements, it is necessary to assume concentration
of the field because of membrane heterogeneity. If this idea is assumed,
then the electromagnetic field could affect the protein channels, leading
to disturbances of intact hydrogen-bond formations of structured channel
water that is reflected in the infrared spectrum. Thus, the protein structure is slighty destabilized[16].

It is well known that fixed surface charges result in the formation of
a diffuse electrical double layer in which the charge at the surface is
balanced by charges of opposite sign in the medium and by the dipole moment
of water molecules adjacent to the surface. Basic information about macromolecular dimensions, size, and shape have been derived from the relaxation
of field-induced changes in optical properties and electrical parameters of
the electrically and optically anisotropic systems. It has been established
that the extremely large dipole moment, which the electric field produced
by the displacement of the counter-ion atmosphere, is responsible for some
of the effects.

244

EFFECTS ON THE CELLULAR LEVEL

Electromagnetic fields effects play an important role in many biological processes of the cellular level. Phenomena, such as nerve excitation, electromagnetic ion transport, and neurostisynthesis of ATP, involve cell functions in which biochemical reactions are coupled to electric field forces[17]. Exogeneously applied weak electrical fields are known to modify or alter normal cellular functions[18,19]. Potentially large numbers of cellular processes could be affected by electromagnetic fields. No obvious linear relationship between the shape of the signal and the effectiveness of a given signal was found[19].

Theoretical considerations of Frohlich[20] have suggested that living objects might contain systems exhibiting large-amplitude oscillations at high frequencies. In particular, Frohlich predicted that there would be a very high frequency oscillation in the frequency range of 100 GHz.

The study of the orientation of erythrocytes of patients suffering from sickle cell anemia was reported by Vienken et al[21]. The suspension of erythrocytes was exposed to an electromagnetic field with a maximum output voltage of 40 V and a frequency range of 1 Hz - 5MHz. Differences in orientation between normal and pathological erythrocytes were found to be frequency dependent. An orientation of sickle cells was also observed in homogeneous magnetic fields due to magnetic anisotropy in the heme molecule in deoxyhemoglobin, whereas normal erythrocytes do not show any orientation in magnetic field.

EFFECTS OF ELECTROMAGNETIC FIELDS ON BIOLOGICAL MEMBRANES

The interest in electromagnetic field effects on macromolecules and biomembranes significantly increased when it was established that electromagnetic fields were capable of producing structural-conformational changes in membrane constituents.

It has become clear that nonionizing electromagnetic fields offer a valuable new tool in studies of membrane transductive functions. The cell membrane surface is considered as a prime site in electromagnetic field transduction. There is good evidence that the coupling of weak, oscillating electrochemical fields occurs on membranes[22]. A minimal sequence of three steps is proposed: (1) the initial events may relate to the vast array of negative charges at terminals on the surface glycoprotein sheet, a step presumed to occur within the area of the membrane; (2) a step involving transmembrane signaling mediated by lipoproteins and glycoproteins; (3) a step coupling the transmembrane signal to intracellular organelles via cytoskeletal elements.

The asymmetric distribution and orientation of membrane lipids and proteins has a great importance in their functions as well as in reactions of biological membranes to applied electromagnetic fields. It has been proposed that one of the mechanisms for alteration of the functional properties of membranes by electromagnetic fields can be their structural rearrangement[23]. Such rearrangements are tightly connected with the modification of the transport properties of membranes. Perturbation of the ionic exchanges through the membrane can be a possible mediator between electromagnetic stimulation and cell response[24].

The influence of electromagnetic field exposure on the mitogenic capability of receptor ligands in human lymphocytes can occur through different mechanisms: (1) ligand electrophoresis and cell receptor electrophoresis along the membrane with possible formation of surface clusters;[25](2) field-induced free energy changes of the ligand-encounter complexes, thus affecting the binding rate constants;[26](3) changes of the mean ligand-encounter lifetime due to local microelectrophoresis; (4) changes in the microstructure of water, which involve hydrated membrane proteins.

When the cell suspension is exposed to an electromagnetic field, cell reactivation proceeds more slowly; pointing out an antagonistic effect of the field with respect to the ligand. An inhibitory effect presumably at the membrane level has recently been described. It is induced by an electromagnetic field signal on the in vitro response of bone cell parathyroid hormone[27].

Strong magnetic fields can be used to obtain highly oriented specimens of biological material which are suitable for structural analysis using optical diffraction and NMR techniques. As magnetic anisotropy is conformational-dependent, structural information can be deduced from the orientational behaviour of studies particles. Also, studies of polymerization and assembly of proteins such as fibrin can be accomplished by monitoring the birefringence in a magnetic field. It has been clearly demonstrated that aggregates of fibrous proteins may be magnetically oriented[28].

The charge in the electric field in a protein could cause a reorientation of the dipoles of nearby water molecules. If a chain of water molecules were involved, the electric-field provoked changes could be partially transferred to a remote part of the protein. Water molecules within the interior of the protein are found to make up elaborate and highly ordered hydrogen bound chains and networks. The transmission of information along the water chain could also be carried out by rotational correlation among the water molecules.

Another approach to the problem is the investigation of the magnetic properties of lipids, the major component of biomembranes. For example, the behaviour of crystals with stacked molecular bilayers (suspended in xylene) was studied in a constant magnetic field of about 5 kG. The lipids showed an orientation whereby the direction of hydrocarbon chains were perpendicular to the bilayer[29].

MECHANISMS OF ELECTROMAGNETIC FIELD ACTION

It has been established that electromagnetic fields play an important role in many biological processes. Cell membranes play a key role in detecting, transforming, and transmitting signals from the cell surface to the interior. There is strong evidence that cell membranes are powerful amplifiers of weak electrochemical events in their vicinity.

The absence of general mechanisms of electromagnetic field actions on biological systems is a major problem in the study of biological effects related to electromagnetic fields.

One of the main aims of investigating electromagnetic field effects on chemical and biological transformations is to determine the reaction mechanism. It is known that the majority of biological processes involve ionic species in aqueous environments. The stability of many biopolymers requires a definite ionic strength.

The mechanisms by which cells sense the application of electromagnetic fields is presently unknown. At least one component of the electrochemical information transfer probably involves changes in the concentration of cations at the cell surface and/or within the cell. One possible mechanism may be that changes in the concentration of protein-bound ions may induce conformational changes in the associated molecules[30]. With regard to molecules that operate in the extracellular environment, alterations that produce conformational changes in plasma membrane components may induce concomitant changes in either the activity or accessibility of these components. With regard to intracellular processes, the glycosylation apparatus of the cell appears to be a candidate target for changes induced by alterations in net ion flux. The conformation of protein susbstrates for various glycosidic enzymes may be changed, signaling the transfer of additional sugar residues. In addition, changes in the conformation of themselves could affect the transferase activity[31].

Some authors[32] propose that the coupling of electromagnetic fields and cells is taking place probably at the level of the plasma membrane. The effect of the field is conspicuous in the presence of activators of inhibitors. Obviously, the very low frequency, low energy pulsating electromagnetic fields are a new compelling tool to investigate: (1) the structural and functional relationship among the biochemical entities regulating cell deviation and differentiation; (2) the therapeutic implications of the field interactions with the receptor-messenger apparatus of the plasma membrane.

The possible participation of the cell membrane in signaling is considered in several observations and hypotheses: A/ The plasma membrane has been attributed to play a role in the spinning of suspended live cells under the action of high frequency electromagnetic fields[33]; B/ According to the electrochemical information transfer hypothesis, the electromagnetically induced currents must couple nonFaradaically with ion transport across the membrane [30]; C/ The same very low frequency fields that stimulated Ca^{++} uptake by the chick tibia also promoted Na^+ flux and ATPase activity across red blood cell membranes; D/ Since the same fields stimulate DNA synthesis and cell proliferation which are regulated by the plasma membrane Na^+K^+ATPase, the role of membrane ATPase in coupling of the signal with the biological processes becomes inevitable; E/ The hydration of plasma membrane and the double electrical layer around the membrane has to be considered when discussing the signal coupling.

A fundamental question is whether the stimulation of the biological processes affected by the electromagnetic fields is electric or magnetic. According to Pilla [30], the stimulation and the response are electric, and non-Faradaic. Two different (by frequency) fields may be coupling with totally different, if only quantitatively different, processes. It is known that AC magnetic fields can induce currents that will flow in close circuits within the body.

The direct interaction of magnetic fields with the biological systems involves energies that are much smaller than the terminal energies [34]. For this reason the possibilities of informational influence have to be considered[2].

Warnke [35]tried to determine the casual mechanism of action of alternating magnetic fields. The parameters in question: (1) induced voltage boosts; (2) induced eddy currents; (3) Lorentz forces; (4) mechanical microvibrations through diamagnetic and paramagnetic substances. Based on induced field forces of about 15 mT, the calculated values would not be energetically sufficient to provoke an nonspecific stimulation. That is why Warnke considers that the information potential of induced voltage and currents plays a decisive role. The eddy currents or electromagnetic forces cause an electric polarization of the membrane and thus, a storage and adding-up of impulses is possible.

These effects are strongly cooperative. The cooperativity is considered as the way in which components of a macromolecule or a complex of macromolecules act together to switch from one stable state to another. Trigger signals to cooperative processes may be weak and amplified response orders of magnitude may be larger. Amplification effects in cooperative events raise questions about thresholds and the minimum size of an effective triggering stimulus.

Adey[8] estimated that even a magnetic field of 1 T gets into the system the energy of the order of 1 K, far below kT, and that the coupling of this field with brain tissue would be very small.

CONCLUSIONS

We would like to note that because of the difference in physical parameters, every particular electromagnetic field is capable of influencing different membrane cellular components as well as cell environments. This is probably the reason that a general mechanism of electromagnetic field action on biological systems has not been proposed yet. In conclusion, we feel that a complete approach including cell interiors, cell membranes, and cell environments (including hydration properties) is necessary for studying the effect of electromagnetic field action.

Further investigations of this action will be developing in several directions:
- effects on population level
- effects on whole tissue organisms
- effects on tissue levels
- effects on cells
- effects on subcellular levels
- effects on membranes

The following general aspects seem to be of great significance:
- Study of basic mechanisms of electromagnetic field action on biological membranes
- Electromagnetic field action on membrane transport
- Medical aspects
  * diagnostics
  * therapy
  * whole body NMR
  * hyperthermia
- Electromanipulation
- Hygienic estimation
  * evaluation of the risks for the population as well as for individuals
  * developing of standards for safety conditions in occupational conditions and everyday life

REFERENCES

1. M. S. Markov, Influence of constant magnetic field on biological systems, in: "Charge and Field Effects in Biosystems", M. J. Allen, P. N. R. Usherwood, eds., Abacus Press, Kent 314 (1984)
2. M. S. Markov, S. I. Todorov, M. R. Ratcheva, Biomagnetic effect of the constant magnetic field action on water and physiological activity, in: "Physical and Chemical Basis of Biological Information Transfer", J. G. Vasileva-Popva, E. V. Jensen, eds., Plenum Press, New York, 441 (1975)
3. V. I. Klassen, "Magnetic Treatment of Water Systems", Chimia, Moscow (1978) (in Russian)

4. R. P. Blakemore, Magnetotactive bacteria, Science, 190:377 (1975)
5. H. A. Lowenstein, Magnetite in denticle capping in recent chitons, Geol. Soc. America Bull., 73:435 (1962)
6. R. R. Baker, J. G. Mather, J. H. Kennaugh, Magnetic bones in human sinuses, Nature, 301:78 (1983)
7. J. L. Kirshvink, D. S. Jones, B. J. MacFadden, eds., "Magnetic Biomineralization and Magnetoreceptors in Organisms: A New Magnetism", Plenum Press, New York (1985)
8. W. R. Adey, Tissue interactions with nonionizing electromagnetic fields, Physiol. Rev., 61:435 (1981)   .
9. M. S. Markov, Direct and indirect action of constant magnetic fields, in: "Proceedings of the Sixth International Conference on Magnet Technology", Bratislava, 384 (1977)
10. M. S. Markov, Biological mechanisms of the magnetic field action, Transactions IEEE, Mag. 17:2334 (1981)
11. S. M. Bawin, W. R. Adey, Sensitivity of calcium binding in cerebral tissue to weak environmental oscillating low frequency electric fields, Proc. Natl. Acad. Sci. USA, 73;1999 (1976)
12. S. M. Bawin, L. K. Kaczmarek, W. R. Adey, Effects of modulated VHF fields on the central nervous system, Ann. N. Y. Acad. Sci., 247:74 (1975)
13. A. F. Lowrence, W. R. Adey, Nonlinear wave mechanisms in interactions between excitable tissue and electromagnetic fields, Neurol. Res., 4:115 (1982)
14. C. V. Byus, R. L. Lundak, R. M. Fletcher, W. R. Adey, Alterations in protein kinase activity following exposure of cultured lymphocytes to modulated microwave fields, Bioelectromagnetics, 5:34 (1984)
15. H. Hayashi, Y. Sakagushi, K. Mochida, Laser-photolysis study of the external magnetic field effect on the dynamic behaviour of radical pairs involving germil radical, Chem. Letters, (1984) 79
16. V. L. Shnyrov, G. G. Zhadan, I. G. Akoev, Calorimetric measurements of the effect of 330-MHz radiofrequency radiation on human erythrocyte ghosts, Bioelectromagnetics, 5:411 (1984)
17. E. Neumann, in: "Modern Bioelectrochemistry", F. Gutmann, H. Keyzer, eds., Plenum Press, New York, (1986) 97
18. R. Goodman, C. A. Bassett, A. Henderson, Pulsing electromagnetic fields induce cellular transcription, Science, 223: 1283 (1983)
19. R. Goodman, A. S. Henderson, Sine waves enhance cellular transcription, Bioelectromagnetics, 7:23 (1986)
20. H. Frohlich, Evidence for Bose-consideration-like excitation of coherent modes in biological systems, Phys. Letters, 51A:21 (1975)
21. J. Vienken, U. Zimmermann, A. Alonso, D. Chapman, Orientation of sickle cells in alternating electric field, Naturwissenschaften, 71:158 (1984)
22. W. R. Adey, Molecular aspects of cell membranes as substrates for interactions with electromagnetic fields, in: "Synergetics of the brain", H. Keitzner ed., (1983)
23. R. B. Olsers, S. Belman, M. Eisenbud, W. W. Mumford, J. R. Rabinowitz, The increased passive efflux of sodium and rubidium from rabbit erythrocyte by microwave radiation, Radiat. Res., 82:244 (1980)
24. M. Hisenkamp, M. Rooze, Morphological effects of electromagnetic stimulation on the skeleton of fetal or newborn mice, Acta Ortop. Scand., 53: suppl. 196:39 (1982)
25. S. McLaughlin, M. M. Poo, The role of electroosmosis in the electric-field-induced moment of charges macromolecules on the surfaces of cells, Biophys. J., 34:35 (1981)
26. A. Chiabrera, M. Grattarola, R. Viviani, C. Braccini, Modelling of the perturbation induced by low frequency electromagnetic fields on the membrane receptors of stimulated human lymphocytes, Studia Biophysica, 91:125 (1981)
27. R. A. Luben, C. D. Cain, M. Chi-Jun, D. M. Rosen, W. R. Adey, Effect

of electromagnetic stimuli on bone and bone cells in vitro, <u>Proc. Natl. Acad. Sci. USA</u>, 79: 4180 (1982)

28. J. Torbet, M. J. Dickens, Orientation of skeletal muscle action in strong magnetic fields, <u>FEBS Letters</u>, 173:403 (1984)

29. J. Sakurai, Y. Kawamura, Magnetic field-induced orientation and bending of the myelin fibers of phosphatidylcholine, BBA, 735:189 (1983)

30. A. A. Pilla, Electrochemical information transfer at cell surfaces and junctions, <u>in</u>: "Bioelectrochemistry", H. Keyzer, F. Gutmann, eds., Plenum Press, New York, (1980)

31. S. J. Fisher, J. Dullings, S. D. Smith, Effect of a pulsed electro-magnetic field on plasma membrane protein glycosylation, <u>J. Bioelectricity,</u> 5: 253 (1986)

32. G. Collacicco, A. A. Pilla, Electromagnetic modulation of biological processes, <u>Z. Naturforschung</u> 38c:468 (1983)

33. U. Zimmermann, J. Vienken, G. Pilwat, Rotation of cells in an alter-nating electric field: the occurence of a resonance frequency, <u>Z. Naturforschung</u>, 36c:173 (1981)

34. A. R. Sheppard, M. Eisenbud, "Biological effects of electric and magnetic fields of extremely low frequency", New York University Press, New York, (1977)

35. U. Warnke, The possible role of pulsating magnetic fields in the reduction of pain, <u>Pain Therapy,</u> 1:229 (1983)

A THEORETICAL EXAMINATION OF HOW AND WHY POWER LINE FIELDS

INTERACT WITH BIOLOGICAL TISSUE

Kent Davey

School of Electrical Engineering
Georgia Institute of Technology
Atlanta, GA 30332-0250

Summary

The objective of this paper is to define and document the mechanism of biological tissue - field interaction at power line frequencies. The basic hypothesis is summarized in the following four theses:

I.   The coupling of the field to the tissue is magnetic not electric.

II.   There is a twofold mechanism of interaction involving both electrokinetic and conductive processes.

III.   Among the major requirements of this coupling is that the field be transient in nature; it is the transient current superimposed on the base sinusoidal field that causes all the biological linkage of interest.

IV.   The orientation of the field is quite important as to the type and magnitude of biological interaction resulting from the field exposure.

The above four theses represent a summary and compilation of a considerable amount of independent research; for the sake of brevity I will not elaborate too heavily on any one point in this paper. After discussing each of the four points, a research strategy is suggested for additional verification. This work has significant implications for a utility planning strategy most likely to mitigate political problems in this area. From this research, it would appear that the utilities are attacking the wrong problem in trying to reduce human field level exposure. The ultimate goal is to reduce current transients. One of the best ways a utility might mitigate deleterious health effects form transmission lines is to push the voltage up higher to minimize current, since current transients constitute the biological coupling mechanism. Of course there are industrial considerations, but if biological coupling were the only issue, high voltage would be the solution. The utilities have supported a barrage of epidemiological studies [1-5] which have turned up conflicting and controversial results. Based on the work, it would appear that these indefinite epidemiological results are expected because of the nature of the actual biomedical - field coupling mechanism.

I. Coupling

The biomedical field coupling is magnetic not electric. The focus in this research is noninvasive coupling; thus, there are no surface skin electrodes.
The electrical relaxation time if found by solving the equation of charge conservation

$$\nabla \cdot \vec{J} = -\frac{\partial \rho}{\partial t} \qquad (1)$$

where
ρ = charge density

t = time

$\vec{J}$ = current density.

The dielectric constant ϵ is used to link the electric field E to the charge density

$$\nabla \cdot (\epsilon \vec{E}) = \rho. \qquad (2)$$

Combining (1) and (2) and using the fact that conductivity links J and E $(\vec{J} = \sigma \vec{E})$ yields

$$\vec{E} \cdot \nabla \sigma + \frac{\sigma \rho}{\epsilon} - \frac{\sigma \nabla \epsilon \cdot \nabla \vec{E}}{\epsilon} = -\frac{\partial \rho}{\partial t}. \qquad (3)$$

This equation simply states that charge in a medium with local conductivity σ and dielectric constant ϵ will decay in time due gradients in either the medium conductivity or dielectric strength and to conductivity itself by the ratio $\frac{\epsilon}{\sigma}$. One figure of merit used to estimate what the representative electrical relaxation time is realized by ignoring all spatial gradients in (3). The temporal description of charge in a homogeneous medium is thus

$$\rho = \rho_0 \exp\left(-\frac{t}{\epsilon/\sigma}\right). \qquad (4)$$

The electrical relaxation time ϵ/σ varies typically from 1 to 100 nanoseconds. This means that if the tissue is exposed to an instantaneous electric field by, for example, two non contacting electrodes, in a few hundred nanoseconds (at most) surface charge will build up on the interface of the tissue to buck out or shield all of the field internal to the tissue. The governing equations have been worked out in considerable detail in a number of references [6,7,8]. Thus at least for frequencies below 100 kHz we must seek another coupling mechanism.

Magnetic fields penetrate biological tissue unimpeded except where high permeable plates or shields exist. The coupling is through eddy currents, but the problem is not a true eddy current problem in that the induced currents have a

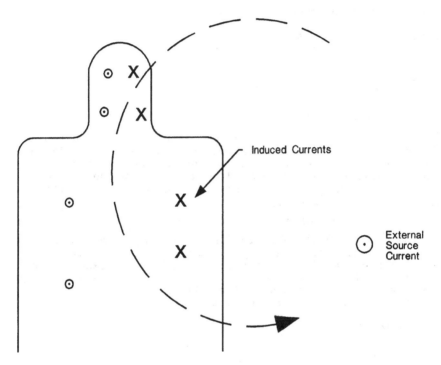

Figure 1. Electric field/current coupling due to an external time varying source current.

negligible effect on the source field. The biological currents are determined by solving Faraday's equation with the
product of local conductivity and electric field substituted
for the induced electric field.

$$\nabla \times \frac{\vec{J}}{\sigma} = -\frac{\partial \vec{B}}{\partial t}.$$
(5)

We have worked out such fields a number of ways and found the
use of a current vector potential or a split vector scalar
potential approach the most expeditious and numerically stable approach for determining these fields.

$$\vec{J} = \nabla \times \vec{A}$$
(6)

which assures the necessary solenoidal condition $\nabla \cdot \vec{J} = 0$.

The second approach is to split E

$$\vec{E} = \vec{T} - \nabla \Phi$$
(7)

where Φ solves the boundary conditions and

$\vec{T}$ equals the time rate of change of the magnetic vector potential.

Fig. 1 gives a rough picture of the nature of the current
induced. Note that the primary induction current is colinear
with the source current. The key point is that it is the
induced electric field generated by a time changing magnetic
field that produces the biological tissue coupling in the
transmission line problem.

II. Mechanism of Coupling

The mechanism of magnetic field coupling is electrokinetic and conductive. The body is made up of proteins, i.e.
polypeptide chains in an electrolytic solution. Except when
the local pH is at the isoelectric point, these chains undergo
charge exchange with the solution. The two primary chemical
exchanges are the carboxyl and ammonium exchange.

$$COOH = COO^- + H^+$$

$$NH_4^+ = NH_3 + H^+$$

Because of the reaction of these side groups on the polypeptide chain, the fixed polypeptide is left with a net charge,
the sign and magnitude of which are based on the local pH.
Consider the case when the chain is positive. The surrounding
polyelectrolytic solution which maintains charge neutrality
on a macroscale loses this neutrality in the vicinity of the
fixed polypeptide. As Fig.2 suggests, the negative ions con-
gregate close to the polypeptide preserving a balance between
electrical attraction and ionic diffusion forces. The length
scale over which the charge imbalance exists, known as the
Debye length, can be calculated exactly in terms of the interfacial surface charge on the chain and the conductivity and
dielectric constants of the surrounding polyelectrolytic
solution [9]. It is typically a few microns in length. When an
external electric field is applied tangential to the solution,
a force (often referred to as the electroosmotic force) is

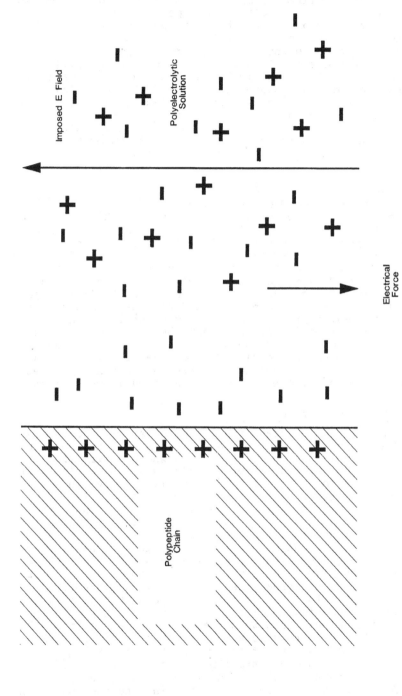

Figure 2. Basic Electrokinetic Field Interaction.

imposed on the surrounding electrolyte. This force is volume coupled only over the Debye region where the zeta potential is nonzero.

The second mechanism, conduction, is more straightforward. Any induced field will result in both positive and negative ion current flow according to their respective ionic mobilities. One example of this application in which there is virtually unanimous agreement of an electrical field - biological coupling is in the area of osteogenesis. Bassett [10,11] pioneered most of the magnetic based coupling (as opposed to direct surface or transcutaneous electrodes), which induces an electric field flushing calcium ions (among others) through the site of the bone nonunion. There are also experimental studies on cartilage which suggest that flushing synovial fluid through the injured site (e.g. by movement or electric field stimulation), will expedite the healing process.

III. The Requirement of a Transient Current

The biological conduction problem is unlike most in that it is the ions that are actually promoting the charge transfer or movement. The characteristic electroinertial time is such that for frequencies above about 5 Hz inertial effects dominate. The ions never have time to migrate in any one direction to accomplish anything for balanced sinusoidal fields above this frequency. This is, of course, why Bassett and others have used non balanced excitation currents in their stimulation coils. The optimum magnetic field time pattern shown in fig.3 is that which produces a constant electric field with time, i.e. monodirectional. Note that the electric field is not truly constant but undergoes by necessity a large negative impulse during each cycle. The ions cannot react to this short time pulse, but 'see' instead the rather constant field over the whole cycle. The commensurate stimulation current is shown in fig.3b.

IV. The Importance of Orientation

It is worthwhile pedagogically to focus on osteogenesis in our discussion of electric field - biological tissue coupling, since it is one of the few areas where a firm link has been established between magnetic fields and tissue; it just so happens that in this case the effect is positive in terms of healing. The question of when a coupling occurs is more germain to the problem under discussion. Fig. 4 shows a nonunion or breakage in a bone with two possible induced electric field orientations. When the electric field is tangential to the break the field can most effectively flush ions (particularly Ca) through the breakage gap via both the electrokinetic and conductive mechanisms discussed above. The fields induced normal to the break are of little value in bone growth. This principle can be applied to a number of examples. Fig. 5 shows the effect of a field oriented along the long axis of a spheroidal organ which has a lower conductivity than the surrounding electrolytic solution. Continuity of current insures that the majority of the current will streamline around the organ. More penetration would be realized if the electric field were normal to the organ's long axis.

Collagen fibrils are often found in a homogeneous mesh the composite of which has a higher resistivity than the surrounding electrolyte (Fig. 6). Unlike the previous case, here

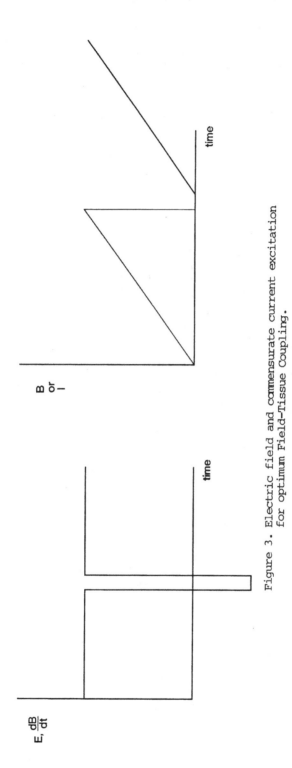

Figure 3. Electric field and commensurate current excitation for optimum Field-Tissue Coupling.

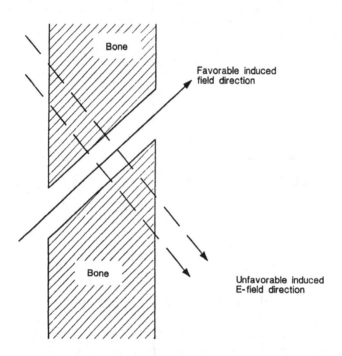

Figure 4. Favorable and unfavorable induced fields
for strong osteogenetic activity.

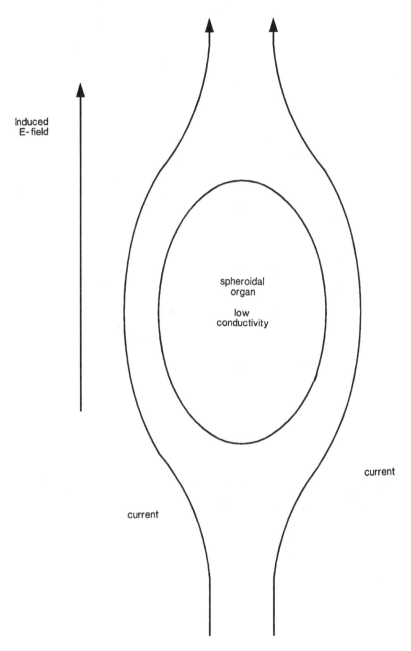

Figure 5. Current flow around a low conductivity spheroidal organ.

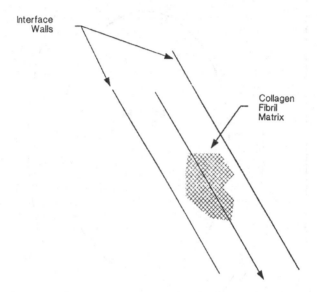

Favorable Induced Field

Figure 6. The desired field in a Collagen Fibril Matrix
for best field coupling is tangential to the wall.

the walls represent a nonconducting medium. An induced electric field normal to these walls will link little magnetic flux and have a small current. The primary coupling direction is tangential to these walls both electrokinectically and conductively.

Conclusion

The arguments in sections III and IV above explain why the epidemological studies have to date been so inconsistent. None have attempted to place an index on transient current field exposure. It is this and not the magnitude of the field itself which has any significance. Second, it matters what orientation a subject spends most of his time in the field. The above argues that there will be significance to the sleeping orientation of the subject to the field. The primary concern from the perspective of the utility should be to minimize the current transient as experienced by the consumer. There are a number of possibilities that could be explored among those being the following:
1) Passive eddy current shielding along the right of way.
2) Use of higher voltages and thus lower absolute currents.
3) Design of electronic buffers at the substation level to compensate actively for current surges.

References

1. W.R. Adey, "Tissue Interactions with Nonionizing Electromagnetic Fields", _Physiological Reviews,_ (1981),61:2,435.
2. "Biological Effects of 60 Hz Power Transmission Lines: A Report of the Florida Electric and Magnetic Fields Advisory Commision", Tallahassee, FL, NTIS no PB 85200871.
3. "Biological Effects of Power Line Fields", New York State Power Lines Scientific Advisory Panel, Final Rept NY Dept of Health, Albany, NY July1, 1987.
4. "Biological and Human Health Effects of Extremely Low Frequency Electromagnetic Fields",Post 1977 Literature Review, Am Inst of Biological Sciences,(1985),Arlington, VA NTIS no AD-A152-731.
5. "A Critical Review of the Scientific Literature on Low Frequency Electric and Magnetic Fields : Assessment of Possible Effects on Human Health and Recommendations for Research", West Associates,Southern Edison Co., Rosemead CA.
6. A.J.Grodzinsky and J.R. Melcher," Electromechanical Transduction with Charged Polyelectrolyte Membranes", _IEEE Trans on Biomedical ENG,_(1976), BME-23:6,421-33.
7. V.G. Levich, _Phsiochemical Hydrodynamics,_ Prentice Hall, Englewood Cliffs, N.J.,(1963),472-93.
8. P. Delahay, Double Layer and Electric Kinetics, _Interscience Publishers,_ New York, (1966), 33-52.
9. J.R. Melcher, Continuum Electromechanics,_MIT Press,_(1981), Cambridge, Mass,10.19-35.
10. C.A. Bassett, R. Pawluck, and A. Pilla,"Augmentation of Bone Repair by Inductively Coupled Electromagnetic Fields",_Science_, 184:(1974),575-77.
11. C.A. Bassett,R. Pawluck, and A. Pilla,"Acceleration of Fracture Repair by Electromagnetic Fields: a surgically non-invasive method",_Ann NY Acad Sci.,_ 238:(1974),242-49.

THE EXTRACELLULAR SPACE AND ENERGETIC HIERARCHIES IN

ELECTROCHEMICAL SIGNALING BETWEEN CELLS

W. Ross Adey

Veterans Administration Medical Center and
University School of Medicine
Loma Linda, California 92357 USA

INTRODUCTION

It is a reasonable assumption that the first living organisms existed
as single cells floating or swimming at the surface of primitive oceans.
This concept of a cell emphasizes the role of a bounding membrane that
delimits an organized interior composed of biomolecules that may have first
existed in the absence of cells, perhaps simply as an unconstrained
molecular "soup" at the surface of primordial seas. Within cells, these
molecular systems mediate processes essential for all terrestrial life in
metabolism, reproduction and responses to environmental stimuli.

This enclosing membrane forms the organism's window on the world
surrounding it. In unicellular organisms swimming freely through large
fluid volumes, cell membranes act as both sensors and effectors. As a
sensor, it detects altered chemistry in the surrounding fluid. It offers a
path for inward signals generated on its surface by a broad spectrum of
stimulating ions and molecules, including hormones, antibodies and
neurotransmitters; and these most elemental signals crossing the membrane
to the cell interior are susceptible to manipulation by a wide range of
natural or imposed electromagnetic (EM) fields that may also pervade the
pericellular fluid. As effectors, cell membranes may induce cell movement
by flagellae or pseudopodia; or secrete substances synthesized internally,
including hormones, antibodies and structural proteins such as collagen.
Many of these effector functions are also sensitive to intrinsic and
imposed EM fields.

This situation is sharply changed in cellular aggregates that form
tissues of higher animals. Unable to move freely in a virtually limitless
ocean, cells are separated by narrow fluid channels that take on special
importance in signaling from cell to cell.

Concepts of an Extracellular Space between Cells in Tissue

These channels act as windows on the electrochemical world surrounding
each cell. Hormones, antibodies and neurotransmitter molecules move along
them to reach binding sites on cell membrane receptors. These narrow fluid
"gutters", typically not more than 150 ^{o}A wide, are also preferred pathways
for intrinsic and environmental EM fields in tissue, since they offer a
much lower electrical impedance than cell membranes.

Numerous stranded protein molecules protrude into these spaces, sensing chemical and electrical signals in surrounding fluid. These strands are external terminals of helical proteins that pass through the enclosing lipid sheet (plasma membrane). On the membrane surface, tips of these strands act as electrically charged receptor sites for hormones and antibodies and for neurotransmitters at synapses. They form an anatomical substrate for the first detection of weak electrochemical oscillations in pericellular fluid, including field potentials arising in activity of adjacent cells or as tissue components of environmental fields.

Research in molecular biology has increasingly emphasized communication between cells that occurs more or less directly due to their mutual proximity. In addition to inward signals that pass down these protein strands to the cell interior (Adey, 1988a and b), there is also a stream of outward signals through cell membranes mediated by these proteins. Many of these signals then pass to neighboring cells through specialized protein plaques that form gap-junctions. Disruption of gap-junction communication is associated with unregulated cell growth (Adey, 1988b and d; Butterworth and Slaga, 1987; Newmark, 1987; Trosko, 1987; Yamasaki, 1987).

In these studies, imposed weak electromagnetic fields have proved unique tools in revealing the sequence and energetics of major steps in cell membrane transductive coupling. Many of these interactions are "windowed" with respect to field frequency and amplitude, and to duration of field exposure. Windowing of these responses points to their nonlinear and nonequilibrium character, and focuses current and future research on physical substrates for these interactions (Adey, 1975, 1977, 1981a and b, 1983, 1984, 1986; Lawrence and Adey, 1982; Maddox, 1986).

Current Flow in Tissue: Role of the Extracellular Space (ECS)

Current flow in tissue from intrinsic or imposed EM fields has a distribution determined on the one hand by high cell membrane resistance, and on the other, by strongly conducting fluid in which cells are bathed (Cole, 1940). Typical cell membrane resistances are in the range 3,000-100,000 $ohm.cm^{-2}$. Extracellular fluid has a much lower specific resistance in the range 50 $ohm.cm^{-1}$. From a consideration of dielectric dispersions, cell membranes behave reactively at frequencies as high as the low gigahertz range (Schwan, 1974). Thus, although the ECS forms only about 10 percent of the conducting cross-section of typical tissue, it is clearly a preferred pathway, carrying at least 90 per cent of any imposed or intrinsic current and directing it along cell membrane surfaces. For this reason, we have found that focal measurements of electrical impedance in small tissue volumes are indicators of changing physiological states. "Electrical impedance changes accompanying physiological responses may arise in perineuronal fluid with a substantial macromolecular content and calcium ions may modulate perineuronal conductivity" (Adey, 1966).

Our studies used a 1.0 kHz impedance measuring current with a density of only 1.0 $uamp/mm^2$ at the electrode surface that minimized stimulation in living tissue (Adey, Kado and Didio, 1962). Coherent detection methods with suitable integration times were used to detect the condition of balance in the impedance measuring bridge. This results in high differential sensitivity and allows detection of perturbations of the order of 0.1 percent of the baseline impedance.

Macromolecular strands derived from intramembranous proteins lie in the ECS and may modulate conductance as a function of cerebral tissue state; thus, there are impedance "transients" accompanying alerting, orienting and visual discriminative responses (Adey et al., 1966) (Fig.1).

A) 100% PERFORMANCE - LIGHT CUE

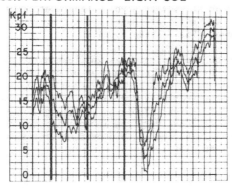

B) IMMEDIATELY AFTER CUE REVERSAL

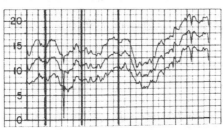

C) RETRAINING TO DARK CUE - 76% PERFORMANCE

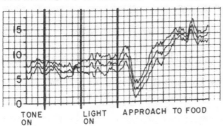

TONE   LIGHT   APPROACH TO FOOD
ON   ON

Fig. 1. Hippocampal impedance averaged over 5-day periods at various
levels of training, with successive presentations of alerting,
orienting and discriminative stimuli to cat. In each graph,
middle trace shows mean. Upper and lower traces show one
standard deviation. Variability was low at 100 percent
performance (A), increased substantially after cue reversal
(B), but decreased again after retraining (C). Calibration 50
pf. (From Adey, Kado, McIlwain and Walter, 1966).

They exhibit differential characteristics in different brain regions. Long
lasting but reversible impedance changes occur with anesthetic and
psychotropic drugs. Asphyxiation sharply reduces the fraction of cerebral
tissue cross-section occupied by the ECS. This leads to a doubling of
tissue impedance, associated with movement of conducting fluid into
neuroglial cells. We have shown that this impedance relates to
extracellular calcium levels (Nicholson, 1965).

OBSERVED BIOLOGICAL SENSITIVITIES TO IMPOSED EM FIELDS

By reason of a much higher conductance in the ECS than in current
pathways that pass through cell membranes, the ECS is the primary route for
induced tissue components of environmental EM fields. Our studies of

current flow along cell membrane surfaces showed its interaction with
anionic fixed charges on strands of glycoprotein that protrude into the ECS
from the lipid plasma membrane, and suggested a functional role for this
cell surface compartment (Elul, 1966, 1967).

As a perspective on the biological significance of this cell surface
current flow, there is evidence from a number of studies that ELF fields
producing tissue gradients as weak as 10^{-7} V/cm are involved in essential
physiological functions in marine vertebrates, birds and mammals, including
man (see Adey, 1981a for review) (Table 1). In vitro studies have also
reported sensitivities at these very low intensities for cerebral Ca^{2+}
efflux (Bawin and Adey, 1976) and in Ca-dependent processes in bone growth
(Fitzsimmons et al., 1986, 1989).

TABLE 1

BIOELECTRIC SENSITIVITIES TO ELF FIELDS

	Function	Tissue Gradient	Imposed Field
Sharks & Rays	Navigation and Predation	10^{-8} V/cm	DC to 10 Hz
Birds	Navigation	10^{-7} V/cm	0.3 gauss
Birds	Circadian Rhythms	10^{-7} V/cm	10 Hz, 2.5 V/m
Monkeys	Subjective Time Estimations	10^{-7} V/cm	7 Hz, 10 V/m
Man	Circadian Rhythms	10^{-7} V/cm	10 Hz, 2.5 V/m

Comparison with Intrinsic Cell and Tissue Neuroelectric Gradients

Membrane Potential	10^{5} V/cm
Synaptic Potential	10^{3} V/cm
Electroencephalogram	10^{-1} V/cm

With RF fields that are amplitude-modulated at ELF frequencies,
induced tissue electric gradients can be substantially higher than with ELF
fields, due to increased coupling between RF fields and tissue (see Adey,
1981a for review). At tissue levels of 10-100 mV/cm cited above, they are
in the same amplitude range as intrinsic oscillations generated
biologically, such as the electroencephalogram (EEG). Induced fields at
these higher levels also produce a wide range of biological interactions.
These responses include entrainment of brain EEG rhythms at the same
frequencies as ELF components of imposed fields, conditioned EEG responses
to imposed fields, and modulation of brain and behavioral states (Bawin et
al., 1973); and in non-nervous tissue, strong effects on cell membrane
functions, including modulation of intercellular communication through
gap-junction mechanisms (Fletcher et al., 1986), reduction of cell-mediated
cytolytic immune responses (Lyle et al., 1983), and modulation of
intracellular enzymes that molecular markers of signals arising at cell
membranes and then coupled to the cell interior (Byus et al., 1984, 1988).

Physiological Benchmarks in Electrical Organization of Cell Membranes

In most cells in the resting state, there is a steady membrane
potential of approximately 0.1 V between the inside and the outside of the
cell, due to differential concentrations of K^+ and Na^+ ions in these two
compartments. The interior of the cell is negative with respect to the

exterior. This membrane potential exists across the extremely thin lipid plasma membrane, typically about 40 A^o thick; a membrane so thin that in consequence there is an enormous electric gradient of 10^5 V/cm across the cell membrane. This large gradient is altered by 10^3 V/cm in synaptic activation in nerve cells (Table 1). In sharp contrast, physiological electric oscillations in fluid surrounding cells are many orders of magnitude weaker than this natural barrier of the membrane potential. For example, the gradient of the EEG measured across the dimensions of a typical brain neuron is a mere 0.1 V/cm in the extracellular fluid, six orders of magnitude less than the electric barrier of the membrane potential.

As a perspective on this disparity between the gradient of the membrane potential and observed cell sensitivities to weak electric gradients in surrounding fluid, we may model the membrane potential as a conducting sheet with a charge of 100 kV and placed only 1 cm above the ground. In sharp contrast, cell sensitivities to gradients as weak as 10^{-7} V/cm may be modeled for surrounding fluid by connecting one terminal of a 1.5 V battery to an electrode in the Pacific Ocean at San Diego and the other terminal to an electrode in the ocean at Seattle 2000 km away.

Thus, it is not surprising that such weak physiological gradients in fluid surrounding cells have been denied a physiological role. Nevertheless, many organisms including man are sensitive to tissue gradients in the range 0.1-100 mV/cm (Adey, 1981a and b). These sensitivities have been confirmed in cell and tissue cultures for many cell types, including lymphocytes (Byus et al., 1984; Lyle et al., 1983, 1988), liver cells (Byus et al., 1987), ovary cells (Byus et al., 1988), bone cells (Luben et al 1982; Luben and Cain, 1984; Cain and Luben, 1987), cartilage cells (Hiraki et al., 1987) and nerve cells (Dixey and Rein, 1981). Embryonic bone matrix formation is increased by exposure to even far weaker gradients down to 10^{-7} V/cm (Fitzsimmons et al., 1986, 1989).

These interactions emphasize the importance of <u>amplification</u> in their ultimate effects on intracellular mechanisms discussed below. Importantly, they have only been seen with oscillating fields and not with static gradients, thus suggesting that integration over time is a required element in the interaction. They typically occur at frequencies below 100 Hz, a region to which we have applied the term <u>biological spectrum</u>.

Biological Responses to Weak Environmental and Intrinsic EM Fields: a Physiological Dilemma

Many biologists have viewed these observations cautiously as beyond the realm of a possible physical reality. As discussed below, it is necessary to view them in the context of the role of cooperative processes and associated nonlinear electrodynamics at cell membranes revealed with imposed EM fields (Adey, 1984, 1986, 1988a b and c; Adey and Lawrence, 1984). These cooperative phenomena are in the realm of nonequilibrium thermodynamics, and are thus far removed from traditional equilibrium models of cellular excitation based on depolarization of the membrane potential and on associated massive changes in ionic equilibria across the cell membrane.

Equilibrium models have also been offered in explanation of the first events in cell membrane transductive coupling coupling of electrochemical stimuli at the cell surface. For nervous tissue, it has been generally accepted that the models of Hodgkin and Huxley (1952) appropriately describe both the sequence and the energetics of excitatory events. On the other hand, the diverse effects of weak imposed EM fields cited above and in Table 1, as well as others described below, make plain that these are inappropriate models.

For example, low-frequency pulsed magnetic fields effective in therapy of ununited fractures (Bassett, 1987) induce tissue gradients of about 3 mV/cm and extracellular current densities around 10^{-6} Amp/cm^2. In cultured bone cells, these fields modify enzyme activity and modulate the secretion of collagen in response to parathyroid hormone; responses at one millionth of threshold transmembrane currents (of the order of 1.0 mA/cm^2) predicted by Hodgkin-Huxley models (Luben et al., 1982). Hodgkin and Huxley originally offered these models only in the context of a mathematical description of major perturbations in Na^+ and K^+ equilibria at certain epochs in the temporal and energetic sequences of excitation of nerve fibers. Their extrapolation by others to address threshold phenomena in cellular systems would appear beyond the scope of their original intent.

To this point, we have focused on evidence that a wide range of tissues and organisms respond to EM fields at intensities many orders of magnitude less than the electric barrier of the membrane potential. Other evidence will be presented relating responses to stimulus characteristics in narrow frequency domains or <u>windows</u>. Jointly, these findings support the view that these interactions are nonlinear and nonequilibrium in character. They are consistent with quantum processes involving long range interactions between electric charges on cell surface macromolecules. They are also consistent with coherent states and phase transitions in populations of charged elements in the biomolecular systems of cell membranes.

These findings point to collective properties of cell membranes as essential in transductive coupling of chemical and electrical stimuli. These collective properties appear to rest on functions of strands of receptor proteins that cross the phospholipid bilayer (plasma membrane) from outside to inside the cell, and importantly, on functional interrelations between these strands and the highly polar phospholipid molecules that surround them.

COLLECTIVE PROPERTIES OF CELL MEMBRANES IN COUPLING OF PHYSIOLOGICAL SIGNALS TO THE CELL INTERIOR

Electromagnetic radiation with wavelengths longer than the ultraviolet region of the spectrum (photon energies less than about 12 eV) does not possess sufficient energy to to cause ionization. Therefore, there has been a persisting view in certain areas of the physical sciences that nonionizing EM fields are incapable of inducing bioeffects other than by heating (Foster and Guy, 1986; Foster and Pickard, 1987). This view overlooks the possibility for cooperativity in biomolecular systems, and the profoundly important role that cooperativity appears to play in detection of tissue components of nonionizing EM fields.

Since our studies have shown similar sensitivities in a wide range of tissues and cell types, we conclude that these electrochemical sensitivities may be a general property of all cells. The nature of these interactions is so far removed from the concepts and models that have guided research in ionizing radiation that expertize in this area can offer little in the search for underlying mechanisms. Equilibrium thermodynamics and the classical models of the statistical mechanics of matter appear equally inappropriate in their applications to most key questions on the biological effects of nonionizing EM fields.

Three Stages in Transmembrane Signaling from Cell Surface to Interior

We may summarize what is known about inward and outward signal paths through cell membranes. Intramembranous protein particles (IMPs) placed within cell membranes provide an essentially direct inward path between the cell surface and intracellular enzymatic systems and organelles. As a functional model, there is a minimal sequence of three steps in this transductive coupling and each is calcium-dependent:

1. The first weak electrochemical events associated with binding of stimulating molecules at their receptor sites and with EM fields are sensed by cell surface glycoproteins that protrude outward from IMPs. These glycoproteins have densely polyanionic terminal strands that form specific molecular receptor sites. They also strongly bind cations, particularly calcium and hydrogen ions.

2. These surface events are amplified, apparently through cooperative modifications in calcium binding, and then signaled to the cell interior by transmembrane portions of IMPs. Lipoproteins and phospholipids of the plasma membrane also participate in this signaling.

3. Inside the membrane, there is coupling of these signals to intracellular enzyme systems, and through the tubes and filaments of the cytoskeleton, to the nucleus and to other organelles.

An outward stream of electrical and chemical signals also passes through cell membranes and links adjacent cells, partly through gap-junctions. Chemical substances synthesized in one cell and essential for the functioning of its neighbors are transported to the adjoining cells in the process of metabolic cooperation. Loss of normal intercellular communication is associated with unregulated growth. Both inward and outward signals are sensitive to a broad spectrum of weak EM fields.

Cooperative Modification of Calcium Binding by EM Fields at Cell Surfaces, with Amplification of Initial Signals

Manipulation of initial events in moecular binding at cell membrane receptor sites by imposed EM fields causes a far greater increase or decrease in Ca^{2+} binding than is accounted for in the events of receptor ligand binding (Kaczmarek and Adey, 1973, 1974; Bawin and Adey, 1976; Bawin et al., 1975; Lin-Liu and Adey, 1982); and the nonequilibrium character of this altered binding is attested by its occurrence in narrow frequency and amplitude windows.

Most of these studies of EM field effects on tissue Ca^{2+} have used cerebral tissue, including cerebral cortex in awake cats (Adey et al., 1982), isolated chick cerebral hemisphere (Bawin and Adey, 1976; Bawin et al., 1975, 1978a and b; Blackman et al., 1979, 1982, 1985a and b), cultured neurons (Dutta et al., 1984), and cerebral synaptosome fractions (Lin-Liu and Adey, 1982). "Tuning curves" (frequency windows) of altered Ca^{2+} efflux from cerebral tissue as a function of low frequencies in imposed EM fields were first seen in pioneering studies by Bawin et al. (1975, 1976), either as a response to simple low-frequency fields, or with low-frequency amplitude-modulation of RF fields. Maximal sensitivities were noted at frequencies around 16 Hz and were less at higher and lower frequencies.

We may note that neither physical size of the test biota nor their geometry are primary determinants of these interactions. This sensitivity of Ca^{2+} binding has been noted in biosystems having an enormous range of physical dimensions, ranging from awake cerebral cortex to cultured neurons, and finally in isolated terminals of cerebral nerve fibers (synaptosomes), with mean diameters around 0.7 um. They have been reported with RF fields amplitude-modulated at low frequencies; with low-frequency electric fields; with low-frequency EM fields; and with combined low-frequency EM fields and static magnetic fields.

These studies have also revealed <u>intensity windows</u> in modification of Ca^{2+} binding by EM fields. Blackman et al. (1979) noted an intensity window in cerebral tissue, confirmed by Bawin et al. (1978a), for RF fields in the range 0.1-1.0 mW/cm^2 (1.0 mW/cm^2 = 61 V/m in air) with sinusoidal amplitude-modulation at 16 Hz. At cellular dimensions, these RF fields typically produce EEG-level gradients (10-100 mV/cm). Ca-dependent processes also exhibit intensity windows. The activity of the enzyme ornithine decarboxylase (ODC), essential for growth in all eukaryotic cells, is modified in cultured liver cells by 60 Hz electric fields in the range 0.10-10.0 mV/cm (Byus et al., 1987).

Concepts of Cooperativity and Stimulus Amplification in Biomolecular Systems

Many functional linkages between participating elements in the dynamic patterns found in biomolecular systems are characterized by <u>cooperativity</u>, defined in this context as ways in which components of a macromolecule, or a system of macromolecules, act together to switch from one stable state to another (see Adey, 1988b for review). These joint actions frequently involve phase transitions, hysteresis, and avalanche effects in input-output relationships (Schmitt et al., 1975; Wyman, 1948; Wyman and Allen, 1951).

Dissipative Processes

There is much evidence that molecular organization in biological systems needed to sense weak stimuli, whether thermal, chemical or electrical, may reside in joint functions of molecular assemblies or their subsets (Katchalsky et al., 1974), with dynamic patterns developing in populations of elements as a result of their complex flow patterns. These flow patterns can undergo sudden transitions to new self-sustaining arrangements that are relatively stable over time. Because these dynamic patterns are initiated by continuing energy inputs, they are classed as <u>dissipative processes</u>. By reason of the continuing energy input, they occur far from equilibrium with respect to at least one important parameter in the system (Katchalsky and Curran, 1965). Moreover, two or more quite distinct mechanisms can give rise to the same dynamic pattern (Othmer and Scriven, 1971).

At cell surfaces, energy levels of fixed charges on polyanionic terminals of glycoprotein strands are determined by energy supplied from intracellular metabolic processes, as seen in receptor movements accompanying the cooperative phenomena of patching and capping (Yahara and Edelman, 1973). Studies of acridine dye binding to biopolymers have shown very long relaxation times in the millisecond range (Schwarz, 1975; Schwarz and Balthasar, 1970; Schwarz and Seelig, 1968; Schwarz, Klose and and Balthasar, 1970). These long relaxation times strongly suggest cooperative interactions that depend on <u>coherent states</u> between neighboring fixed charges on polymer sheets. All charge sites are then at the same higher energy level above the ground state, thus creating a <u>coherent domain</u> over a significant area at the membrane surface. These domains may persist for a finite time until returning to ground state, releasing energy cooperatively in response to a weak trigger.

Stimulus Amplification in Cooperative Systems

Initial triggers to cooperative processes may be weak and the amplified responses many orders of magnitude larger, as in the sharply nonlinear release of $^{45}Ca^{2+}$ from binding sites in cerebral tissue by added Ca ions (Kaczmarek and Adey, 1973) and by weak EM fields (Bawin and Adey, 1976; Bawin et al., 1975); and in a series of Ca-dependent processes at cell membranes that include the large generation of cAMP by glucagon binding to membrane receptors (Rodbell et al., 1974); the generation of cAMP by binding of parathyroid hormone to its membrane receptors and the modulation of this process by weak EM fields (Luben et al., 1982; Cain et al., 1987); the amplification of immune responses in patching and capping at cell membranes (Yahara and Edelman, 1973) and in the modulation of cell-mediated cytotoxicity of lymphocytes by weak EM fields (Lyle et al., 1983, 1988); and in swimming behavior of bacteria elicited by small concentrations of an attractant (Koshland, 1975).

There are enormous amplifications between initial triggers in these systems and the ensuing responses, raising questions about thresholds and the minimum size of an effective triggering stimulus. Studies of cooperativity in biological systems (Schwarz, 1975) and in biomembranes (Blank, 1976) have usually focused on effects of a change in an external parameter on the equilibrium constant of a specific reaction. Although a sharp transition from one highly stable state to another such state can also be achieved by noncooperative means, much larger transition energies would be required and the transition would occur more slowly. Sharp and fast transitions characteristic of many biological systems thus involve cooperative interactions, such as the individually weak forces in a series of hydrogen bonds or in hydrophobic reactions (Engel and Schwarz, 1970; Schwarz, 1975; Schwarz and Balthasar, 1970; Schwarz et al., 1970).

However, when compared with tissue electric gradients induced by imposed EM fields cited above, the requisite electric gradients found in some experimental molecular transitions are very large. For example, long-lasting conformation changes occur in poly(A).poly(U) and in ribosomal RNA with pulsed electric fields of 20 kV/cm and with a decay time of 10 usec (Neumann and Katchalsky, 1972). The helix-coil conformational change in poly(gamma-benzyl L-glutamate) can be induced by a gradient of 260 kV/cm (Schwarz and Seelig, 1968).

These sensitivities for nucleic acid chains in pure solution contrast sharply with effects of low-frequency trasins of EM pulses on DNA synthesis in cultured cells, where significant effects occur at field intensities in the range of 10^{-8}-10^{-4} T (Takahashi, et al., 1986). Observed field interactions with cells and tissues based on oscillating low-frequency electric gradients between 10^{-7} and 10^{-1} V/cm noted above would involve degrees of cooperativity many orders of magnitude greater than in the example just cited (Adey, 1988d). These differences may relate in part to far greater sensitivities of cellular systems to low frequency oscillating EM fields than to imposed step functions or DC gradients (Blank, 1972) that have been used in many experiments and models to test levels of cooperativity in biological systems.

The thermal Boltzmann (kT) noise in the system is a most important factor in determining this threshold for a low-level coherent oscillation to elicit a cooperative response. Boltzmann noise is 0.02 eV at room temperature and is the basis of molecular collisional interactions. If modeled on this thermal threshold, the sensing of a gradient of 10^{-7} V/cm would require a cooperative molecular system extending over 300 m. The abundant evidence that extracellular gradients from 10^{-1} V/cm down to 10^{-7} V/cm are biologically significant in systems of cellular dimensions is a salutary reminder of the importance of better understanding molecular and morphological substrates of this transductive coupling. Relevant factors may include temporal entrainment of activity in large systems of random generators by coherent oscillations far weaker than this random activity (Nicolis et al., 1973, 1974).

Based on a model of dielectric dispersion at the charged surface of micron-sized particles (Einolf and Carstensen, 1971), we have hypothesized that cell surfaces may act as extremely narrow-bandwidth low-pass filters in the transfer of thermal noise along the surface of micron-sized spheres and tubes (Adey and Bawin, 1976), thus enhancing the signal-to-noise ratio of ELF oscillations in the pass band of the filtering system. Transfer functions of 10^{-8} V/cm along cell surfaces at frequencies in the range 10-50 Hz calculated from this model agree well with observed biological ELF thresholds in the range 10^{-7}-10^{-8} V/cm.

CELL MEMBRANE RECEPTOR PROTEINS AS SUBSTRATES FOR INWARD SIGNALING AND ENERGY TRANSFER

Much attention now focuses on the long strands of membrane receptor proteins as models of coupling proteins in studies of the nature of transmembrane signals. Although still incomplete, their classification has identified two major structural types. They share important common features. All have an external amino terminus, negatively charged and located on a terminal glycoprotein strand. Their inner ends are characterized by a carboxyl group. All have a short hydrophobic segment placed within the lipid bilayer (plasma membrane), contrasting sharply with strings of hydrophilic amino acids that form the much longer external and internal segments of the strand. Their principal differences are in the number of consecutive crossings by which the strand threads itself in and out of the plasma membrane.

In their simplest forms, membrane receptor proteins appear to cross the plasma membrane only once. Those in this category include the human epidermal growth factor (EGF), the nerve growth factor (NGF), and the insulin receptor. In a second group, the strand crosses the membrane seven times. Receptors in this group include M1- and M2-muscarinic receptors, alpha- and beta-adrenergic receptors, rhodopsin, the substance K receptor, and probably the parathyroid hormone (PTH) receptor. As with single-crossing receptors, each amino acid domain lying within the plasma membrane appears to form a hydrophobic region. Two members in this latter category exhibit sensitivity to EM fields (rhodopsin and the PTH receptor)>

Structure of Receptor Proteins Crossing the Plasma Membrane Once

The sequence of the entire 1210 amino acid chain of the EGF receptor protein has been deduced by Ullrich et al. (1985), with striking findings on the sequences that make up the extracellular, intramembranous and cytoplasmic portions of the chain. Extracellular and intracellular segments are each composed of about 600 hydrophilic amino acids. The most striking feature of the entire sequence is the extremely short length of thepresumed intramembranous segment of 23 amino acids, predominantly hydrophobic, and only a single amino acid with a sidechain capable of hydrogen bonding.

Subsequent studies of the NGF receptor protein have also shown a strikingly similar segment of 23 amino acids within the membrane (Radeke et al., 1987), suggesting that this configuration plays a fundamental role in processes of transmembrane signaling. This view is strengthened by studies with a chimaeric protein constructed of the extracellular portion of the insulin receptor protein joined to the transmembrane and intracellular domains of the EGF receptor protein (Riedel et al., 1986). In this

272

molecule, the EGF receptor kinase domain of the chimaeric protein is activated by insulin binding. The authors conclude that insulin receptors and EGF receptors employ closely related or identical mechanisms for signal transduction across the plasma membrane.

Functional Organization of Receptor Proteins Making Seven Crossings of the Plasma Membrane

In a paper entitled "The Return of the Magnificent Seven", Hanley and Jackson (1987) discuss the work of Masu et al. (1987) in cloning the substance K receptor and revealing a possoble arrangement with four extracellular and four intracellular domains connected by seven membrane-spanning hydrophobic segments. This study did not reveal an expected complementarity at the nucleotide level between the peptide and its cognate receptor, and thus the site of ligand binding to the receptor remained unclear.

Later studies by Lefkowitz and his colleagues (Kobilka et al., 1988) have examined chimaeric alpha$_2$- and beta$_2$-adrenergic receptors to determine domains involved in effector coupling and ligand binding specificity. They identified seven hydrophobic domains consistent with transmembrane spanning segments (Fig. 2). Their effector coupling is to guanine nucleotide regulatory proteins (G proteins). <u>The seventh hydrophobic domain (counted away from the external amino terminus) appears to be a major determinant of both agonist and antagonist ligand binding specificity</u>. However, the authors caution that it cannot be concluded that this seventh hydrophobic domain forms the ligand binding pocket, since it may confer ligand binding specificity by interacting with the domains directly involved in formation of the binding site.

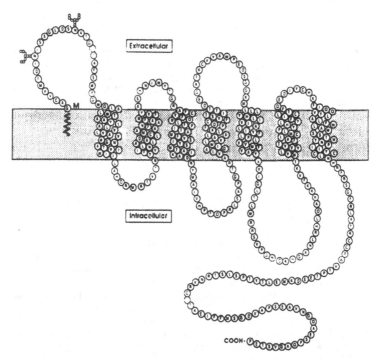

Fig. 2. A model of substance-K receptor in the plasma membrane. Segments spanning the lipid bilayer (plasma membrane) are a sequence of seven hydrophobic domains. (From Hanley and Jackson, 1987, with permission).

Configuration of Hydrophobic Domains

These hydrophobic domains may form membrane-spanning alpha-helices, as suggested by electron diffraction studies (Henderson and Unwin, 1975). The less hydrophobic amino acids of these alpha-helices are likely to project toward the interior of the molecule, while the more hydrophobic may form a boundary with the plasma membrane. The alpha-helices lying adjacent to one another are presumed to have evolved in ways that minimize steric and electrostatic repulsive forces between them.

Implications from Experimental Data on the Nature of Transmembrane Signals

Ullrich et al. (1985) concluded that a hydrophobic segment of the receptor protein as short as 23 amino acids is probably too short to be involved in conformation changes; and that its hydrophobic character makes unlikely its participation in either ionic or proton movement by Coulombic forces. As an alternative, they have suggested that an EGF-induced conformation change in the extracellular segment of the receptor protein strand may be transmitted to the cytoplasmic domain by movement of this short intramembranous segment in and out of the lipid bilayer, or by receptor aggregation.

The hydrophobic character of this intramembranous receptor protein segment and the generally hydrophobic nature of the surrounding plasma membrane both suggest that charged molecules and ions would be stripped of their hydration shells in this domain. In the context of ion transport through channels formed by certain intramembranous proteins, is there evidence that transmembrane movement of ions is mediated by these receptor protein segments, despite their hydrophobicity and their location in a highly hydrophobic environment?

Addition of epidermal growth factor (EGF) to human epidermal cell cultures causes a 2 to 4-fold increase in cytoplasmic free Ca^{2+} within 30-60 sec (Moolenaar et al., 1986). This EGF-induced signal appears to result from Ca^{2+} entry via a voltage-independent protein channel, since it is not accompanied by changes in membrane potential. It is completely dependent on extracellular Ca^{2+} moving into the cell interior. This action is inhibited by cancer-promoting phorbol esters which have a specific membrane receptor (Ca^{2+}-dependent protein kinase C), an enzyme involved in a cascade of growth regulating mechanisms, including synthesis by ornithine decarboxylase (ODC) of polyamines required in DNA synthesis. As discussed below, this sequence leading to stimulation of ODC by phorbol esters is sensitive to both ELF and ELF-modulated RF fields (Byus et al., 1987, 1988).

Transmembrane Signaling in Hydrophobic Cell Membrane Domains: Cooperative Phospholipid-Protein Interactions

Using electron spin resonance labeling of the tails of phospholipid molecules in an artificial phospholipid bilayer, McConnell (1975) noted that intrusion of a protein strand into this bilayer induces coherent states between charges on the tails of adjoining phospholipid molecules for considerable distances away from the protein strand. At the same time, the motions of the phospholipid tails are constrained and they behave more rigidly. These interactions suggest establishment of energetic domains determined by joint states of intramembranous proteins and surrounding phospholipid molecules. Moreover, states of dielectric strain associated with these lipoprotein interactions are likely to determine optical properties within their cooperative domains. As in fiber optic systems, dependence of these optical properties on membrane excitation states may determine stability of dark soliton propagation as a means of transmembrane signaling, as discussed below.

What role may this hydrophobic intramembranous portion of a receptor protein play in transduction of EM fields? The evidence favors transduction of visible light in these regions of rhodopsin molecules. For ELF magnetic fields, Luben et al. (1982, 1984) concluded that the receptor protein for the parathyroid (PTH) hormone is a probable site of field transduction, based on studies of PTH liganding to its receptor and on differentisal actions of these ELF fields on collagen synthesis in bone cells by PTH or by 1,25-dihydroxy-vitamin D_3. In continuing studies, Luben and Duong (1989) have incompletely sequenced the PTH receptor in approximately half the molecule, with evidence from a sequence of four consecutive hydrophobic segments that its molecular configuration is consistent with other receptors having seven hydrophobic domains.

Thus, experimental evidence that we have discussed so far points to two probable sites of EM field interactions with the mechanisms of transductive coupling at cell membranes. The first involves highly cooperative interactions by weak fields with cationic binding to the polyanionic fixed charges on cell surface glycoproteins, particularly for Ca^{2+}. A second site of interaction is apparently in proteins in hydrophobic regions of the plasma membrane. They couple signals from surface receptor sites to intracellular structures, including enzymes located internally at juxta-membrane sites, such as adenylate cyclase.

A Possible Role for Soliton Waves in Ionic Movements in Hydrophobic Domains at Cell Membranes: Chaotic Models of Nonlinear Oscillating Systems

In ionic movements through these hydrophobic domains, physical models offer several options.

We have hypothesized that this transmembrane signaling may involve nonlinear vibration modes in helical proteins and generation of Davydov-Scott soliton waves (Lawrence and Adey, 1982), moving in sequence down the length of glycoprotein and lipoprotein molecules (Adey and Lawrence, 1984). These solitons may arise in interactions of phonons and excitons along linear molecules, resulting in nonlinear molecular vibrations. It is proposed that nonlinear interatomic forces (specifically in the hydrogen bond) can lead to robust solitary waves with greatly increased radiative lifetimes (Davydov, 1979; Hyman et al., 1981). Davydov concludes that this would correspondingly increase the tendency for molecular vibrations to be the vehicle for energy transfer over long molecular chains, specifically over the amide "spines" in alpha-helix proteins and DNA, where there is the bond sequence - - -HNC=O- - - HNC=O- - -, etc. Davydov's nonlinear analysis shows that propagation of amide-I vibrations can couple to nonlinear sound waves in the alpha-helix, and the coupled vibration propagates as a localized and dynamically stable wave.

It is important to set a perspective on the sequence of events that might lead to soliton waves in biomolecular systems as a manifestation of self-sustained oscillations. These oscillations may be modeled on requirements for interaction of regular external perturbations with internal oscillations, resulting in synchronization of the system to the external drives (Kaiser, 1984). The observed phenomena depend on increasing levels of external driving energy. With increasing driving energy, the first shift from free internal oscillations to a self-organized state involves limit cycles, with periodic patterns developed and maintained by nonlinear processes and the influx and efflux of energy. At higher input levels, entrainment occurs, associated with sharp resonances, windows and threshold effects. These interactions are all athermal. A further increase in energy of external driving fields, both static and periodic, leads to sequences of period-doubling bifurcations, alternating with quasiperiodic and irregular epochs (quasiperiodicity, chaos). As a consequence, a regularly driven self-oscillating system may exhibit intrinsic chaotic behavior, even when the underlying dynamic is strongly deterministic. Finally, still higher levels of energy input destabilize the system (collapse), leading to the onset of propagating pulses (solitary waves or solitons). A nonlinear temporal structure is thus replaced by a nonlinear spatiotemporal structure.

However, the search for soliton waves of the Davydov-Scott type in DNA and helical proteins has been inconclusive and the concept has been criticized on theoretical grounds (Lawrence et al., 1987). They have not been found in extensive experimental studies using appropriately sensitive physical techniques (Layne et al., 1985). Computer modeling by Lawrence et al. show soliton scattering immediately upon formation in amide spines without propagation. It appears that the 23 amino acid hydrophobic segment is too short for their propagation. On the other hand, there is strong evidence for nonlinear, nonequilibrium processes at critical steps in transmembrane coupling, based on windows of sensitivity in EM field frequency and amplitude, and in duration of exposure (Adey, 1988b).

Dark Solitons and Analogies with Electrooptical Properties of Cell Membranes

Therefore, we may speculate that, if solitonic phenomena are involved in transmembrane signaling, they may be of a quite different type from the Davydov-Scott soliton. A model relevant to organization of cell membranes is suggested by studies of dark solitons in single-mode optic fibers (Christiansen, 1989). Due to nonlinear interactions between light and the transmitting medium in these fibers (the refractive index varies with light intensity), the normal tendency of light pulses to disperse can be negated. As a result, optical solitons can propagate without distortion. Dark pulses, short breaks in an otherwise continuous laser beam, can be made to propagate stably as far as 4000 km. In water waves, wave breaking occurs because the group velocity of the wave is greater at the peak than at the edges; the wave overtakes itself. The nonlinear Schrodinger equation also predicts wave-breaking effects in solitons, and determining a wide range of frequencies within the pulse (chirping).

These dark solitons suggest analogies with the sharp changes in optical properties of living vertebrate and invertebrate axons accompanying polarizing currents (Tobias et al., 1950), and the highly cooperative movements of about 18 $^{\circ}$A reported with laser interferometry at the axon surface within 1 msec of excitation (Hill et al., 1977). These electrooptical properties of cell membranes suggest an important avenue for future research in exploring the cooperative domains that exist between protein strands and their mantle of phospholipids within the plasma membrane, as discussed above.

Cyclotron Resonance Models of Ionic Movements in Hydrophobic Domains at Cell Membranes

Polk (1984) first suggested that free (unhydrated) Ca^{2+} ions in the earth's geomagnetic field would exhibit cyclotron resonances around 10 Hz; and that these cyclotron currents would be as much as five orders of magnitude greater than the Faraday currents if the Ca^{2+} ions exhibited nearest neighbor coherence. Interactions between the earth's geomagnetic field and a weak low-frequency EM field (40 V/m peak-to-peak in air, estimated tissue components 10^{-7} V/cm) modify Ca^{2+} efflux from chick cerebral tissue (Blackman, 1985b). For example, halving the local geomagnetic field with a Helmholtz coil rendered ineffective a previously interactive 15 Hz field; and doubling the geomagnetic field caused a 30 Hz signal that had been ineffective in an unmodified geomagnetic field to elicit a response.

In the cyclotron resonance model, most of the singly and doubly charged ions of biological interest have gyrofrequencies in a vacuum in the range 10-100 Hz for a mean value of the earth's geomagnetic field of 0.5 gauss (Liboff, 1985). Liboff hypothesizes that imposed EM fields at frequencies close to a given resonance may couple to the corresponding ionic species in such a way as to selectively transfer energy to these ions. He proposes that data from Blackman's experiments cited above may relate to cyclotron resonance in singly ionized K^+, with secondary effects on Ca^{2+} efflux.

OUTWARD SIGNALS THROUGH CELL MEMBRANES: ROLE OF GAP-JUNCTIONS

Specialized regions of contact between membranes of adjacent cells (Robertson, 1963) couple cells electrically (Furshpan and Furakawa, 1962) and chemically. High resolution electron microscopy with lanthanum staining rveals a 2-3 nm cleft containing protein to which Revel and Karnovsky (1967) assigned the term gap-junction. Gap-junctions are perforated by numerous tiny tubes (connexons) 1.5 nm in diameter that span the entire membrane; so that when connexons of adjacent cells come into register, the interior of adjacent cells are effectively in continuity.

Connexons thus provide a physical substrate for ionic coupling and transfer of essential metabolic substances between cells in the process of metabolic cooperation (see Fletcher et al., 1987b for review). Studies by Pitts and coworkers (Pitts and Finbow, 1986) demonstrated that intimately contacting cells could exchange products of metabolism. For example, mutant cells having a defective purine salvage pathway which prevented them from utilizing a specific nucleotide precursor could, nevertheless, incorporate a metabolite of that molecule into their nucleic acid as long as they contacted cells with a functional metabolic pathway.

However, it remains for future research to establish the exact nature of this metabolic cooperation, since the identity of material in recipient cell nuclei remains unknown, and no precise change has so far been identified in metabolic events as a result of metabolite transfer; but the fact that dansylated amino acids can pass from one cell of a contacting pair to its partner (Johnson and Sheridan, 1971) supports its physiological significance. Cell density-dependent suppression of transformation (Herschman and Brankow, 1986, 1987) further supports this concept.

Oncogenes may also interrupt intercellular pathways via gap-junctions, possibly by expression of peptides that act at cell membranes as spurious growth factors (Castagna, 1987). In experiments using normal rat kidney, (NRK) cells that bear a proviral insert from a temperature-sensitive mutant of Rous sarcoma virus (LA25), Atkinson et al. (1981, 1986) demonstrated an unequivocal relationship between gap-junctions and intercellular communication. They used the temperature sensitivity of the viral gene to show that intercellular communication was fully effective at 39C, but that both communication channels and identifiable gap-junctions were lost at 33C, the virus permissive temperature.

Disruption of Intercellular Communication through Gap-Junctions by Phorbol Esters and EM Fields

Fletcher et al. (1987a) noted that blockage of the entry of natural cytolytic substances, alpha-lymphotoxin (LT) and recombinant tumor necrosis factor (TNF), into Chinese hamster ovary cells depends on their ability to form gap-junctions, a function that varies between different strains of these cells. Fletcher found that the phorbol ester cancer promoter TPA opens gap-junctions to permit entry of LT, leading to cell death (lysis) in a dose-dependent fashion (Fig. 3).

Weak radiofrequency fields (450 MHz, 1.0-1.5 mW/cm^2 incident energy) with 16 Hz sinusoidal amplitude-modulation enhanced this ability of phorbol ester tumor promoters to impair gap-junction communication. Moreover, the enhanced response required the presence of 16 Hz modulation and did not occur with an unmodulated carrier wave having the same incident energy (Fig 4).

INTRACELLULAR ENZYMES AS MOLECULAR MARKERS OF EM FIELD INTERACTIONS WITH HORMONES AND CANCER PROMOTERS AT CELL MEMBRANES

Metabolic, messenger and growth enzymes are all initiated or stimulated by molecules with specific receptor sites on cell membranes. Activation of these enzymes and the reactions in which they participate involve energies millions of times greater than in the cell surface triggering events initiated by the EM fields, emphasizing the membrane amplification inherent in this transmembrane signaling sequence.

We have found three groups of intracellular enzymes that respond to signals initiated at cell membranes as a response to athermal EM field exposure. These responses occur with or without concurrent cell membrane

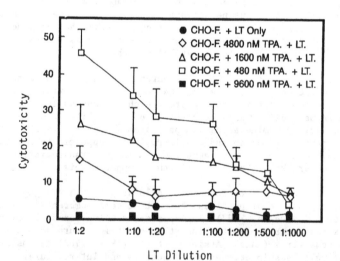

Fig. 3 Effects of cancer-promoting phorbol ester (TPA) on alpha-lymphotoxin (LT)-mediated cytolysis of Chinese hamster ovary cells treated with increasing doses of TPA in the presence of graduated concentrations of LT. (From Fletcher et al., 1987a).

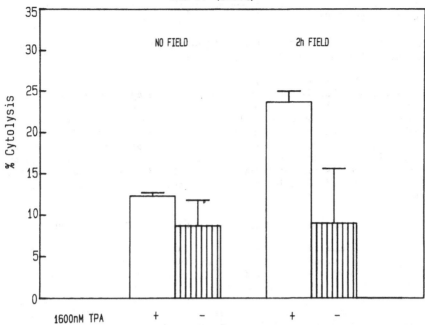

Fig. 4 Separate and combined actions of phorbol ester tumor
promoter TPA and 450 MHz RF field that impair gap-junction
communication between hamster ovary cells, allowing entry
of alpha-lymphotoxin (LT). LT action was enhanced by
fields with 16 Hz sinusoidal amplitude-modulation
(2h-FIELD), but there was no effect of the unmodulated
carrier wave (NO FIELD) at the same incident energy
(1.0 mW/cm^2). Tests with (+) and without (-) TPA (2400nM)
were compared with effects of modulated and unmodulated
fields. (From Fletcher et al., 1986).

stimulation initiated chemically by physiological molecules (hormones,
antibodies, etc.), and by cancer promoting substances. These enzymes are:

1) membrane-bound adenylate cyclase involved in activation of
   protein kinase enzymes through conversion of adenosine
   triphosphate (ATP) to cyclic-adenosine monophosphate (cAMP),
   as seen in bone cells and chondrocytes exposed to low-
   frequency pulsed magnetic fields (Hiraki et al., 1987; Luben
   et al., 1982; Luben and Cain, 1984; Cain et al., 1987).

2) cAMP-independent protein kinases that perform messenger
   functions (Byus et al., 1984).

3) ornithine decarboxylase (ODC), essential for growth in all
   cells by its participation in synthesis of polyamines
   essential for DNA formation (Byus et al., 1987, 1988). All
   are Ca-dependent and their actions have been reviewed in
   detail elsewhere (Adey, 1986).

We shall focus here on EM field interactions at cell membranes on activity of protein kinase enzymes and on the growth enzyme ornithine decarboxylase (ODC).

Lymphocyte Protein Kinase Responses to RF Fields Amplitude-Modulated at Low Frequencies: Windows in Frequency and Time

Some intracellular protein kinases are activated by signals arising in cell membranes that do not involve the adenylate cyclase/cAMP pathway discussed above. This group includes membrane protein kinases related to actions of cancer promoting phorbol esters (see below). In human tonsil lymphocytes exposed to a 450 Mhz field (1.0 mW/cm^2), cAMP-independent protein kinases showed windowed activity with respect to exposure duration and modulation frequency (Byus et al., 1984). Reduced enzyme activity only occurred at modulation fequencies between 16 and 60 Hz, and only for the first 15-30 min of RF field exposure. Unmodulated fields elicited no responses.

Protein Kinase C and Membrane Effects of Cancer Promoting Phorbol Esters

Amongst these membrane-related cAMP-independent protein kinases is Ca-dependent phosphatidylserine protein kinase (protein kinase C, PKC). PKC is the major membrane receptor for the cancer promoting phorbol ester TPA (Castagna et al., 1982; Nishizuka et al., 1983, 1984). It is normally activated by diacylglycerol formed transiently from breakdown of inositol phospholipids in response to extracellular signals. TPA activates PKC irreversibly. From these findings, a series of bridges have been built, uniting research on tumor formation, growth factors, signal transduction, and the action of specific oncogenes.

PKC is transferred into cell membranes from the cytosol as an essential step in its activation by TPA. This translocation involves synergism between Ca^{2+} and the phorbol ester (Wolf et al., 1985). TPA has a diacylglycerol-like structure and is able to substitute for diacylglycerol at extremely low concentrations. Like diacylglycerol, TPA dramatically increases affinity of the enzyme for Ca^{2+} to the 10^{-7} M range, resulting in its full activation without detectable mobilization of Ca^{2+} (Nishizuka, 1984).

Ornithine Decarboxylase (ODC) as a Marker of EM Field Interactions with Cancer Promoters at Cell Membranes

The enzyme ornithine decarboxylase (ODC) is required for growth in all eukaryotic cells. PKC plays a key role in its activation, since ODC activation by the phorbol ester TPA is inhibited staurosporine, a potent inhibitor of PKC (Yamada et al., 1988). PKC may also be involved in ODC gene expression. TPA-increased synthesis of steady-state levels of ODC RNA may be mediated by PKC and is regulated at the transcriptional level (Hsieh and Verma, 1988).

ODC removes carboxyl (-COOH) groups (decarboxylation) from the amino acid ornithine, initiating synthesis of long chain polyamine molecules of putrescine, spermatidine and spermine. These polyamines are utilized in DNA and protein synthesis. All agents that stimulate ODC are not cancer-promoting, but all cancer promoters stimulate ODC. Its activation accompanies increased cell numbers characteristic of both normal tissue repair and tumor formation. Thus, increased ODC activity is not indicative in itself that cells have either become transformed or are in the process of tumor formation.

ODC activity is induced by chemical cancer promoters with cell membrane binding sites, including phorbol esters, and by ELF electromagnetic fields and by RF fields with ELF amplitude-modulation that are also sensed at cell membranes. 60 Hz electric fields (10 mV/cm, 1 h exposure) increased ODC activity 5-fold in human lymphoma CEM cells, and 2- to 3-fold in mouse myeloma cells (P3) relative to unexposed cells (Byus et al., 1987). Byus et al. (1988) noted a similar sensitivity in Reuber H35 hepatoma cells in an athermal 1 h exposure (temperature rise < 0.1 °C) to a 450 MHz field (1.0 mW/cm^2 peak-envelope incident energy) with sinusoidal 16 Hz amplitude modulation. There were no responses to 60 and 100 Hz modulation.

Increased ODC activity in Reuber H35 hepatoma cells stimulated with TPA was further increased in this study by prior exposure to the 16 Hz-modulated 450 MHz field. Similar 1 h exposures of Chinese hamster ovary (CHO) cells and 294T melanoma cells also increased their ODC activity; and when these CHO cells were first exposed for 1 h, they then responded to TPA phorbol ester with a further increase in ODC activity (Fig. 5).

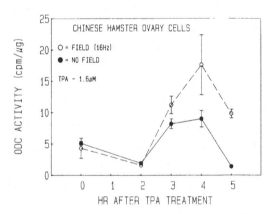

Fig. 5 Ornithine decarboxylase (ODC) activity in Chinese hamster ovary cells stimulated by the tumor promoter TPA, with and without exposure to a 450 MHz field (1.0 mW/cm^2) amplitude-modulated at 16 Hz. (From Byus et al., 1988).

EPIGENETIC CARCINOGENESIS; CELL MEMBRANE DYSFUNCTIONS AND DISRUPTED INTERCELLULAR COMMUNICATION IN TUMOR FORMATION

Tumor formation as a manifestation of unregulated growth involves a sequence of defined stages: initiation, promotion and progression. Development of a fully malignant tumor involves complex interactions between environmental (chemical, ionizing and nonionizing radiation, viruses) and endogenous (genetic, hormonal factors (Weinstein, 1988).

281

The overall process can occupy a major fraction of the lifetime of the individual (Foulds, 1954; Slaga et al., 1978; Weinstein et al., 1984).

Initiation appears to involve a change in genetic stores of DNA and may arise from a single event. That change is not expressed and a tumor does not result unless one or more promoting agents act repeatedly at a later time. Following initiation, the time between exposure to a promoter and appearance of a tumor may be brief. Initiated cells may remain quiescent if not stimulated by a promoter, and a tumor may never develop if sufficient exposures to promoters do not occur.

Promotion occurs in previously initiated cells by action of agents having very weak or no carcinogenetic activity when tested alone, but they markedly enhance tumor yield when applied repeatedly following a low or suboptimal dose of a carcinogen (Berenblum, 1982; Slaga et al., 1978; Nishizuka, 1983, 1984; Berridge, 1987). Skin tumor promoters include croton oil and its phorbol ester derivatives, extracts of unburned tobacco, tobacco smoke condensate, dinitrobenzene and benzpyrene. In contrast to initiators, phorbol esters do not bind to DNA but act by binding to cell membrane associated receptors. They thus act at the epigenetic level.

Nonionizing EM fields interact with TPA at cell membranes to modulate its actions on inward and outward signal streams. Available evidence indicates that nonionizing EM fields do not function as classical initiators in the etiology of cancer by causing DNA damage and gene mutation (Adey, 1986, 1987a and b, 1988a and b). On the other hand, there is evidence that they act, separately and jointly with chemical cancer promoters, to modulate transmembrane signaling and disrupt intercellular communication.

Trosko has hypothesized that two major types of intercellular communication help to maintain "normal orchestration" of proliferation and differentiation during development and between quiescent stem-progenitor and differentiated cells in the adult (Fletcher et al., 1987b; Trosko, 1987; Trosko and Chang, 1986; Yotti et al., 1979); one involving transfer of molecular signals from cells of one differentiation or tissue type to another across an extracellular space, and often over a significant distance (for example, in actions of hormones, growth factors and neurotransmitters); and the other mediated by transfer of relatively small molecular weight molecules and ions to neighboring cells via gap-junctions.

Gap-junctional communication in Trosko's model would involve a minimum of four steps: recognition by neighboring cells of one another through the action of adhesion molecules; functional gap-junctions; small regulatory or signaling molecules; and transducing protein receptors for these signaling molecules.

Disruption of intercellular communication through gap-junctions leads to serious disorders in growth control, including eefects on tissue repair and neoplastic transformation (Loewenstein, 1977, 1979, 1981). Controlled growth occurs in the presence of gap junctions, but in their absence, growthmay be unregulated. In vitro, this can be reversed if cancer cells make contact with normal cells (Herschmann and Brankow, 1986, 1987; Newmark, 1987). We have noted above that the phorbol ester tumor promoter TPA opens gap-junctions, permitting entry of natural cytolytic substances, alpha-lymphotoxin (LT) and tumor necrosis factor (TNF); and that this action of TPA is enhanced by weak RF fields with low-frequency amplitude-modulation (Fletcher et al., 1986, 1987a) (Fig. 3).

Oncogenes may also interrupt intercellular pathways via gap-junctions, possibly by expression of peptides that act at cell membranes as spurious growth factors (Castagna, 1987). In ewxperiments with normal rat kidney (NRK) cells that bear a proviral insert from a temperature-sensitive-mutant of Rous sarcoma virus (LA25), Atkinson et al., (1981, 1986) de,onstrated an unequivocal relationship between gap-junctions and intercellular communication. Using the temperature sensitivity of the viral gene, they showed that intercellular communication was fully effective at 39C, but that both communication channels and identifiable gap-junctions were lost at 33C, the virus permissive temperature.

Transfer of fluorescent dyes between neighboring cells allows assessment of the integrity of their gap-junction communication. Progressive changes in homologous and heterologous gap-junction communication were observed in cell lines derived from selected stages of SENCAR mouse skin carcinogenesis (Klann et al., 1989). There was a progressive loss of homologous communication as the neoplastic process increased. Tests of heterologous communication between these cells and their normal counterparts were made in cocultures and showed functional communication without selectivity; suggesting that progressive loss of homologous but not heterologous communication accompanies neoplastic development.

SUMMARY

This review of structural and functional aspects of communication between neighboring cells emphasizes the role of the cell membrane in detection and transductive coupling of weak oscillating electromagnetic (EM) fields in the pericellular environment. Cerebral extracellular conductance changes accompany physiological responses. Perineuronal fluid has a substantial macromolecular content and calcium ions may modulate perineuronal conductivity. Protruding from the plasma membrane, stranded glycoproteins with strongly polyanionic terminals attract cations, particularly calcium and hydrogen, forming a cell surface compartment.

In their interactions in this cell surface compartment, weak imposed EM fields have been found unique tools in identifying the sequence and energetics of key events that couple chemical and electric stimuli from the cell surface to the interior. These fields are typically in the spectrum below 100 Hz and at intensities in the range of 10^{-1}-10^{-7} V/cm in pericellular fluid. Cell membranes are primary sites of interactions with these fields. Regulation of cell surface chemical events by these fields involves a major amplification of initial weak triggers associated with binding of hormones, antibodies and neurotransmitters to their specific binding sites. Calcium ions play a key role in this stimulus amplification. The evidence supports nonlinear, nonequilibrium processes at critical steps in transmembrane coupling.

Inward signals from surface receptors to the cell interior are conveyed by receptor proteins. Short intramembranous segments of these protein strands are highly hydrophobic, with evidence for existence of cooperative domains between these strands and surrounding phospholipids of the plasma membrane. Signaling along these hydrophobic segments may involve nonlinear vibration modes and altered electrooptical properties in cooperative interactions with surrounding phospholipids.

Communication between cells through gap-junctions is also sensitive to low-frequency EM fields at athermal intensities. We hypothesize that cancer promotion with tumor formation may involve dysfunctions at cell membranes, disrupting inward and outward signal streams.

There is much evidence that the molecular organization in biological systems essential for sensing weak stimuli, whether thermal, chemical or electrical, resides in joint functions of molecular assemblies, or in subsets of these assemblies. On the other hand, it is at the atomic level within these molecular systems that physical, rather than chemical events now appear to shape the flow of signals and the transmission of energy.

ACKNOWLEDGMENTS

We gratefully acknowledge support for studies in our laboratory from the US Department of Energy, the US Environmental Protection Agency, the US Bureau of Devices and Radiological Health (FDA), the US Office of Naval Research, the US Veterans Administration, the Southern California Edison Company, the Health Foundation and the General Motors Medical Research Institute.

REFERENCES

Adey, W.R., 1966, Intrinsic organization of cerebral tissue in alerting, orienting and discriminative responses, page 615 in: "The Neurosciences. First Study Program," Quarton, G.C., Melnechuk, T., and Schmitt, F.O., eds., Rockefeller University, New York.

Adey, W.R., 1975, Effects of electromagnetic radiation on the nervous system, Ann. NY Acad. Sci. 247:15.

Adey, W.R., 1977, Models of membranes of cerebral cells as substrates for information storage, BioSystems 8:163.

Adey, W.R., 1981a, Tissue interactions with nonionizing electromagnetic fields, Physiol. Rev., 61:435.

Adey, W.R., 1981b, Ionic nonequilibrium phenomena in tissue interactions with nonionizing electromagnetic fields, page 271 in: "Biological Effects of Nonionizing Radiation", Illinger, K.H., ed., Am. Chem. Soc., Washington, D.C. Symposium Ser. No. 157.

Adey, W.R., 1983, Molecular aspects of cell membranes as substrates for interactions with electromagnetic fields, page 201 in: "Synergetics of the Brain," Basar, E., Flohr, H., Haken H., and Mandell, A.J., eds., Springer, Berlin.

Adey, W.R., 1984, Nonlinear, nonequilibrium aspects of electromagnetic field interactions at cell membranes, page 3 in: "Nonlinear Electrodynamics in Biological Systems," Adey, W.R. and Lawrence, A.F., eds., Plenum, New York.

Adey, W.R., 1986, The sequence and energetics of cell membrane transductive coupling to intracellular enzyme systems, Bioelectrochem. Bioenergetics. 15:447.

Adey, W.R. 1987a. Evidence for tissue interactions with microwave and other nonionizing electromagnetic fields in cancer promotion, page 147 in: "Biophysical Aspects of Cancer," Fiala, J. and Pokorny, J., eds. Charles University, Prague.

Adey, W.R. 1987b. Cell membranes, electromagnetic fields and intercellular communication, in Dynamics of Sensory and Cognitive Processing in the Brain, Second Symposium, Basar, E., ed., Springer Verlag, Heidelberg, in press.

Adey, W.R., 1988a, Cell membranes: the electromagnetic environment and cancer promotion, Neurochem. Res. 13:671.

Adey, W.R., 1988b, Physiological signalling across cell membranes and cooperative influences of extremely low frequency electromagnetic fields, page 148 in: "Biological Coherence and Response to External Stimuli," Frohlich, H., ed., Springer Verlag, Heidelberg.

Adey. W.R., 1988c, Biological effects of radio frequency radiation, page 109 in: "Interaction of Electromagnetic Waves with Biological Systems," Lin, J.C., ed., Plenum, New York.

Adey, W.R., 1988d, Effects of microwaves on cells and molecules, Nature London 333:401.

Adey, W.R., Bawin, S.M., and Lawrence, A.F. 1982. "Effects of weak amplitude-modulated microwave fields on calcium efflux from awake cat cerebral cortex, Bioelectromagnetics 3:295.

Adey, W.R. Kado, R.T., and Didio, J., 1962, Impedance measurements in brain tissue using microvolt signals. Exper. Neurol., 5:47.

Adey, W.R., Kado, R.T., McIlwain, J.T., and Walter, D.O., 1966, The role of neuronal elements in regional cerebral impedance changes in alerting, orienting and discriminative responses, Exper. Neurol. 15:490.

Adey, W.R., and Lawrence, A.F. (Eds.) 1984. Nonlinear Electrodynamics in Biological Systems. Plenum, New York.

Atkinson, M.M., Menko, A.S., Johnson, R.G., Sheppard, J.R. and Sheridan, J.D., 1981, Rapid and reversible reduction of junctional permeability in cells infected with a temperature-sensitive mutant of Rous sarcoma virus, J. Cell. Biol. 91:573.

Atkinson, M.M., Anderson, S.K., and Sheridan, J.D., 1986, Modification of gap junctions in transformed cells by a temperature-sensitive mutant of Rous sarcoma virus, J. Membrane Biol. 91:53.

Balcer-Kubiczek, E.K., and Harrison, G.H, 1985, Evidence for microwave carcinogenesis, in vitro, Carcinogenesis 6:859.

Balcer-Kubiczek, E.K., and Harrison, G.H., 1989, Induction of neoplastic transformation in C3H10T1/2 cells by 2.45 GHz microwaves and phorbol ester, Radiation Res. 117:531.

Bassett, C.A.L., 1987, Low energy pulsing electromagnetic fields modify biomedical processes, BioEssays 6:36.

Bawin, S.M., and Adey, W.R., 1976, Sensitivity of calcium binding in cerebral tissue to weak electric fields oscillating at low frequency, Proc. Nat. Acad. Sci. USA 73:1999.

Bawin, S.M., Adey, W.R., and Sabbot, I.M., 1978b, Ionic factors in release of $^{45}Ca^{2+}$ from chick cerebral tissue by electromagnetic fields, Proc. Natl. Acad. Sci. USA 75:6314.

Bawin, S.M., Kaczmarek, L.K., and Adey, W.R., 1975, Effects of modulated VHF fields on the central nervous system, Ann. NY Acad. Sci. 247:74.

Bawin, S.M., Sheppard, A.R., and Adey, W.R., 1978a, Possible mechanisms of weak electromagnetic field coupling in brain tissue, Bioelectrochem. Bioenergetics 45:67.

Berenblum, I., 1982, Sequential aspects of chemical carcinogenesis: skin, page 237 in: "Cancer: A Comprehensive Treatise," Vol. 1, Ed. 2, Becker, F.F. ed., Plenum Press, New York.

Berridge, M.J., 1987, Inositol triphosphate and diacylglycerol: two interacting second messengers, Annual Rev. Biochem. 56:159.

Blackman, C.F., Elder, J.A., Weil, C.M., Benane, S.G., Eichinger, D.C., and House, D.E., 1979, Induction of calcium ion efflux from brain tissue by radio frequency radiation, Radio Sci. 14:93.

Blackman, C.F., Benane, S.G., Kinney, L.S., Joines, W.T., and House, D.E., 1982, Effects of ELF fields on calcium-ion efflux from brain tissue, in vitro, Radiation Res. 92:510.

Blackman, C.F., Benane, S.G., House, D.E., and Joines, W.T., 1985a, Effects of ELF (1-120 HZ) and modulated (50 Hz) RF fields on the efflux of calcium ions from brain tissue, in vitro, Bioelectromagnetics 6:1.

Blackman, C.F., Benane, S.G., Rabinowitz, J.R., House, D.E., and Joines, W.T., 1985b, A role for the magnetic field in the radiation-induced efflux of calcium ions from brain tissue in vitro, Bioelectromagnetics 6:327.

Blank, M., 1972, Cooperative effects in membrane reactions, J. Colloid Interface Sci. 41:97.

Blank, M. 1976. Hemoglobin reactions as interfacial phenomena, J. Electrochem. Soc. 123:1653.

Byus, C.V., Lundak, R.L., Fletcher, R.M., and Adey, W.R., 1984,

Alterations in protein kinase activity following exposure of cultured lymphocytes to modulated microwave fields, Bioelectromagnetics 5:34.

Byus, C.V., Pieper, S.E., and Adey, W.R., 1987, The effects of low-energy 60-Hz environmental electromagnetic fields upon the growth-related enzyme ornithine decarboxylase, Carcinogenesis 8:1385.

Byus, C.V. Kartun, K., Pieper, S., and Adey, W.R., 1988. Increased ornithine decarboxylase activity in cultured cells exposed to low energy modulated microwave fields and phorbol ester tumor promoters, Cancer Res. 48:4222.

Cain, C.D. and Luben, R.A., 1987, Pulsed electromagnetic field effects on PTH-stimulated cAMP accumulation and bone resorption in mouse calvaria, page 269 in: "Interactions of Biological Systems with Static and ELF Electric and Magnetic Fields," 23rd Hanford Life Sciences Symposium. US Dept of Energy Symposium Series, No. 60. Anderson, L.E., Kelmar, B.J., and Weigel R.J., eds. US Dept of Energy, Washington, D.C.

Cain, C.D., Adey, W.R., and Luben, R.A., 1987, Evidence that pulsed electromagnetic fields inhibit coupling of adenylate cyclase by parathyroid hormone in bone cells, J. Bone Mineral Res. 2:437.

Castagna, M. 1987. Phorbol esters as signal transducers and tumor promoters, Biology of the Cell 59:3.

Castagna, M. Takai, Y., Kaibuchi, K., Sano, K., Kikkawa, U., and Nishizuka, Y., 1982, Direct activation of calcium-activated phospholipid-dependent protein kinase by tumor-promoting phorbol esters, J. Biol. Chem. 257:7847.

Christiansen, P.L., 1989, Shocking optical solitons, Nature, London 339:17.

Cole, K.S., 1940, Permeability and impermeability of cell membranes for ions, Cold Spring Harbor Symp. Quant. Biol. 4:110.

Davydov, A.S., 1979, Solitons in molecular systems, Physica Scripta 20:387.

Dixey, R., and Rein, G. 1982. ^{3}H-noradrenaline release potentiated in a clonal nerve cell line by low-intensity pulsed magnetic fields, Nature London 296:253.

Dutta, S.K., Subramoniam,A., Ghosh, B., and Parshad, R., 1984, microwave radiation-induced calcium efflux from brain tissue, in vitro, Bioelectromagnetics 5:71.

Einolf, C.W., and Carstensen, E.L., 1971, Low-frequency dielectric dispersion in suspensions of ion-exchange resins, J. Phys. Chem. 75:1091.

Elul, R., 1966, Applications of non-uniform electric fields. Part 1. Electrophoretic evaluation of adsorption. Trans. Faraday Soc. London 528:3484.

Elul, R., 1967, Fixed charge in the cell membrane, J. Physiol. London 189:351.

Engel, J., and Schwarz, G., 1970, Cooperative conformational transitions of linear biopolymers, Angew. Chem. Int. Ed. 9:389.

Fitzsimmons, R.J., Farley, J., Adey, W.R., and Baylink, D.J., 1986, Embryonic bone matrix formation is increased after exposure to a low-amplitude capacitively-coupled electric field, in vitro, Biochim. Biophys. Acta 882:51.

Fitzsimmons, R.J., Farley, J.R., Adey, W.R., and Baylink, D.J., 1989, Frequency dependence of increased cell proliferation, in vitro, in exposures to a low-amplitude, low-frequency electric field: evidence for dependence on increased mitogen activity released into the culture medium, J. Cell. Physiol. 139:586.

Fletcher, W.H., Shiu, W.W., Haviland, D.A., Ware, C.F., and Adey, W.R., 1986, A modulated-microwave field and tumor promoters similarly enhance the action of alpha-lymphotoxin (aLT), Bioelectromagnetic

Soc., 8th Annual Meeting, Madison WI, Proceedings, p. 12.

Fletcher, W.H., Shiu, W.W., Ishida, T.A., Haviland, D.L., and Ware, C.F., 1987a, Resistance to the cytolytic action of lymphotoxin and tumor necrosis factor coincides with the presence of gap junctions uniting target cells, J. Immunol. 139:956.

Fletcher, W.H., Byus, C.V., and Walsh, D.A., 1987b, Receptor-mediated action without receptor occupancy: a function for cell-cell communication in ovarian follicles, Adv. in Exper. Med. Biol. 219:299.

Foster, K.R., and Guy, A.W., 1986, The microwave problem, Sci. Am. 255(3):32.

Foster, K.R., and Pickard, W.F., 1987, Microwaves: the risks of risk research, Nature London 330:531.

Foulds, L., 1954. The experimental study of tumor progression, Cancer Res. 14:327.

Goodman, R., and Henderson, A.S., 1988, Exposure of salivary gland cells to low-frequency electromagnetic fields alters polypeptide synthesis, Proc. Natl. Acad. Sci. USA 85:3928.

Hanley, M.R., and Jackson, T, 1987, Return of the magnificent seven, Nature London 329:766.

Henderson, R., and Unwin, P.N.T., 1975, Three-dimensional model of purple membrane obtained by electron microscopy, Nature London 257:28.

Herschman, H.R., and Brankow, D.W., 1986, Ultraviolet irradiation transforms C3H10T1/2 cells to a unique suppressible phenotype, Science 234:1385.

Herschman, H.R. and Brankow, D.W., 1987, Cell size, cell density and nature of the tumor promoter are critical variables in expression of a transformed phenotype (focus formation) in co-cultures of UV-TDTx and C3H10T1/2 cells," Carcinogenesis 8:993.

Hill, B.C., Schubert, E.D., Nokes, M.A., and Michelson, R.P., 1977, Laser interferometer measurements of charges in crayfish axon diameter concurrent with axon potential, Science 196:426.

Hiraki, Y., Endo, N., Takigawa, M., Asada, A., Takahashi, H., and Suzuki, F., 1987, Enhanced responsiveness to parathyroid hormone and induction of functional differentiation of cultured rabbit costal chondrocytes by a pulsed electromagnetic field, Biochim. Biophys. Acta 931:94.

Hodgkin, A.L., and Huxley, A.F., 1952, A quantitative description of membrane current and its application to conduction and excitation in nerve, J. Physiol. London 117,500.

Housey, G.M., Johnson, M.D., Hsiao, W.-L., O'Brian, C.A., Murphy, J.P., Kirschmeier, P., and Weinstein, I.B., 1988, Overproduction of protein kinase C causes disordered growth control in rat fibroblasts, Cell 52:343.

Johnson, R.G., and Sheridan, J.D., 1971, Junctions between cancer cells in culture: ultrastructure and permeability, Science 174:717.

Kaczmarek, L.K., and Adey, W.R., 1973, The efflux of $^{45}Ca^{2+}$ and ^{3}H-gamma-aminobutyric acid from cat cerebral cortex, Brain Res. 63:331.

Kaczmarek, L.K., and Adey, W.R., 1974, Weak electric gradients change ionic and transmitter fluxes in cortex, Brain Res. 66:537.

Katchalsky, A., 1974, Concepts of dynamic patterns. Early history and philosophy, Neurosci. Res. Program Bull. 12:30.

Katchalsky, A., and Curran, P.F., 1965, "Nonequilibrium Thermodynamics in Biophysics," Harvard University Press, Cambridge, MA.

Klann, R.C., Fitzgerald, D.J., Piccoli, C., Slaga, T.J., and Yamasaki, H., 1989, Gap-junctional intercellular communication in epidermal cell lines from selected stages of SENCAR mouse skin carcinogenesis, Cancer Res. 49:699.

Kobilka, B.K., Kobilka, T.S., Daniel, K., Regan, J.W., Caron, M.G. and Lefkowitz, R.J., 1988, Chimeric alpha$_2$-, beta$_2$-adrenergic receptors: delineation of domains involved in effector coupling and ligand binding specificity, Science 240:1310.

Koshland, D.E. 1975. Transductive coupling in chemotactic processes: chemoreceptor-flagellar coupling in bacteria, page 273 in: "Functional Linkage in Biomolecular Systems," Schmitt, F.O., Schneider, D.M. and Crothers, D.M., eds., Raven Press, New York.

Lawrence, A.F., and Adey, W.R., 1982, Nonlinear wave mechanisms in interactions between excitable tissue and electromagnetic fields, Neurol. Res. 4:115.

Lawrence, A.F., McDaniel, J.C., Chang, D.B., and Birge, R.R., 1987, The nature of phonons and solitary waves in alpha-helical proteins, Biophys. J. 51:785.

Layne, S.P., Bigio, I.J., Scott, A.C., and Lomdahl, P.S., 1985, Transient fluorescence in synchronously dividing Eschericia coli, Proc. Nat. Acad. Sci. USA 82:7599.

Liboff, A.R., 1985, Cyclotron resonance in membrane transport, page 281 in: "Interactions Between Electromagnetic Fields and Cells," Chiabrera, A., Nicolini, C., and Schwan, H.P., eds., Plenum Press, New York.

Lin-Liu, S., and Adey, W.R., 1982, Low frequency amplitude-modulated microwave fields change calcium efflux rates from synaptosomes, Bioelectromagnetics 3:309.

Loewenstein, W.R., 1977, Permeability of membrane junctions, Ann. NY Acad. Sci. 137:441.

Loewenstein, W.R., 1979, Junctional intercellular communication and the control of growth, Biochim. Biophys. Acta 560:1.

Loewenstein, W.R., 1981, Junctional intercellular communication: the cell-to-cell membrane channel, Physiol. Rev. 61:829.

Luben, R.A., Cain, C.D., Chen, M.-Y., Rosen, D.M., and Adey, W.R., 1982, Effects of electromagnetic stimuli on bone and bone cells, in vitro: inhibition of responses to parathyroid hormone by low-energy, low-frequency fields, Proc. Natl. Acad. Sci. USA 79:4180.

Luben, R.A., and Cain, C.D., 1984, Use of bone cell hormone responses to investigate bioelectromagnetic effects on membranes, in vitro, page 23 in: "Nonlinear Electrodynamics in Biological Systems," Adey, W.R. and Lawrence, A.F., eds., Plenum Press, New York.

Luben, R.A. and Duong, H.P., 1989, A candidate sequence for the mouse 70 kDa mouse osteoblast PTH receptor (PTHR) is homologous to the rhodopsin family of G-protein linked receptors (GPLR), Amer. Soc. Cell Biol., Proc. Annual Meeting, San Francisco, Abstract No. 348.

Lyle, D.B., Schechter, P., Adey, W.R. and Lundak, R.L., 1983, Suppression of T Lymphocyte cytotoxicity following exposure to sinusoidally amplitude-modulated fields, Bioelectromagnetics 4:281.

Lyle, D.B., Ayotte, R.D., Sheppard, A.R., and Adey, W.R., 1988, Suppression of T lymphocyte cytotoxicity following exposure to 60 Hz sinusoidal electric fields, Bioelectromagnetics 9:229.

Masu, Y., Nakayama, K., Tamaki, H., Harada, Y., Kuno, M., and Nakanishi, S., 1987, cDNA cloning of bovine substance-K receptor through oocyte expression system, Nature London 329:836.

McConnell, H.M., 1975, Coupling between lateral and perpendicular motion in biological membranes, page 123 in: "Functional Linkage in Biomolecular Systems," Schmitt, F.O., Schneider, D.M. and Crothers, D.M., eds., Raven Press, New York.

Moolenaar, W.H., Aerts, R.J., Tertoolen, L.G.J., and DeLast, S.W., 1986, The epidermal growth factor-induced calcium signal in A431 Cells, J. Biol. Chem. 261:279.

Newmark, P., 1987, Oncogenes and cell growth, Nature, London 327:101.

Nishizuka, Y., 1983, Protein kinase C as a possible receptor protein of tumor-promoting phorbol esters, <u>J. Biol. Chem</u>. 258:11442.

Nishizuka, Y., 1984 The role of protein kinase C in cell surface signal transduction and tumour promotion, <u>Nature</u> London 308:693.

Othmer, H.G., and Scriven, L.E., 1971, Instability and dynamic pattern in cellular networks, <u>J. Theor. Biol</u>. 32:507.

Pitts, J.D., and Finbow, M.E., 1986, The gap junction, <u>J. Cell. Sci.</u>, Suppl. 4:239.

Polk, C., 1984, Time-varying magnetic fields and DNA synthesis: magnitude of forces due to magnetic fields on surface-bound couterions. Proc. Bioelectromagnetics Soc., 6th Annual Meeting, p 77 (Abstract).

Revel, J.-P. and Karnovsky, M.J., 1967, Hexagonal array of subunits in intercellular junctions of the mouse heart and liver, <u>J. Cell Biol.</u> 33:C7.

Riedel, H., Schlessinger, J., and Ullrich, A, 1986, A chimeric, ligand binding v-erbB/EGF receptor retains transforming potential, <u>Science</u> 236:197.

Robertson, J.D., 1963, The occurrence of a subunit pattern in the unit membranes of club endings in Mauthner cell synapses in goldfish brains, <u>J. Cell Biol</u>. 19:201.

Rodbell, M., Lin, M.C., and Salomon, Y., 1974, Evidence for interdependent action of glucagon and nucleotides on the hepatic adenylate cyclase system, <u>J. Biol. Chem</u>. 249:59.

Schmitt, F.O., Schneider, D.M., and Crothers, D.M., eds., 1975, "Functional Linkage in Biomolecular Systems," Raven Press, New York.

Schwarz, G., 1975, Sharpness and kinetics of cooperative transitions, page 32 <u>in</u>: "Functional Linkage in Biomolecular Systems, Schmitt, F.O., Schneider, D.M. and Crothers, D.M., eds., Raven Press, New York.

Schwarz, G., and Balthasar, W., 1970, Cooperative bindingof linear biopolymers. 3. Thermodynamic and kinetic analysis of the acridine orange-poly (L-glutamic acid) system. <u>Eur. J. Biochem</u>. 12:461.

Schwarz, G., Klose, S., and Balthasar, W., 1970, Cooperative binding to linear biopolymers 2. Thermodynamic analysis of the proflavine-poly (L-glutamic acid) system. <u>Eur. J. Bioche</u>. 12:454.

Schwarz, G., and Seelig, J., 1968, Kinetic properties and electric field effect of the helix-coil transition of poly (gamma-benzyl L-glutamate) determined from dielectric relaxation measurements, <u>Biopolymers</u> 6:1263.

Slaga, T.J., and Butterworth,, B.B., eds., 1987, "Nongenotoxic Mechanisms in Carcinogenesis," 25th Banbury Report. Cold Spring Harbor Laboratory, New York 11724.

Slaga, T.J., Sivak, A., and Boutwell, R.K., eds., 1978, "Mechanisms of Tumor Promotion and Cocarcinogenesis," Vol. 2. Raven Press, New York.

Takahashi, K., Kaneko, I., Date, M., and Fukada, E., 1986, Effect of pulsing electromagnetic fields om DNA synthesis in mammalian cells in culture, <u>Experientia</u> 42:185.

Tobias, J.M., and Solomon, S., 1950, Opacity and diameter changes in polarized nerve, <u>J. Cell. Compar. Physiol</u>. 35:25.

Trosko, J.E., 1987, Mechanisms of tumor promotion: possible role of inhibited intercellular communication, <u>Eur. J. Cancer Clin. Oncol</u>. 23:599.

Trosko, J.E., and Chang, C.C., 1986, Oncogene and chemical inhibition of gap-junctional intercellular communication: implications for teratogenesis and carcinogenesis, page 21 <u>in</u>: "Genetic Toxicology of Environmental Chemicals. Part B: Genetic Effects and Applied Mutagenesis," Liss, New York.

Ullrich, A., Coussens, L., Hayflick, J.S., Dull, T.J., Gray, A., Tam, A.W., Lee, J., Yarden, Y., Libermann, T.A., Schlessinger, J., Downard, J., Mayes, E.L.V., Whittle, N., Waterfield, M.D., and

Seeburg, P.H., 1985, Human epidermal growth factor receptor cDNA sequence and aberrant expression of the amplified gene in A431 epidermoid carcinoma cells, _Nature_ London 309:428.

Weinstein, I.B., 1988, The origins of human cancer: molecular mechanisms of carcinogenesis and their implications for cancer prevention and treatment, _Cancer Res._ 48:4135.

Weinstein, I.B., Gattoni-Celli, S., Kirschmeier, P., Lambert, M., Hsiao, W.-L., Backer, J., and Jeffrey, A., 1984, Multistage carcinogenesis involves multiple genes and multiple mechanisms, page 225 _in_: "Cancer Cells I. The Transformed Phenotype, Levine, A., Van de Woude, G., Watson, J.D. and Topp, W.C., eds. Cold Spring Harbor Laboratory Press, New York 11724.

Wolf, M., Levine, H., May, W.S., Cuatrecasas, P., and Sayhoun, N., 1985, A model for intracellular translocation of protein kinase C involving synergism between Ca^{2+} and phorbol esters, _Nature_ London 317:346.

Wyman, J., 1948, Heme Proteins, _Adv. Protein Chem._ 4:407.

Wyman, J., and Allen, D.W., 1951, The problems of the heme interactions in hemoglobin and the basis of the Bohr effect, _J. Polymer Sci._ 7:491.

Yahara, I., and Edelman, G.M., 1972, Restriction of the mobility of lymphocyte immunoglobulin receptors by concanavalin A, _Proc. Nat. Acad. Sci. USA_ 69:608.

Yamasaki, H., 1987, The role of cell-to-cell communication in tumor promotion, page 297 _in_: "Nongenotoxic Mechanisms in Carcinogenesis," Butterworth, T.E. and Slaga, T.J., eds., 25th Banbury Report. Cold Spring Harbor Laboratory, New York 11724.

Yotti, L.P., Chang, C.C., and Trosko, J.E., 1979, Elimination of metabolic cooperation in Chinese hamster ovary cells by tumor promoter, _Science_ 206:1069.

PHOTO-INDUCED BIOELECTROCHEMICAL PROCESSES

DIAZOLUMINOMELANIN: A SYNTHETIC ELECTRON AND NONRADIATIVE

TRANSFER BIOPOLYMER

Johnathan L. Kiel, Gerald J. O'Brien, David M. Simmons
and David N. Erwin
Radiation Sciences Division (RZP)
USAF School of Aerospace Medicine (HSD/AFSC)
Brooks AFB, Texas 78235-5000, U.S.A.

INTRODUCTION

The green hemoprotein (GHP) of erythrocytes is a type-b cytochrome
with no known function except for the incidental finding of its
ferroactivation of phosphoenolpyruvate carboxykinase which is not
present in erythrocytes (Chee and Lardy, 1981). We have previously
postulated that GHP can transfer superoxide to methemoglobin, converting
it to oxyhemoglobin (Kiel et al., 1988). Although no artificial
substrate has been available for measuring the electron transfers
mediated by GHP, previous data has indicated that the peroxidations
putatively mediated by GHP and hemoglobin in erythrocytes can be
inhibited by 3-amino-L-tyrosine (Kiel and Erwin, 1986; Kiel et al.,
1988). This derivatized tyrosine is an inhibitor of peroxidases (Kiel,
1988) and inhibits the oxidative burst (superoxide and hydrogen peroxide
production) of mouse peritoneal macrophages (Lefkowitz et al., 1988).
The latter is associated with an electron transport chain containing a
type-b cytochrome (Rossi et al., 1986). Furthermore, when
lactoperoxidase is reduced by thiol, its binding of tyrosine derivatives
becomes significantly enhanced (Pommier and Cahnmann, 1979). Therefore,
a potential existed for the development of a tyrosine-derived substrate
that specifically bound to GHP.

Chemiluminescence is a sensitive indicator of electron transfers
and peroxidations that hemoproteins are capable of mediating (Ewetz and
Thore, 1976). Luminol (5-amino- 2,3-dihydrophthalazine-1,4-dione) is
highly efficient in this role (Prichard and Courmier, 1968). However,
hemoproteins such as horseradish peroxidase do not bind luminol, and the
luminescent reaction is an interaction between unbound luminol radical
and hydrogen peroxide (Prichard and Courmier, 1968). However, luminol
has been linked to other fluorescent compounds and effected nonradiative energy transfers (Roswell and White, 1978). Furthermore, large
polymers have been constructed that demonstrate this nonradiative
transfer amongst donor and acceptor positions in the polymer molecule
(Morawetz, 1988).

Based on the above information, we attempted to formulate a
compound of 3-amino-L-tyrosine and luminol that would specifically bind
to GHP and produce chemiluminescence indicative of GHP-catalyzed
peroxidation. Since both starting compounds are amino-aromatics, we

chose diazo coupling as a convenient method for forming the heterodimer (March, 1977).

MATERIALS AND METHODS

Luminol (5-amino-2,3-dihydro-1,4-phthalazinedione), 3-amino-L-tyrosine hydrochloride (3AT), ferrous sulfate, ferric sulfate, sodium oleate, hydrochloric acid, dimethyl sulfoxide (DMSO), sodium hydroxide, sodium nitrite, and phosphate buffer salts were obtained from Sigma Chemical Company (St. Louis, Missouri). Hydrogen peroxide (3%) was from Hydrox Chemical Company (Elk Grove Village, Illinois). Acetone and sodium carbonate were from Fisher Scientific (Houston, Texas). Green hemoprotein was obtained from fresh human hemolysate from one of the investigators and was prepared in partially purified form as previously described (Kiel et al., 1988).

Chemiluminescence Measurements

GHP-mediated chemiluminescence was measured using a Turner Designs 20e Luminometer (Mountain View, California). The reactions were run at room temperature (23-25 C).

Chemiluminescence was measured in the radiofrequency radiation field (2450 MHz, 25 W, continuous wave) using the Quantitative Luminescence Imaging System (QLIS) built for USAF by the Electro-Optic Systems Section, Automation and Measurement Sciences Department, Pacific Northwest Laboratories (Richland, Washington) of the Department of Energy (U.S. Patent pending). Briefly, the QLIS is constructed from a coherent fiber optic image guide contained in a circularly polarized (wire mesh) microwave guide powered by a MCL (La Grange, Illinois) RF power generator (Model #15222), a Newvicon video camera with a 35 mm f/1.4 lens, and an image processor with VIOS software drivers and Datacube boards serviced by a microprocessor (MicroVax II). The QLIS was calibrated with a TLS Systems Model #40108-4 solid-state scintillator with phosphor #5000 (520-nm emission peak) and a carbon-14 activator of 17.5 mCi. The calibration source was held in an adapter which allowed the source to be coupled to the input window of the fiber optic bundle.

Chemiluminescent solutions were held in 1.0 X 1.0 X 4.5-cm polyacrylate cuvettes which were, in turn, held in a polyacrylate holder that connected to the fiber optic bundle input face of the QLIS. Temperature was measured during microwave radiation in the center of the solution in each cuvette with a nonperturbing electrothermia probe and monitor (Vitek, Model #101).

Chemiluminescent emission spectra were measured in a Photo Research PR-713 Spectro Radiometer. Fluorescent spectra were measured in an Aminco-Bowman Spectrophotofluorometer.

Preparation of Chemiluminescent Material

The basic chemiluminescent compound was made by combining a solution of 3-amino-L-tyrosine hydrochloride and sodium nitrite (10 mM each) in water, after a 5 min preincubation, with a 10 mM solution of luminol in DMSO and allowing the material to react for 50 min. The reaction was stopped by adding an equal volume of acetone followed by addition of saturated sodium hydroxide solution (10% total volume of reaction mixture). This mixture turned dark brown. After several days, crystals formed on the surface and bottom of the reaction mixture. These crystals were removed from the reaction liquor by filtration on

filter paper washed with acetone, and dried. The flat multilayered
transparent crystals were purple in color. This material was assigned
the trivial name diazoluminomelanin (DALM). When these crystals were
heated on a hot plate to temperatures that char paper, the crystals left
a white, ceramic-like brittle residue.

An alternate form of the above compound was prepared by allowing
the reaction to proceed in the presence of magnetic fluid in place of
the water solution in the above reaction mixture. The magnetic fluid was
prepared from ferrous sulfate and ferric sulfate as described by
Shimoiizaka et al. (1976). Black crystals formed in this preparation
and were treated as above. The crystals were ground into a powder. The
material readily attached to a magnetic stirring bar, demonstrating its
magnetic property.

Manganese chloride (5 mM) was added to some of the sodium salt in
order to prepare an alternate form. The crystals were needle-like and
red-brown.

Chemiluminescent Reactions

The reactions between GHP and DALM were carried out under two sets
of conditions. First, 100 ul GHP solution with an optical absorbance of
0.2 at 416 nm were mixed with 37.5 ug DALM in 620 ul of 1 mM, pH 7.4,
sodium phosphate buffer containing 0.0003% hydrogen peroxide. The
luminescence of this reaction was monitored with the 20e Luminometer and
compared to the reaction without GHP. Second, 300 ul of 1 mM, pH 7.4,
sodium phosphate buffer containing 10 ug/ml DALM was activated with 100
ul 0.001% hydrogen peroxide, followed by the addition of two 10-ul
aliquots of GHP solution (0.10 O.D. at 412 nm and 0.27 O.D. at 414 nm,
respectively).

In some instances, the crystals of DALM were placed on filter
paper, dissolved with 3% hydrogen peroxide and the luminescence was
visually observed in a dark room. A long-wavelength ultraviolet lamp
(UVL.56, Ultraviolet Products, Inc., San Gabriel, California) was used
to visually observe fluorescence and phosphorescence.

Reaction mixtures of DALM for microwave radiation exposure were
composed of 1.5 ml 1 M sodium carbonate in water solution containing 1
mg/ml luminol, 0.75 ml of the first solution with 0.75 ml sodium salt of
DALM (1 mg/ml) in water solution, or 1.0 ml of the second mixture with
0.5 ml of 2 mg/ml magnetite DALM in water solution. Hydrogen peroxide,
100 ul 3%, was added to each of the previous reaction mixtures. These
reaction mixtures were placed in polyacrylate cuvettes in the QLIS and
activated with microwave radiation heating.

RESULTS AND DISCUSSION

The chemiluminescent and fluorescent emission peaks of
3-amino-L-tyrosine, luminol, and diazoluminomelanin are shown in Table
1. Whereas, the luminol chemiluminescence and fluorescence are usually
at the same wavelength, both being shifted alike dependent on the
solvent (Roswell and White, 1978), the corresponding emissions of DALM
are distinctly different. The chemiluminescent emissions of DALM at 519
nm (major peak) and 600 nm (minor peak) resemble the emissions of
luminol in DMSO (510 nm peak). The chemiluminescence of luminol in
water and DMSO (50:50) peaks at 484 nm and resembles the fluorescence
peak at 480 nm of DALM in dilute aqueous solution (10 to 100 uM). In
more concentrated solutions, DALM fluorescence is severely quenched.

Table 1. Peak Wavelengths (nm) of Excitation and Chemiluminescent and Fluorescent Emissions of 3-Amino-L-tyrosine (3AT), Luminol, and Diazoluminomelanin (DALM)

Compound	Excitation	Emission	
		Chemiluminescence	Fluorescence
3AT	256	NA[a]	365
	360	NA	420
Luminol	275	NA	425 (water)
			510 (DMSO)
	NA	425 (water)	NA
		510 (DMSO)	NA
		484 (water/DMSO)	NA
DALM	365	NA	480
	NA	519 (major peak)	NA
		600 (minor peak)	NA

[a]Not applicable.

The source of such anomalous behavior in DALM was revealed by examining its chemiluminescence and fluorescence on filter paper. When DALM was dissolved in 3% hydrogen peroxide, the solution applied to filter paper, and observed in a dark room, the luminescence started at the edges of the applied solution and increased toward the center of the damp spot upon partially drying or blowing of exhaled air across the surface. Examination of the spots under a long-wavelength ultraviolet lamp revealed UV absorption of the fresh solution at the center of each spot with fluorescence at the edges, and with time, toward the center. There appeared to be a release of fluorescent material from the original compound following peroxidation. Like natural melanin, the DALM has a broad wavelength absorption, particularly in the UV portion of the spectrum. The peroxidation evidently allowed for the release of material excited by nonradiative transfer of energy from chemically excited states of luminol contained within the DALM. Even though the DALM is readily soluble in water and the aforementioned reactions occurred in aqueous solution, the microenvironment of luminol probably resembles that provided by DMSO, accounting for the shift to longer wavelengths of emission for DALM. This form of luminol is probably converted by peroxidation into the quininoid form of aminophthalate free ion, the iminent emitter of light (Roswell and White, 1978) in organic solvents.

The obervation that the human breath can activate DALM luminescence suggested that either moving air containing oxygen was needed or air containing carbon dioxide. Forced air blown across the DALM solutions on filter paper did not observably increase the luminescence. However, air with 5% carbon dioxide or brief pulses of 100% carbon dioxide did increase chemiluminescence. The 100% carbon dioxide briefly quenched the luminescence during its application.

The maganese salt of DALM luminesced when activated in the same manner as the sodium salt. However, it luminesced more brightly and for shorter periods of time (data not shown).

When heated on a laboratory hot plate, the purple crystalline DALM converted to a white opaque ceramic-like material and did not char like filter paper placed on the hot plate for the same length of time of heating. This white material was capable of being activated to yield chemiluminescence with hydrogen peroxide. However, it obtained an additional property, phosphorescence. After being exposed to the long-wavelength ultraviolet lamp for a few seconds, the ceramic-like material phosphoresced with a green light for several seconds.

Fig. 1 and 2 show that DALM is capable of acting as a substrate for green hemoprotein. Fig. 1. shows a preparation of GHP and DALM activated with a low concentration of hydrogen peroxide (88 uM) that yielded a 5 to 20-fold increase of the baseline luminescence of uncatalyzed DALM peroxidation.

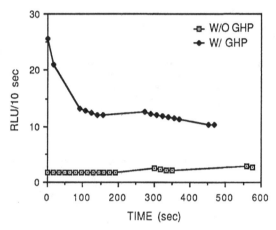

Fig. 1. Luminescence produced from diazoluminomelanin (52 ug/ml) in 1 mM pH 7.4 phosphate buffer containing 0.0003% H_2O_2 with (upper curve) or without (lower curve) human erythrocyte green hemoprotein. RLU, relative light units of a 20e Luminometer.

Fig. 2. shows another preparation of GHP that increased the baseline chemiluminescence of DALM by over 500-fold with about 70 uM hydrogen peroxide. Therefore, DALM has shown to be an excellent substrate for GHP-mediated peroxidation. The failure of the heme optical absorptions to be proportional to the activation by the two GHP preparations is because they also indicate the presence of other contaminating hemoproteins, especially hemoglobin, which does not substantially increase DALM chemiluminescence (data not shown).

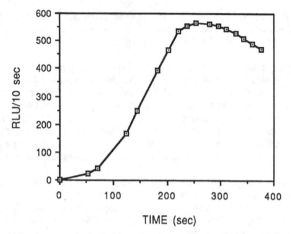

Fig. 2. Chemiluminescence of diazoluminomelanin (10 ug/ml) activated
with 0.00025% H_2O_2 in 1 mM pH 7.4 phosphate buffer and human
green hemoprotein. GHP (10 ul) was added after the first 30
sec and again 18 sec later.

Fig. 3 and 4 display the thermochemiluminescent responses of
various chemiluminescent preparations of luminol and DALM. Note that
the thermal response in Fig. 3 of the luminol, hydrogen peroxide, and
sodium carbonate solution is less than that of the same combination with
purple crystalline DALM solution added, and considerably less than that
of the solution with both purple DALM and magnetite DALM solutions
added. From these results, we conclude that DALM not only catalyzes its
own peroxidation, but can also facilitate the peroxidation of free
luminol. Therefore, DALM is probably capable of co-oxidizing with a
number of reductants. Furthermore, magnetite acted as an inorganic
catalyst for this peroxidation.

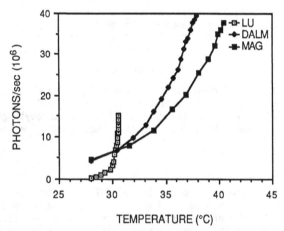

Fig. 3. Thermochemiluminescence of luminol (Lu, 1 mg/ml) in sodium
carbonate (1 M) aqueous solution; sodium salt of
diazoluminomelanin (DALM, 500 ug/ml) and luminol (500 ug/ml)
in 0.5 M sodium carbonate; and sodium salt of
diazoluminomelanin (333 ug/ml), magnetite diazoluminomelanin
(Mag, 667 ug/ml), and luminol (333 ug/ml) in 0.33 M sodium
carbonate. The solutions were activated with 100 ul 3% H_2O_2,
respectively. Microwave radiation was at 2450 MHz, continuous
wave, 25 W input power.

Fig. 4 shows that the three preparations of luminol and DALM did not increase in luminescence with time in the same way as they did with temperature. These results indicate that the DALM solutions were not only more inherently active in peroxidatively driven chemiluminescence, but also were more effective absorbers of microwave radiation, even in dilute solutions (333 or 667 ug/ml). This same experimental approach will be used with GHP, when pure preparations of the enzyme are available, in order to determine if any enhanced absorption and activation by microwave radiation is mediated through GHP.

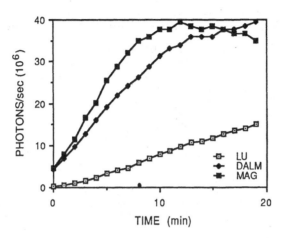

Fig. 4. Thermchemiluminescence versus time of luminol (Lu) in sodium carbonate solution; luminol, sodium salt of diazoluminomelanin (DALM), and sodium carbonate solution; and luminol, sodium salt of diazoluminomelanin, magnetite diazoluminomelanin (Mag), and sodium carbonate solution. The preparations were the same as in Fig. 3.

In conclusion, diazoluminomelanin is a very versatile electron and nonradiative energy transfer biopolymer with potential as a substrate for erythrocyte green hemoprotein and, perhaps, related cytochrome b's. Furthermore, because it can yield steady-state metastable chemiluminescent solutions with high thermal sensitivity, it may prove to be a useful microwave radiation dosimetric material in the future.

ACKNOWLEDGEMENTS

This work was sponsored in part by the U.S. Air Force Office of Scientific Research.

REFERENCES

Chee, P. P., and Lardy, H. A., 1981, Isolation from erythrocytes of a green hemoprotein with ferroactivator activity, J. Biol. Chem., 256:3865.
Ewetz, L., and Thore, A., 1976, Factors affecting the specificity of the luminol reaction with hematin compounds, Analyt. Biochem., 71:564.
Kiel, J. L., 1988,"Method for Immunosuppression," U.S. Patent 4,766,150.

Kiel, J. L., and Erwin, D. N., 1986, Physiologic aging of mature porcine erythrocytes: Effects of various metabolites, antimetabolites, and physical stressors, Am. J. Vet. Res., 47:2155.

Kiel, J. L., McQueen, C., and Erwin, D. N., 1988, Green hemoprotein of erythrocytes: Methemoglobin superoxide transferase, Physiol. Chem. Phys. Med. NMR, 20:123.

Lefkowitz, D. L., Lefkowitz, S. S., Mone, J., and Everse, J., 1988, Peroxidase-induced enhancement of chemiluminescence by murine peritoneal macrophages, Life Sci., 43:739.

March, J., 1977, "Advanced Organic Chemistry: Reactions, Mechanisms, and Structure," McGraw-Hill, New York.

Morawetz, H., 1988, Studies of synthetic polymers by nonradiative energy transfer, Science, 240:172.

Pommier, J., and Cahnmann, H. J., 1979, Interaction of lactoperoxidase with thiols and diiodotyrosine, J. Biol. Chem., 254:3006.

Prichard, P. M., and Cormier, M. J., 1968, Studies on the mechanism of the horseradish peroxidase catalyzed luminescent peroxidation of luminol, Biochem. Biophys. Res. Comm., 31:131.

Rossi, F., Bellavite, P., and Papini, E., 1986, Respiratory response of phagocytes: Terminal NADPH oxidase and the mechanisms of its activation, in: "Biochemistry of Macrophages," Pitman, London.

Roswell, D. F., and White, E. H., 1978, The chemiluminescence of luminol and related hydrazides, in: "Methods in Enzymology LVII," M. A. DeLuca, ed., Academic Press, New York.

Shimoiizaka, J., Nakatsuka, K., Chubachi, R., and Sato, Y., 1976, Stabilization of aqueous magnetic suspension. Preparation of water-based magnetic fluid, Nippon Kagaku Kaishi, 1976:6.

APPLICATIONS OF BIOELECTROCHEMICAL TECHNOLOGY

ELECTROSTIMULATION OF CELL METABOLISM AND PROLIFERATION

Hermann Berg

Acad. of Sci. of the GDR, Central Institute of
Microbiology and Exptl. Therapy, Dept. of Biophysical Chemistry, DDR-6900 Jena (GDR)

ABSTRACT

Besides electroporation and electrofusion of cells by
single pulses of high field strength a lot of effects on
cell metabolism and proliferation have been detected caused
by pulsating (pulsatile) electromagnetic induced currents
(PEMIC) or direct influence of alternating currents (DAC) on
cell suspension or tissues. These electrostimulations mostly
increase the basic syntheses of ATP, DNA, m-RNA, proteins
and the membrane transport. However, a systematic research
and theoretical explanations are still in the beginning and
some contradictions can be found in literature, too.

In order to extend applications of electrostimulations
of metabolic reactions we started PEMIC and DAC experiments
on yeast and Streptomyces strains under variable conditions
will be presented on the
- CO_2 release from Sacch. cerev.,
- proliferation rate of Sacch. cerev.,
- fermentation processes of Streptomyces,
- formation of anthracycline chromophores.
Finally some relations between PEMIC, DAC and single pulses
will be discussed showing not only the powerful tool of electric fields in basic cell biology on one hand, but mentioning
also the potentially dangerous influence on human beings on
the other.

INTRODUCTION

Since at the end of the seventieth many electromagnetic
and magnetic field effects are widely used in cell biology
for electroporation and electrofusion of membranes, for elec-
trolesion of cell walls and for incorporation of drugs. Nowadays well known single high pulses for those purposes are
now a valuable tool extending the possibilities of classical
genetic methods /1-3/. For the processes leading to reversible pore formation rather good models are developed /4-6/
and the technique is more or less a matter of routine, because commercial apparates are available. Up to now more than

500 contributions to this break-through branch of bioelectrochemistry have been published demonstrating nearly universal
feasibilities in genetics, biology in general and medicine.
The results were discussed on dozens of international Symposia of the Bioelectrochem. Society, the Biomagnetic Society
etc. especially on the Jena Symposia : IX (1982) /7/, XI
(1986) /8/ and XII (1988) /9/. Detailed descriptions can be
found mostly in the journals : studia biophysica (Berlin),
Bioelectrochemistry and Bioenergetics (Elsevier), Journal of
Bioelectricity, Bioelectromagnetics etc. and in such mono-
graphs as
"Bioelectrochemistry II"   /10/
"Modern Bioelectrochemistry" /11/
"Modern Bioelectricity"   /12/
"Comprehensive Treatise of Electrochemistry" (Volume 10) /13/
"Electromagnetic Fields and Biomembranes" /14/.
Besides this treatment another experimental possibility becomes important in cell physiology, namely the influence on
metabolism by the application of pulsating electromagnetically induced currents (PEMIC) etc. or on cells and tissues
lasting for longer times. Whereas Galvanic effects on microorganisms were detected 100 years ago /15/, the systematic
research began 1964 with the PEMIC electrostimulation of calcium transport between bone cells /16/. Nowadays stimulations
caused by electroinduction /17/ have been detected as morphological changes, membrane permeations, biopolymer syntheses,
enzymatic actions, antibiotics productions etc. It seems
possible that all primary and secondary biochemical reactions
can be influenced in such a way, not to forget the effects
of static magnetic fields /18/ and also currents with direct
contact between electrodes and cell suspension.
    The aim of this paper is to review mainly electrostimu-
lation on the cellular level by pulsating electromagnetically induced currents (PEMIC, PEMF) that means extremely low
frequency electric fields (ELF EF) on the one hand and to
compare with own results on the other hand. Finally some conclusions should be drawn regarding the influence of electric
fields in the atmosphere (electromagnetic pollution) on living
beings.

EXPERIMENTAL

    1. The electrostimulation techniques.

Nowadays mainly three techniques are applied /19-21/ :

    1.1. The direct current, either constant (Dc) or alter-
         nating (Dac) of 20 µA and about 1 V electrode po-
         tential. Platinum or stainless steel electrodes
         are us used /16/.
    1.2. Capacitively coupled electrical field (Cc) : sine
         wave 60 kHz, 0.33 V/cm, 5 µA/cm^2 with stainless
         steel electrodes /12, 21/.
    1.3. Inductively coupled electromagnetic field (PEMIC,
         PEMF) by Helmholtz coils (of about 200 µT as a
         mean value, corresponding to a 10 times higher mag-
         netic field than that of the earth /19, 20/,
         - either sine wave : about 100 Hz, 10 mV/cm,
           200 µT
         - or rectangular pulses, e.g. a 5 ms burst repeti-

tion rate 15 Hz of having 200 /us main and 20 /us
opposite polarity /20/. Resulting field strength
in the medium : 1-15 mV/cm, (up to 100 mV/cm
reached) 200 /uT.

The technique 1.3. (PEMIC) is widely applied using a fast
driving voltage step of about V_o = 25 V for coils of low in-
ductivity L_c and resistance R_c. Then the relation for the
rectangular-type induced waveform dI_c/dt is approximately
valid /23/

$$dI_c/dt = V_o/L_c(1-t \; R_c/L_c) \tag{1}$$

2. The cell suspensions.

Our own PEMIC experiments were performed with yeast (Sacch.
cerev.) or fungi (Streptom. noursei or Streptom. griseus).

2.1. PEMIC electrostimulation of the CO_2 production of
yeast.
The cells $(3x10^7/ml)$ were transfered from their
dry state (by lyophilisation) to a yeast-extract
peptone-glucose starvation medium of low pH value
(<5). The number n of CO_2 bubbles of stimulated
and control suspensions were measured by an optical
bubble-counting device according to E.Bauer /22/.
2.2. The electrostimulation of yeast proliferation.
Stimulation of a yeast suspension in a nutrient
medium occured with Dac of 15 Hz, 72 Hz and 1100 Hz
(mA range) at 28°C for 6 h using platinum electrodes. The cells in the control and in these sine
fields were counted by the image-analizer "Qanti-
met" (+10% reproducibility).
2.3. The electrostimulation of nourseothricin production.
Streptomyces noursei ZIMET 43716 was inoculated into a micro-fermenter (6.2 l glass cylinder) from a
large pilot fermenter. The micro-fermenter was
equipped outside with two Helmholtz coils (65 turns
copper wire) and inside with two stirres for rapid
mixing and oxygen transport.
PEMIC conditions: 5 ms bursts consisting of single
220 /us pulses and 60 ms burst intervals, 1.5 mV/cm
induced field strength.
Fermentation conditions: to the complex medium (for
details compare /60/) ammonium hydroxyd, KH_2PO_4
and glucose were feeded.
Analytical determinations (see /60/):
- nourseothricin,
- acid soluble phosphate,
- extracellulare phosphate,
- glucose, and the biomass dry weight.
2.4. The electrostimulation for formation of chromophores of anthracycline antibiotics.
For the reason of the high oxygen consumption of
Streptomyces griseus the PEMIC equipment was
mounted together with the cultivation flash on a
rapid shaking mashine.

3. Theoretical mechanisms.

In contrast to the energy of single high pulses for electroporation and electrofusion (10^5 J) the PEMIC energy transduction during 1 hour is extremely low /23/ e.g. in the order of 10^{-4} J for a current density of 10 uA/cm^2. Moreover the electric field strength E of the PEMIC signal is also about 9 orders of magnitude lower than E of the membrane interface (10^6 V/cm).

For such reasons there must be an amplification mechanisms or perturbation effects on metastable equilibria triggering highly cooperative processes, which were detected e.g. for Ca^{++} release from cerebral tissue /24/. It is well known that similar effects in tissues require many orders of magnitude lower energy than in solution, e.g. for conformational changes of nucleic acids.

Another possibility is the resonance of a PEMIC perturbation of a sensitive intrinsic self-oscillating system exhibiting a chaotic destabilisation which may lead to a collapse propagating solitons moving down the length of lipoproteins resulting in nonlinear vibrations /25/. A third type of main effects of imposed ELF EF (100 Hz) are slow shifts of charges especially on glycoproteins, e.g. Ca^{++} ions play a key role in this mechanism for amplification. I such a way cell membranes can be primary sites for Fröhlich's coherent oscillations in proteins of about 10^{11} Hz /26/.

In this context the ELF EF propagated redistributions of ions and microelectrophoresis of ligands near surface receptors perturbate /27/ the complex equilibrium in an appreciable manner (+20%) above the threshold field strength : $E^* < 25$ mV/cm. A model based mainly on Coulombic coupling between transmembrane potential $\Delta\phi_m$ and conformational equilibria of membrane proteins involves partially already mentioned ideas /27, 28/. $\Delta\phi_m$, which can be changed rapidly and dramatically by opening or closing of ion channels between -10 to -250 mV, imposes $-20 < E < -500$ kV/cm for a membrane thickness of 6 nm. An applied electric field E on a spherical cell induces - according to Maxwell - :

$$\Delta\phi_m = 1.5 \; r \; E \cos \theta \hspace{3cm} (2)$$

(θ is the angle between the field vector and a line from the cell center to a point on the surface.)
According to $\Delta\phi_m/(\Delta E)$ the amplification experienced by a membrane protein is approximately 1000 times! Especially charged amino acids and prosthetic groups are targets of such field interactions forming helical macro dipoles of 100 Debye. If in an equilibrium conformations of an enzyme differ in the electric moment $D \approx 200$ Debye and $\Delta\phi_m$ reaches about 200 mV than a free energy absorption of about 4 kcal/mol from the field occurs during such transfer process, with other words a frequency dependent turnover between two conformations with different affinities to a substrate is possible, if resonance takes place.

The electroconformational coupling works as an energy transducer. This phenomenon should be valid also for channel proteins regulating the permeation of membranes not to forget their possible phase transitions. In spite of such a weak energy input there exists a so strong amplification that an increasing amount of examples in cell biology has been detected. Typical electrostimulations of metabolic reactions, of membrane behaviour and cell division will be listed now, whereas morphological changes and behaviour effects must be omitted.

4. Selected characteristic PEMIC applications.

Nowadays some hundreds of observations are published, however, there are also contradictory results because too many parameters must be identical - especially frequency and field strength. Consequently the reproduction is sometimes difficult. Therefore only similar results are involved into the following tables, which have been detected by several authors.

4.1. Electrostimulation of biopolymer syntheses.
The most important PEMIC effect is the stimulation of the DNA synthesis possibly via plasma membrane ATPases /29/, because subsequent reaction products also an increase as shown in table 1.

Table 1. Electrostimulation of biopolymer syntheses

Substrate	Effect	References
-DNA (^{3}H-thymidine incorporation)		
embryonic bone cells (rat)	increase	30 (1984)
embryonic chondrocytes	increase	31 (1978)
fibroblasts (human)	increase	32 (1984)
V79 cells (Chinese hamster) 4×10^{-5} T	increase	33 (1986)
4×10^{-4} T	decrease	
-m-RNA (^{3}H-uridine incorporation)		
X chromosome (salivary gland, Diptera)	increase	34 (1988)
leucemic cells (human)	increase	35 (1988)
-Proteins (labelled aminoacid incorp.)		
skin cells (rat)	increase	36 (1984)
bone fibroblasts : collagen	increase	37 (1985)
ovar cells (Chinese hamster)	increase	38 (1985)
fibroblasts (human) : glycoprotein	increase	39 (1986)

4.2. For fermentation purposes the electrostimulation of enzymes is of considerable interest (Table 2). Especially the ATP synthesis is essentially required for many cellular processes. Membrane bound enzymes seem to be energetically favoured.

Table 2. Electrostimulation of enzyme activity.

Enzyme	Reaction	Effect	References
ATPase (beef heart)	ATP synthesis	increase	40 (1988)
ATPase (E. coli)	ATP synthesis	increase	41 (1986)
Na, K-ATPase (erythrocytes)	K^+, Rb^+ uptake $25^{\circ}C$	increase	42 (1984)
Adenylcyclase (erythrocytes)	c-AMP	increase	43 (1984)
Adenylcyclase (fibroblast)	c-AMP (longterm)	inhibition	44 (1986)
Tyrosinase (melanoma cells)	c-AMP, melanin	increase	45 (1986)
Ornithine decarb- oxylase (fibro- blast)	4-6 h +phorbol- ester	decrease	46 (1988)
	10-12h "	increase	

4.3. Strongly coupled with enzyme reactions and $\Delta\phi_m$ is the enhancement of membrane transport (Table 3) of ions and substrates. Especially the Ca^{++} flux is essential for bone growth and repair but also for electrofusion and a lot of cell reactions.

Table 3. Electrostimulations of membrane transport and $\Delta\phi_m$

Process	Cell	Effect	References
Na^+ release	erythrocytes (human)	increase	47 (1983)
Ca^{++} influx	bone cells (chicken)	increase	29,48 (1983)
Ca^{++} efflux	cerebral cells	increase	49 (1984)
Ca^{++} influx	lymphocytes (human)	increase	50 (1987)
$\Delta\phi_m$	protoplasts (mesophyll)	changes	51 (1987)

4.4. Not only for the increase of biomass but also for the production of certain cell ingredients (e.g. antibodies and other proteins for therapy) a faster cell division (Table 4) is useful.

Table 4. Electrostimulation of proliferation

Cells	Conditions	Effect	References
Lymphocyte (human)	lectin induced	increase	52 (1986)
Lymphocyte (human)	lectin induced	reduced	53 (1983)
Ganglia (chicken)	PEMIC or Dc : neurite outgrowth	enhanced	54 (1984)
Leukocyte (human)	60 Hz, 10^{-4} T	no signif.	55 (1986)
Corneal fibroblasts (rabbit)	69 Hz,s, 10^{-4} T ^{3}H-thymidine incorporation	no evidence	56 (1986)
Calvaria cells (chicken)	6×10^4 Hz (Cc.) 0.1 mV/cm +Ca-calmodulin	increase	57 (1987) 58 (1986)

The reason why of different results with the lymphocyte-lectin model in table 4 is the following /59/ : the lymphocyte response depends on immunological functional conditions that cannot be foreseen.

5. Recent results and discussion of the Department Biophysical Chemistry / ZIMET.

Besides our experiments on electroincorporation and electrofusion since 1978 we started 1982 with electrostimulation after the IXth Jena Symposium /7/, where we discussed technical prerequisites with A.Pilla /20/.
Up to now we studied only PEMIC effects on the increase of the productivity of cellular metabolism.

5.1. A PEMIC electrostimulation of the CO_2 production of yeast.
The electromagnetic field stimulates the glycolytic CO_2 production of Sacch. cerev. in such a way that these cells reach their full metabolic activity in a shorter time than the control, with other words their initial rate :
$n/t = 12 \ h^{-1}$ is twice of the rate of the con-
$n/t = \ 6 \ h^{-1}$. trol
After about 2 h both attained nearly the same maximal rate. This kind of stimulation is most effective under starvation conditions, where the cells need some time to regenerate their optimal metabolism.

5.2. The electrostimulation of yeast proliferation.
As fig. 1 shows no effect occurs in the low frequency range, whereas at high frequency a significant electrostimulation by a direct sinusoidal alternating current was found. This result confirmed again the well known basic rule that the most time consuming work is to detect the suitable "electric window", where the cell reacts sensitively.

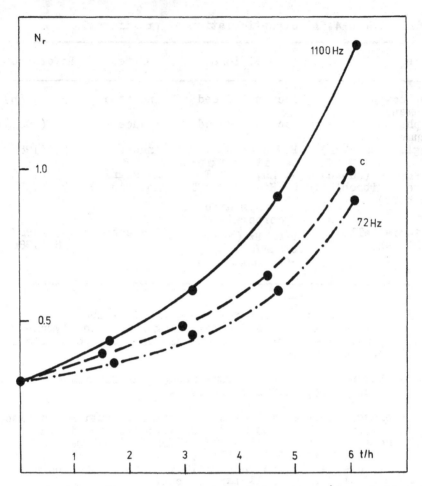

Fig. 1 PEMIC influence on yeast proliferation (description in
the text), N_r - relative number of colonies, c - control

5.3. The electrostimulation of nourseothricin production.
During the initial phase (up to the 50th h) the e-
lectrostimulation production of nourseothricin is
higher than in the control fermenter (Fig. 2). Af-
terwards a retardation in the former takes place,
whereas the rate of the control production is rather stable until the end of fermentation. Moreover
electrostimulation in this initial phase leads consequently also to more biomass, higher O_2 consumption, and glucose utilization. The reason why this
retardation was found is the linkage between the
nourseothricin formation and the phosphate metabolism /60/.
Under electrostimulation the phosphate consumption
is lower than the phosphate release from potato
starch by extracellular phosphatase leading to a
higher free inorganic phosphate concentration
(than in the control fermenter), which inhibits the
nourseothricin formation. As a consequence of such
a linkage in this particular case an intermittent
electrostimulation is required in order to avoid
the inhibition effect of phosphate.

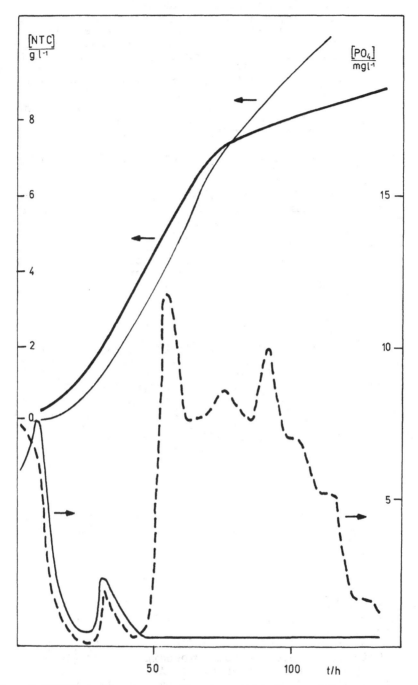

Fig. 2 PEMIC influence on the concentration of fermentation
products : nourseothricin (NTC) and free phosphate
(description in the text; control: thin lines)

5.4. The electrostimulation of formation of chromophores of anthracycline antibiotics.
Preliminary experiments have been indicated that also in the case of Streptomyces griseus an increase of the concentration of anthraquinone dyes of more than 20% was found spectrophotometrically under PEMIC conditions. Future experiments are necessary to determine the parameters of fermentation as in the case of 5.3. and to ascertain the mechanism. It may be possible that proliferation, substrate transport and enzyme activation will be stimulated synergetically.

6. Conclusions from electromagnetic field effects in general.

As can be concluded from the examples in the tables 1-4 and from our own experiments that PEMIC effects seem to be of an universal significance in cell biology and they are comparable with excitation of molecules by light energies. Therefore they are to be taken into account also for living beings and in literature a steadily increasing amount of observations are described, which may be dangerous for the population especially in highly industrialized countries. To mention two examples :
- 100 V/cm were measured near 60 Hz high voltage power lines, which may produce about 0.1 mV/cm tissue gradients in man.
- 1 V/cm of 100-400 MHz radio fields (1 mW/cm^2) may produce 100 mV/cm tissue gradients, depending on body orientation with respect to the field.
These environmental fields are comparable with PEMIC influences of 1-100 mV/cm! This can be a reason why physiological changes and even a higher risk for health are possible /61, 62, 63/. In this respect the interest in electric field effects increases markedly and will be discussed more and more on bioelectrochemical meetings and journals.

ACKNOWLEDGEMENT

The cooperation of E. Bauer, H.-H. Große, K. Harz, H.-E. Jacob, U. Katenkamp, U. Luthardt and P. Mühlig is highly appreciated.

REFERENCES

1. H. Berg, K. Augsten, E. Bauer, W. Förster, H.-E. Jacob, A. Kurischko, P. Mühlig, and H. Weber, Bioelectrochem. Bioenerg. 12:119 (1984)
2. U. Zimmermann, Trends in Biotechnology 1:149 (1983)
3. J. Teissie, Bioelectrochem. Bioenerg. 20:133 (1988)
4. L. Chernomordik, S. Sukharev, I. Abidor, and Yu. Chismadjew, Biochim. Biophys. Acta 736:203 (1983)
5. J. Weaver, R. Mintzer, H. Ling, and S. Sloan, Bioelectrochem. Bioenerg. 15:229 (1986)
6. I. Shugar, and E. Neumann, Biophys. Chem. 19:221 (1984)
7. "Photodynamic and Electric Effects on Biopolymers and Membranes", eds. H. Berg, H.-E. Jacob, studia biophysica (Berlin) Vol. 94 (1983)

8. "Bioelectrochemistry in Biotechnology", eds. H. Berg, H.-E. Jacob, studia biophysica (Berlin) Vol.119 (1987)
9. "Trends in Bioelectrochemistry of Biopolymers and Membranes", ed. H. Berg, studia biophysica (Berlin) Vol.130 (1989)
10. "Bioelectrochemistry II" (eds. G. Milazzo, M. Blank), Plenum Press, N.Y. (1987)
11. "Modern Bioelectrochemistry" (eds. F. Gutmann, H. Keyzer), Plenum Press, N.Y. (1986)
12. "Modern Bioelectricity" (ed. A. Marino), Marcel Dekker Inc., N.Y. (1988)
13. "Comprehensive Treatise of Electrochemistry",(eds. S. Srinivasan, Yu. A. Chizmadzhev, J. O'M. Bockris, B.E. Conway, E. Yeager), Vol. 10, Bioelectrochem. (1985)
14. "Electromagnetic Fields and Biomembranes", (ed. M.Markov), Plenum Press, N.Y. (1988)
15. M. Apostoli, Laquerierre Comp. Rend. Acad. Sci., Paris, 110:918 (1890)
16. C. Bassett, R. Pawluk, and R. Becker, Nature 204:652 (1964)
17. H. Berg, studia biophysica (Berlin) 119:17 (1987)
18. M. Markov, studia biophysica (Berlin) 119:147 (1987)
19. C. Bassett, R. Pawluk, A. Pilla, Science 184   (1974)
20. A. Pilla in "Modern Bioelectricity" (ed. A. Marino), Marcel Dekker (1988) p. 427
21. S. Singh, and J. Katz, J. Bioelectricity 5:285 (1986)
22. E. Bauer, H.-E. Jacob, and H. Berg, studia biophysica 119:137 (1987)
23. G. Colacicco, and A. Pilla, Biochem. Bioenerg. 12:239 (1984)
24. W. Adey, Bioelectrochem. Bioenerg. 15:447 (1986)
25. A. Davydov, Physica scripta 20:387 (1979)
26. H. Fröhlich, Proc. Natl. Acad. Sci. 72:4211 (1975)
27. T. Y. Tsong, R. D. Astumian, Progr. Biophys. Mol. Biol. 50:1 (1987)
28. T. Y. Tsong, and R. D. Astumian, Ann. Rev. Physiol. 50:273 (1988)
29. G. Colacicco, and A. Pilla, Z. Naturforsch. 38c:468 (1983)
30. R. Korenstein, D. Somjen, H. Fischler, and I. Bindermann, Biochim. Biophys. Acta 803:302 (1984)
31. G. Rodan, L. Bourret, and L. Norton, Science 199:690 (1978)
32. A. Pilla, P. Sechand, and B. MC Leod, J. Biol. Physics 11:51 (1983)
33. K. Takahashi, I. Kaneko, M. Date, and E. Fukada, Experientia 42:185 (1986)
34. M. Blank, and R. Goodman, Bioelectrochem. Bioenerg. 19: 569 (1988)
35. J. Phillips, and L. Mc Chesney, Abstr. 10. Ann. Meeting, Bioelectromagn.Soc., Stamford CT, 1982
36. P. Delport, N. Cheng, J. Mulier, W. Sansen, and W. De Loecker, Bioelectrochem. Bioenerg. 14:93 (1985)
37. R. Farndale, and J. Murray, Bioelectrochem. Bioenerg. 14:83 (1985)
38. R. Goodman, A. Abott, A. Krim, and A. Henderson, J. Bioelectricity 4:565 (1985)
39. S. Fisher, J. Dulling, and St. Smith, J. Bioelectricity 5:253 (1986)
40. F. Chauvin, R. Astumian, and T. Tsong, (1988) submitted
41. J. Teissie, Biochemistry 25:368 (1986)

42. E. Serpesu, and T. Tsong, J. Biol. Chem. 259:7155 (1984)
43. D. Jones, J. Bioelectricity 3:427 (1984)
44. R. Farndale, and J. Murray, Biochim. Biophys. Acta 881: 46 (1986)
45. D. Jones, R. Pedley, and J. Ryaby, J. Bioelectricity 5: 145 (1986)
46. C. Cain, E. Salvador, and W. Adey, Abstr. 10. Ann. Meeting, Bioelectromagn. Soc., Stamford CT, 1982
47. K. Gary, A. Pilla, and C. Mayaud, J. Electrochem. Soc. 130:120 (1983)
48. G. Colaccico, A. Pilla, Bioelectrochem. Bioenerg. 10:119 (1983)
49. J. Bond, Bioelectrochem. Bioenerg. 12:177 (1984)
50. A. Liboff, R. Rozek, M. Sherman, B. Mc Leod, and St. Smith, J. Bioelectricity 6:13 (1987)
51. G. Fuhr, R. Hagedorn, R. Glaser, J. Gimsa, and T. Müller, J. Bioelectricity 6:49 (1987)
52. M. Cantini, F. Cossarizza, R. Bersani et al., J. Bio- electricity 5 (1986) 91
53. P. Conti, G. Gigante, M. Cifone, E. Alesse, G. Ianni, M. Reale, and P. Angeletti, FEBS Letters 162:156 (1983)
54. B. Sisken, B. Mc Leod, and A. Pilla, J. Bioelectricity 3:81 (1984)
55. W. Winters, N.Y. State Power Lines Project (1986)
56. P. Basu, N.Y. State Power Lines Project (1986)
57. R. Fitzsimmons, J. Farley, W. Adey, and D. Baylink, Clin. Res. 35:149A (1987)
58. W. Adey, Bioelectrochem. Bioenerg. 15:447 (1986)
59. R. Cadossi, G. Emilia, G. Ceccherelli, and G. Torelli in: "Modern Bioelectricity" (ed. A. Marino), Marcel Dekker (1988) p. 451
60. H.-H. Große, E. Bauer, and H. Berg, Bioelectrochem. Bio- energ. 20:279 (1988)
61. A. Marino in: "Modern Bioelectricity" (ed. A. Marino), Marcel Dekker (1988) p. 965
62. R. Becker, J. Bioelectricity 7:103 (1988)
63. A. Marino, and J. Ray : The Electric Wilderness, San Francisco Press Inc. 1985

THE STUDY OF MEMBRANE ELECTROFUSION

AND ELECTROPORATION MECHANISMS

Arthur E. Sowers

The Jerome H. Holland Laboratory
American Red Cross
Rockville, M.D.

INTRODUCTION

Electrofusion, the induction of membrane fusion with an electric field pulse, was reported independently in four laboratories (Neumann et al., 1980; Senda et al., 1979; Teissie et al., 1982; Zimmermann et al., 1981). Although most papers which report on the use of electrofusion have genetic manipulation purposes, electrofusion has been shown to have special advantages for producing monoclonal antibody-secreting hybridomas (Karsten et al., 1988; Lo and Tsong, 1989; Glassy and Pratt, 1989). Also, as will be explained below, electrofusion is a promising new way to study membrane fusion mechanisms.

Electroporation, the induction of increases in membrane permeability with an electric field pulse, has been known since the 1960s, but within the last five years has become an enormously popular technique for genetic transfection of a wide variety of cells. Electric pulses with approximately the same characteristics will induce both electrofusion and electroporation. This has given rise to the commonly accepted but unproven notion that electropores may be a fusion intermediate structure. While both electrofusion and electroporation have been well characterized, they are very poorly understood from the point of view of fundamental principles.

Our studies have been primarily directed at elucidating the mechanism of membrane electrofusion. In addition to this goal, we have been secondarily interested in: i) determining whether electropores are directly, indirectly, or not involved in this mechanism, ii) determining what parts of the electrofusion mechanism may be relevant to membrane fusion as it occurs in nature, and iii) attempting to identify factors and phenomena which have primary or controlling effects on the electrofusion. Progress over the last few years has been encouraging and there is now evidence that biologically relevant factors play a role in membrane electrofusion.

An electric pulse, as an artificial fusogen has three unique advantages: it is non-chemical, it induces fusion simultaneously with the pulse, and it can induce fusion in high yields. When dielectrophoresis is used to induce close membrane-membrane contact, then the experiments benefit from three more advantages because dielectrophoresis is: non-chemical, reversible, and mild.

High fusion yields make it practicable to study single fusion events. Since all fusion events are simultaneous with the pulse, data recording instruments can be easily coordinated with the application of the pulse. Most importantly, the means of inducing fusion, inducing membrane-membrane contact, and the chemical conditions in the medium can be manipulated independently of one another.

Electrofusion has been extensively reviewed. Numerous highly overlapping reviews have been authored primarily by Dr. Ulrich Zimmermann and coworkers. The most popular review appears to be Zimmermann (1982) while the most recent of which the present author is aware is Zimmermann (1986). Excellent reviews of electrofusion have also been written by Bates et al. (1987), Berg (1987), Hoffmann and Evans (1986), and Pohl et al. (1984). These reviews have emphasized primarily applications and demonstrate the widespread applicability of the method. In contrast, our research and reviews on electrofusion have both emphasized the mechanism of electrofusion. There has been considerable progress in understanding the electrofusion mechanism since the appearance of our first review (Sowers and Kapoor, 1987). The most recent and comprehensive treatments are in Sowers (1989) and this chapter. A monograph is now available which is devoted to electroporation and electrofusion (Neumann et al., 1989). While there is considerable overlap among these reviews, no single review is all-encompassing.

In comparison with electrofusion, electroporation has been more extensively reviewed by many more authors (see for example: Chassy et al., 1988; Shigakawa and Dower, 1988; Potter, 1988; Knight and Scrutton, 1986). All of these recent reviews emphasize applications. It is unfortunate that an excellent earlier comprehensive review by Tsong (1983) was overlooked by many of the recent electroporation users who were primarily interested in genetics. The monograph on electroporation and electrofusion (Neumann et al., 1989) should help fill the gaps in the uneven developments in these fields as well as provide access to earlier literature. Many of the electroporation papers otherwise accept the limitations of the method but do not mention that the mechanism is poorly understood or suggest that an understanding of the mechanism may lead to methodological improvements. There are also misconceptions about electrofusion. Some of these are mentioned below but reviewed more completely in Sowers (1989).

APPARATUS

The vast majority of our work involves a light microscope equipped with optics for phase contrast and epifluorescence imaging. Some of our experiments utilize a low light level video camera connected to a video tape recorder through a time alpha-numerics generator (for documentation) and a video analyzer for semi-quantitative (analog) analysis of image intensity which are within the limitations (Inoue, 1986) of these systems.

Our pulse has in almost all cases had the shape of an exponentially decaying wave. The other major pulse waveform is the square wave. While different effects from these waveforms have been reported (eg. Liang et al., 1988; Sowers, 1989), there are as yet no definitive studies identifying a crucial fundamental difference between the two waveforms and hence this variable is yet to be critically evaluated. Our pulse generator depends on a discharging capacitor and is homemade and described in detail in Sowers (1989).

All electroporation and electrofusion experiments must take place in a chamber which permits two electrodes to come into contact with the aqueous medium containing the membranes. The remainder of the chamber must be insulating and, if microscopy or other monitoring devices are to be used, be provided with necessary features. The geometry of the chamber, and the shape of the electric field caused by this geometry, requires careful thought. A variety of chambers have been used and are described in the earlier literature.

TWO ELECTROFUSION PROTOCOLS

The usual procedure used in electrofusion is to use dielectrophoresis to align membranes into close contact and then apply the fusion-inducing pulse. We refer to this as the Contact First (CF) protocol. It should be pointed out that it has been observed that application of pulses before the induction of close membrane-membrane contact will also induce fusion (Bates et al., 1987; Sowers, 1983; Sowers, 1984; Sowers, 1989; Teissie and Rols, 1986; Zimmermann et al., 1984). We refer to this as the Pulse First (PF) protocol. Both of these protocols are illustrated in Fig. 1. Fusion obtained by this protocol is more poorly understood than the normal fusion protocol. However, this observation may be very significant for understanding fusion mechanisms.

Almost all of our work has utilized an alternating current frequency of 60 Hz (cycles per sec) because adequate alignment can be obtained to produce fusion, and it is available directly from the utility power lines. We have found with our chamber (Fig. 1 in Sowers, 1984) that there is no difference in fusion yield for alternating current with a frequency as high as 100,000 Hz (Sowers, unpublished). Our chamber produces a bulk homogeneous electric field.

DEFINITION OF MEMBRANE FUSION

Fig. 2 shows that if the membranes in question are the plasma membranes of cells and the cells are large enough to be seen by light microscopy, then observation of the rightmost two morphologies (steps C and D) is good evidence for fusion. However, it must be understood that fusion actually occurred at step B. This has been rigorously discussed previously (Gingell and Ginsberg, 1978; Knutton and Pasternak, 1979) and must be taken into consideration in interpreting unexpected experimental results such as found by Wojciesyyn et

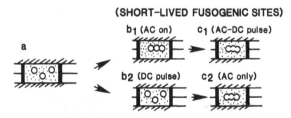

Fig. 1.   Two protocols for electrofusion. Three membranes in chamber (a) can be aligned by dielectrophoresis into contact first (b_1) and then treated with a pulse(c_1) or treated with several pulses (b_2) first and then aligned into contact by dielectrophoresis (c_2).

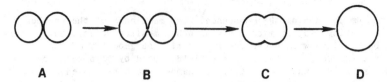

Fig. 2.   Changes in gross morphology when two spherical-shaped
          membranes undergo fusion: A, two unfused membranes in close
          contact; B, single membrane obtained by fusion of membranes
          in A (note that overall morphology does not change, but
          that originally unfused membranes and enclosed spaces are
          now connected to each other); C, partial expansion of the
          diameter of the lumen; D, complete expansion of lumen
          diameter to single large sphere.

al., (1983). There is also now evidence that, in a cellular secretion system,
the step A - B may be reversible (Breckenridge and Almers, 1987; Fernandez et
al., 1984; Zimmerberg et al., 1987).

Regardless of what fusogen is used, membrane fusion must be defined.
This must include the criteria for fusion and the procedure for measuring
fusion events and calculating the fusion yield. While cytoplasmic events may
influence membrane fusion, there are, from a rigorous point of view, only two
fundamental and measurable criteria. They are: i) connection of two
originally unfused (but close-spaced) and separate membranes into one fused
but continuous fusion product (showing, therefore, membrane mixing), and ii)
connection of the two originally separate spaces enclosed by the unfused
membranes into one continuous space (showing, therefore, contents mixing)(Fig.
3). It is extremely important to appreciate the fundamental significance of
this since the many papers which report fusion yield as based on only the
morphologies shown at stages C and D in Fig. 2 may be invalid because fusion
events which go only to stage B, but not beyond, will be missed.

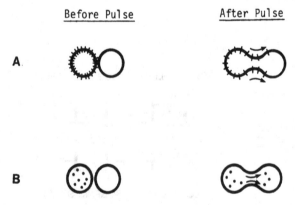

Fig. 3.   Measuring fusion in terms of membrane mixing (A) and
          contents mixing (B). Visual marker molecules, which are
          either lipid soluble (A) or water soluble (B), are used to
          separately monitor membrane mixing or contents mixing as
          unambiguous event monitors.

Fusion yield can be calculated in many ways. When cultured cells are used, sometimes a "fusion index" is used. This essentially compares the number of large (i.e. fused plasma membranes) cells which contain more than one nucleus with the number of cells which contain only one nucleus. This, again, will detect only stages C and D in Fig. 1.

CHOICE OF FUSION PARTNERS AND CALCULATION OF FUSION YIELD

A given membrane type may be fused with another membrane of the same type (homofusions) or a membrane of one type is fused with a membrane of a different type (heterofusions). An example of a homofusion would be a fusion of a human erythrocyte ghost with another human erythrocyte ghost. In contrast, a fusion of a human erythrocyte ghost with, for example, a rabbit erythrocyte ghost would be a heterofusion. These two categories of fusion will be discussed below in further detail.

The most convenient way to detect fusion is via the membrane connection (stage B of Fig. 2). This is done using a fluorescent lipid soluble label such as 1,1'-dihexadecyl-3,3,3',3'-tetra-methylindocarbocyanine perchlorate (DiI = stock number D-384 from Molecular Probes, Eugene, OR). This is a convenient label for such systems as erythrocyte ghost membranes because it has a high quantum yield and low bleaching. However, it is not commonly used with cultured cells because it can be toxic to cell metabolism and certain other cellular and physiological functions (Montecucco et al., 1979; Smith et al., 1981).

Calculation of fusion yield is accomplished by counting the number of fluorescent membranes which remain unfused (lone fluorescent membranes) and counting the number of fluorescent doublets or higher numbers of membranes immediately adjacent to one another which bear fluorescence (Fig. 4). The following additional steps are required: i) adjustment of ratio of fluorescent-labeled to unlabeled membranes, ii) adjustment of the concentration of membranes, and iii) the sample field must be examined before the pulse to detect happenstance alignment of two labeled membranes to monitor bias in experimental results. In the case of membrane labels, lateral diffusion takes place on a time scale which permits visual perception to detect fusion events as early as 7-15 sec after a pulse (Sowers, 1984; 1985; 1989). After a half to one minute, the label concentration at all points on the originally unlabeled membrane comes to lateral equilibrium and thus cannot be easily distinguished from two unfused membranes which were originally both labeled but happenstance aligned together. This approach is used both for lipid soluble membrane labels as well as soluble contents mixing labels.

REPRODUCIBILITY OF DATA AND EXPECTED ERROR

Our methodology involves a visual count of a relatively low number of labeled membranes. A sample population of 100 labeled membranes is always selected from the center of the fusion slide (Fig. 1 in Sowers, 1984). Each source of fluorescence is categorized as either a single membrane (i.e. unfused) or two or more adjacent fluorescent membranes (i.e. fused). When repetitions are performed on the same sample of membranes, either by using a new fusion slide or by counting in a different sample area on the same fusion slide, the standard deviation for each data set as a function of the average fusion yield for the same data set is relatively low (Fig. 5).

Fusion yield in ghost membranes from donated human blood may commonly vary 30% around a mean value to, on rare occasions, as much as a factor of

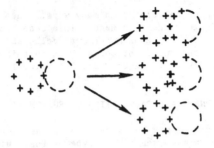

Fig. 4.   Example calculation of fusion yield. Left: Note that the
left member of the membrane pair is labeled with a
fluorescent lipid analog (+) while the right member is
unlabeled (---). While both are observable only with a
light microscope equipped with phase optics, a microscope
equipped with fluorescence optics will detect the left
member. Right: Three fates are possible after a fusion
inducing pulse. Right-upper, lateral diffusion of the
membrane mixing indicator is accompanied by lumen expansion. Right-middle, lateral diffusion of the membrane
mixing indicator is not accompanied by lumen expansion.
Right-bottom, membrane mixing indicator does not move into
originally unlabeled membrane. If right side of figure
represents a sample population then fusion yield is
calculated as number of membrane mixing events divided by
total number of originally labeled membranes multiplied by
100 and rounded off (i.e. 66.6%).

ten depending on blood donor. The source of this variability has not yet
been identified, but when the ghost membranes are made from erythrocytes
from rabbits housed and maintained under a controlled diet, then day to
day variability drops to within experimental error which is quite
acceptable for a very wide range of experiments. Actual data and the
standard deviations calculated from that data (plotted as a function of
fusion yield) shows a quality which is quite acceptable for a very wide
range of experiments (Fig. 5).

From one experiment to another, there are additional sources of
error. The meter setting for pulse voltage and especially the length of
the chamber, since it is hand cut from glass slide, are subject to an
estimated combined error of 4-7%. Error in measuring pulse decay half-
time is limited by analog oscilloscope time base accuracy, but should be
highly reproducible since electrical components were used which had
considerable excess power handling capacity. Routine checks suggest that
there are no major changes from one experiment to the next. Other sources
of variability are described below.

ELECTROFUSION FINDINGS WITH ERYTHROCYTE GHOSTS

Erythrocyte ghosts are preferred over intact erythrocytes as a model
fusion system because they are completely free of cytoplasmic components
and osmotic differences across the membrane are nonexistent or insignificant. They are made from intact erythrocytes by lysis of the plasma
membrane which allows hemoglobin and other residual soluble proteins to

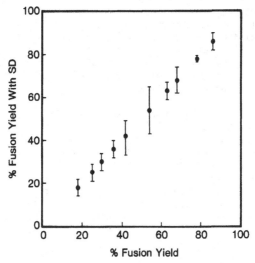

Fig. 5.   Standard deviations calculated from randomly chosen data
           sets containing replications of fusion assays on same
           membranes under same conditions and plotted as function of
           average fusion yield for each set. The number of
           replications range from 4 to 11 per average fusion yield.

escape to the medium (Dodge et al., 1963). After several centrifugations,
both the hemoglobin and the other proteins are removed, thus leaving be-
hind a plasma membrane with the same buffer on both sides of the membrane.
If intact erythrocytes are used, then the electropores, which are also induced by the same fusogenic electric pulse, will allow ion exchange between the internal (cytoplasmic) mileau and the external medium. This
contributes additional non-equilibrium factors to the system and this increases the complexity of the experiment. Also, erythrocyte ghosts can be
loaded with a much broader range of soluble labels. We have found that
when ghost membranes are made and stored at 4^0C, fusion yields and other
properties are relatively unchanged for three to four days after preparation.

    Electrofusion in intact erythrocytes and erythrocyte ghosts has
qualitative aspects which are controlled by buffer strength. Intact
erythrocytes in isotonic phosphate buffer (pH 7.4) all lose their
hemoglobin upon fusion and go through stage C and approach Stage D in Fig.
2 (Sowers, 1985). They do this by first converting from discocytes to
echinocytes and then end as spherocytes before fusing. Erythrocyte ghosts
in low strength (2 - 60 mM) sodium phosphate buffer (pH 8.5), however,
show a more complex behavior. Below a buffer strength of 15-20 mM, human
erythrocyte ghosts membranes go to Stage B, but rarely go to stage C or D
in Fig. 2. However, above a buffer strength of 25-30 mM, most fusions go
to Stage C or D. However, DiI will detect all Stage B, C, and D events
(Sowers, 1988). Fusion yields are roughly proportional to buffer strength
up to strengths of about 30-60 mM (but see below). There is almost no
fusion at 2 mM or below.

Zimmermann (1982, 1986) has recommended an electrofusion strategy which uses very low strength buffers (to enhance dielectrophoretic cell alignment and minimize joule heating in the chamber by the alternating current) and large molecular weight sucrose (for osmotic strength control). However, we have found no evidence of excessive heating. The Zimmermann strategy not only seems to limit experimental possibilities but also has not yet been properly analyzed.

Studies using the CF and PF protocols (Fig. 1) imply, respectively, that in the former the fusogenic change in the membrane is short-lived while in the latter the change is long-lived. When we used the PF protocol to map out the fusogenicity as a function of location on homofused human erythrocyte ghosts we found that only a highly localized area became fusogenic. We also found that this fusogenic change was laterally immobile (Sowers, 1986; Sowers, 1987). This finding implies that whatever the nature of this fusogenic state, it is: a) not completely made of lipids, or b) it is somehow linked to the cytoskeletal system in erythrocytes which restrict the lateral mobility of some plasma membrane components.

Electric pulse-induced electropores are now also recognized to be long-lived entities which have effective diameters that decrease from large sizes relatively quickly. However, after the immediate decrease takes place, further decreases in effective diameter take place at progressively slower rates (Chernomordik et al., 1987; Schwister and Deuticke, 1985; Serpersu et al., 1985).

Further study of the induction of membrane mixing (MM) events and contents mixing (CM) events led to new fundamental knowledge (Sowers, 1988; 1989). For example, discrepancies between measurements of MM events and CM events led to the discovery of CM events which are unrelated to fusion events (Fig. 6). This was made possible by the fact that the induction of close membrane-membrane contact was reversible. Thus, the dielectrophoretic force could be turned off with a switch. This made it possible for Brownian motion to cause unfused membranes to separate and thus actually show that they were unfused.

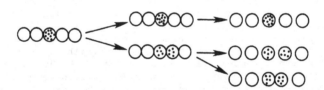

Fig. 6.   Contents mixing events are either associated with fusion or are an artifact. Left: One membrane labeled with a contents mixing indicator aligned into contact with four unlabeled membranes. Middle: Two fates of aligned mixtures of labeled and unlabeled membranes when the dielectropho-resis is maintained after the fusogenic pulse is applied. Middle-upper: No contents mixing. Middle-lower: Movement of some of fluorescence to originally unlabeled membrane to right of the originally labeled membrane. Right: Post-pulse fate when the dielectrophoretic force is removed and enough time is allowed to pass for Brownian motion to cause unfused membranes to separate. Right-upper: No contents mixing, no membrane fusion. Right-middle: Contents mixing but no membrane attachment through a lumen (artifact). Right-bottom: Contents mixing plus membrane attachment through a lumen (valid fusion event).

It has also been observed that erythrocytes (Tsong and Kingsley, 1975) and erythrocyte ghosts (Sowers, 1983; 1984; 1989) undergo a reversible shape change, which always ends in a perfect sphere, when they are exposed to fusion-inducing pulses. Although not quantitated, the shape change occurs as a step change which coincides with the application of a pulse. Conversion of a discocyte-shaped erythrocyte ghost to a spherocytic ghost nearly always requires many pulses (i.e. 3 - 10). It is possible that electroosmosis (see below) may play a role in this phenomenon.

PROPER CONDITIONS FOR ALIGNMENT OF MEMBRANES INTO CONTACT

It is known that the force developed by dielectrophoresis is dependent, among other variables, upon the field strength of the alternating current in the chamber (Pohl, 1978; Hofmann and Evans, 1986). When it is necessary to change the strength of the buffer to examine the effect of ionic strength on membrane fusion, then this affects the amount of voltage needed to generate a given field strength needed to align membranes. This is because the increase in buffer strength causes an increase in conductivity which loads down our alternating current signal source (schematic diagram given in Fig. 3 in Sowers, 1989). For example, when the sodium phosphate buffer (pH 8.5) strength is increased from 20 mM to 60 mM, the alternating field strength must also be increased before alignment will happen as shown in Fig. 7. It is necessary to perform separate experiments to demonstrate that the fusion yield is independent of alternating field strength. Fig. 7 shows that fusion yield is relatively independent of the alternating current field strength at 10 and 20 mM, but much less so at 60 mM.

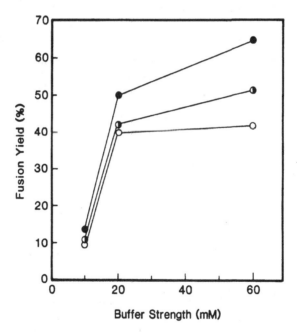

Fig. 7.   Fusion yield in human erythrocyte ghosts in sodium phosphate buffer at pH 8.5 as a function of buffer strength and alternating electric field (60 Hz) used for dielectrophoresis. Alignment field strengths (in V/mm): (O), 6.0-6.75; (◐), 7-7.75; (●), 11.5-12.5.

Fusion yield will also be abnormally low if an insufficient amount of time is allowed for complete alignment to take place. Care must be used in performing these procedures.

Interpretation of Fig. 7 also requires the data in Table 1. Unlike the data for 10 and 20 mM, the data for 60 mM shows a fairly strong fusion yield dependence on alignment field strength. There is very little difference between the field strengths needed for alignment between 10 mM and 20 mM because the load presented by the solution in the chamber is a much smaller fraction of the total internal resistance of the source of the alternating current. Can the difference in the data when the buffer strength is changed from 20 mM to 60 mM be due to the extra heating that takes place at 60 mM? It is clear that at 20 mM a doubling of alignment voltage (from 10 V to 20 V) will also double current and thus quadruple energy input. Since the time to align the membranes is 1/3 as much (1 min compared to 3 mins), the actual effect due, for example, to heating will be 4/3 of that at 20 mM. Yet fusion yields do not change significantly except at just the upper end of the alternating current range and at 60 mM. However, at 60 mM, about 50% more alignment field strength (15.4 compared to 10.5) is needed to just barely bring about alignment and it takes about twice as long (7 mins compared to 3 mins) to occur. Alignment at 60 mM can take place in only 3 mins if the field strength is about 150% of what was needed to barely bring about alignment. This extra voltage will lead to a rate of power input of about 2.2 greater, but in this case the power is transmitted to the chamber over less than half the time. Hence joule heating levels are nearly identical at both ends of the range. The only difference left is alignment force or some other alignment field effect which is accentuated at a buffer strength of 60 mM.

Table 1. Alignment Time and Field Strengths for Different Buffer Strengths

Buffer strength	Range of AC setting* which leads to chains	Corresponding voltage and minimum time for complete alignment
10 mM	20 - 42	10.6 V - 21.0 V 3 mins 0.7 min
20 mM	20 - 42	10.5 V - 21.5 V 3 mins 0.7-1 min
60 mM	26 - 42	15.4 V - 23.2 V 7 mins 3 mins

*Pulse generator circuit given in Fig. 3 of Sowers, 1989.

EFFECT OF PULSE CHARACTERISTICS ON FUSION YIELD

DiI-based fusion yield is presently considered as the most rigorous and unambiguous indicator of membrane electrofusion. As has been previously observed (see reviews), there is a reciprocity between fusion yield and pulse strength or duration. It has been observed previously in a number of cell types that optimum fusion in each cell type requires certain specific pulse strength and duration characteristics (Zimmermann et al., 1981). However, the converse observation that there may be a relationship between fusion yield and biological factors in the cells or membranes when assays are conducted in the presence of identical solution compositions and electric pulse treatments has only recently been presented (Fig. 8).

Fig. 8 shows the homofusion yield for rabbit and human erythrocyte ghost partners in sodium phosphate buffer (20 mM, pH 8.5) (Sowers, 1988; 1989; and unpublished) and for chicken embryo erythrocyte ghosts in 120 mM ammonium acetate buffer (pH 7.4) (Sowers and Kapoor, 1988; Sowers, unpublished) as a function of pulse characteristics. Isofusion contours, plotted after interpolation of data from plots of both fusion yield against pulse field strength and fusion yields against pulse duration permit the relationship between three variables (fusion yield, pulse strength, and pulse duration) to be more clearly visualized. These plots show that use of the same buffer and use of a single 450 V/mm, 0.6 msec decay half-time pulse produces approximately a one order of magnitude difference in fusion yield between human and rabbit erythrocyte ghosts. Furthermore, the rabbit erythrocyte ghosts are much more fusible than the human erythrocyte ghosts and this difference must be due to specific biological factors. Second, at low fusion yields (25% for the rabbit erythrocyte ghosts and 10% for the chicken embryo erythrocyte ghosts), it can be seen that in the former, fusion yield is nearly independent of pulse duration, but, conversely, in the latter, fusion yield is nearly independent of pulse voltage. This indicates qualitatively different effects of the pulse. Third, it also shows the value in exploring the effects of a wider range of pulse characteristics and making a plot of data such as Fig. 8.

Plots of estimated net loss in soluble fluorescent due to electroporation for larger ranges in pulse voltage and duration suggest (Fig. 9): a) a non-linear relationship between electroporation and pulse characteristics, b) an absolute requirement for high voltage before high fusion yields will be obtained, and c) a linear and reciprocal relationship between membrane fragmentation and the product of pulse voltage and pulse duration.

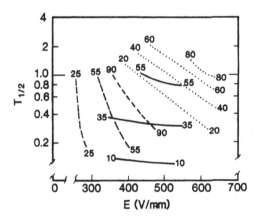

Fig. 8.   DiI-based homofusion yield induced by a single pulse in pearl chains of human, rabbit, and chicken erythrocyte ghosts as a function of pulse strength (E) and pulse decay half-time ($T_{1/2}$) in msec. Membranes are aligned into close contact with AC before a single pulse is applied. Chicken embryo erythrocyte ghosts were in 120 mM ammonium acetate at pH 7.4 (Sowers and Kapoor, 1988). Human and rabbit erythrocyte ghosts were in 20 mM sodium phosphate (pH 8.5).

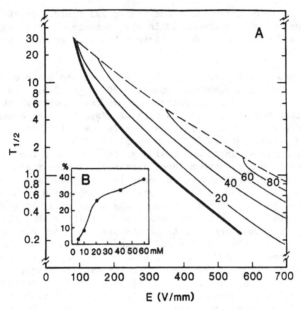

Fig. 9. DiI-based homofusion yield in human erythrocyte ghosts as
induced with a single pulse after alignment into contact
with an alternating (60 Hz) field strength of 10-13 V/mm.
A: Thin solid lines represent isofusion contours interpolated from data obtained as function of pulse strength,
E, and pulse decay half-time, $T_{1/2}$, in msec. Buffer is 20
mM sodium phosphate at pH 8.5. Thick solid line represents
isoloss line for a visually estimated 30% average loss of
MW = 10 kD FITC-dextran from randomly positioned human
erythrocyte ghosts after a single pulse. Dashed line represents threshold combination of pulse strength and duration above which membranes fragment into small vesicles.
B: Fusion yield as function of buffer strength using a
pulse of 600 V/mm strength and 0.6 msec decay half-time
(other conditions same as A). Reproduced from Sowers
(1989).

We have observed that exponentially-decaying pulses with long decay
half-times ($T_{1/2} \gtrsim 2$ msec) and much lower voltages (Fig. 9) can induce
fusion, but they also cause movement of membranes by electrophoresis.
This results in image blur when video data is being recorded and represents a methodological shortcoming associated with this waveform.
Future planned work with square waveform pulses may be able to reduce such
effects.

In human erythrocyte ghosts, pulses which are stronger (in strength
or duration) than the upper fusion yield limit cause fragmentation of the
membrane (Fig. 10). As viewed directly or by playback of recorded video
images, the membrane fragments simultaneously within one or two video
frames (about 17 - 33 msec). The envelope for the pulse regime which
leads to fragmentation is independent of the storage interval between
blood donation and the production of ghost membranes, independent of the
storage interval between production of ghost membranes and assay, and
independent of blood donor (Fig. 10 and see above). When the buffer
strength is changed from 20 mM to 60 mM, the linearly reciprocal

relationship between pulse strength and pulse duration is parallel to that at 20 mM but shifts in position. This indicates that membrane capacitance and solution conductivity may be the sole factors in controlling membrane fragmentation. Biological factors and individual variability must come into play at combinations of pulse strength and duration which are below the fragmentation threshold because individual human blood donors can sometimes show a variability in fusion yield of up to a factor of 3 or more for a given set of solution chemistry and pulse characteristics. When we use blood from rabbits maintained on a constant diet to make the erythrocyte ghosts then this variability shrinks to approximately +/- 10%.

ELECTROOSMOSIS IN ELECTROPORES

There is now good evidence that electroosmosis occurs in electropores during the passage of an electric current. This finding arose out of observations of "asymmetrical permeabilization" of spherical cells when they were exposed to one or more pulses. Mehrle et al. (1985) and Rosignol et al. (1983) showed evidence that a higher degree of permeabilization occurred in the hemisphere facing the positive electrode while our work (Sowers and Lieber, 1986) showed evidence that the increased permeabilization occurred in the hemisphere facing the negative electrode. A later paper by Mehrle et al. (1989) showed evidence that the asymmetry was not as high as originally thought but explained it as due to effects from the field pulse and the intrinsic membrane potential. However, this is not likely to be the case as the pulse-induced membrane potential would be expected to be on the volt level and the intrinsic membrane potential would be expected to be 30 millivolts or less (Fig. 4 in McLaughlin, 1989) compared to the 0.5 volt membrane potential included by a pulse.

However, our hypothesis that electroosmosis plays a role in electropore properties resolves the hemisphere permeabilization discrepancy since it predicts that, as in the Mehrle et al. (1985) and the Rossignol et al. (1983) experiments, a tracer molecule will enter a cell through electropores at the hemisphere facing the positive pole. It also predicts that a tracer molecule already inside a cell will exit from the cell through electropores in the hemisphere facing the negative pole (Sowers and Lieber, 1986).

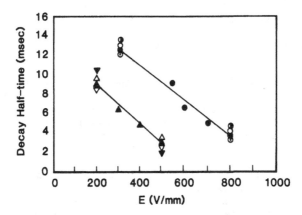

Fig. 10. Membrane fragmentation thresholds in randomly positioned human erythrocyte ghosts in 20 mM (circles) and 60 mM (triangles) sodium phosphate buffer (pH 8.5) as a function of pulse parameters. Different symbol details represent units of blood from different donors.

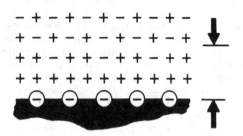

Fig. 11. Diffuse double layer at solid-liquid interface in which
solid surface has fixed net negative charges (McLaughlin
and Mathias, 1985; McLaughlin, 1989). Arrows indicates
thickness of layer. Note that an electric field generated
parallel to surface by a positive pole at left and negative
pole at right will cause all cations in liquid in the
double layer to move to right and thus generate a net
hydrodynamic fluid flow to right.

Electroosmosis, a physical-chemical phenomenon known for over a
century (McLaughlin and Mathias, 1985; McLaughlin, 1989), occurs when an
electric field is induced parallel to a charged surface in contact with an
aqueous solution. If the surface is negatively charged, then charge
neutralization must occur at a layer some distance above the surface as
the cations and anions in the solution redistribute themselves in
accordance with electrostatic requirements. This redistribution will
result in an excess of cations over anions in a layer (Fig. 11) which has
a thickness of 1 nm to 10 nm, for monovalent ion concentrations of 0.1 M
to 0.001 M, respectively. The electric field will cause a net hydrodynamic flow in the direction favored by the charge of the ion in excess and
the polarity of the field.

Since electropores must have a membrane surface in contact with the
medium and since the membrane surface is made up of charged amino acids
and phospholipid head groups with a net negative charge, electroosmosis
would be expected to occur as shown in Fig. 12.

Electroosmosis is expected to be reduced at higher ionic strengths
because of a reduced diffuse double layer thickness (arrow in Fig. 11).
Furthermore, divalent cations will reduce the layer thickness even more
(McLaughlin, 1977). Thus, it would be predicted that at higher ionic
strengths and in the presence of divalent cations the motion of the layer
during electroosmosis would be reduced. Indeed, this conclusion is
reinforced by the results of our experiments on the effect of buffer
strength on electroporation (Sowers, 1986) and two papers which utilized
electroporation to move genetic elements into cells in the presence of
various concentrations of divalent cations (Neumann et al., 1982; Miller
et al., 1988).

An additional test of the hypothesis that the hemisphere permeability
asymmetry is actually an artifact was shown in the experiment in which two
spherical erythrocyte ghosts were separated by a 2-3 um space and on an
axis with the field pulse (Sowers, 1988). One erythrocyte ghost released
its load of FITC-dextran through the hemisphere facing the negative electrode, but the empty erythrocyte ghosts in front of this released load

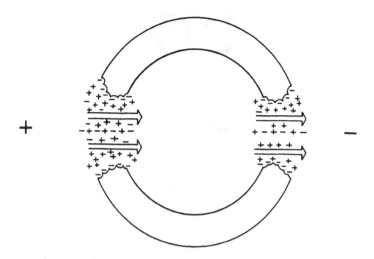

Fig. 12. Electropores cause electroosmosis in each of two hemis-
          pheres which induces a flow into cavity in the positive
          electrode-facing hemisphere and a flow out of cavity in the
          negative electrode-facing hemisphere. Thus markers flow
          into cells in the data from Mehrle et al. (1985) and
          Rossignol et al. (1983) and out of cells in the data from
          Sowers and Lieber (1986).

actually captured a significant amount of these fluorescent molecules
(Fig. 13). The only way this could have happened is if there was a significant amount of electropores in both hemispheres. There may be yet some
other difference in the electropores because of other membrane properties
or system asymmetries, but the permeabilization is generally substantial
in both hemispheres.

FUSION YIELD DEPENDENCE ON IONIC STRENGTH

     In previous cases we have observed that an increase in buffer
strength over the ranges used in our studies will increase absolute (DiI-
based) fusion yield (Sowers, 1984; 1986; 1988; 1989). We have since noted
(see Fig. 7) that at higher ionic strength (i.e. 60 mM sodium phosphate
buffer, pH 8.5) there is a moderate dependence of fusion yield on alignment voltage which could account for our earlier observations. Conversely, at very low buffer strengths, little or no fusion is observed
(Sowers, 1989). Our choice of 20 mM and 60 mM for electrofusion studies
in human erythrocyte ghosts was based on two facts. First, the dielectrophoretic forces in our homogeneous field chambers (Fig. 1 in Sowers, 1984)
obtained with 60 Hz alternating current becomes too weak above 60 mM to
effect alignment. Conversely, below 20 mM not only does the cytoskeletal
network begins to break down after several hours at room temperature, but
fusion yields decrease to the point of degrading the statistics. The one
complicating factor in the use of 20 and 60 mM concentrations of this
buffer on erythrocyte ghosts is that the ghosts are generally spherical or
near spherical at 20 mM but are discocyte-like at 60 mM. The discocyte-
shaped ghosts generally align edge-to-edge in an alternating field. This
has the effect of making their effective diameter slightly larger which
may result, from theoretical studies, in slightly higher induced transmembrane voltages (Fricke, 1953). Beyond that, there may be effects from a
more complex geometry. Our comparisons up to now have generally assumed

Fig. 13. Results of experiment to show similar levels of permeabilization in each hemisphere of spherical membrane exposed to
an electric field pulse. A mixture of human erythroctye
ghosts were used. Some membranes (1 out of 15) were
labeled with MW = 10 kD FITC-dextran while the remainder
(14 out of 15) were faintly labeled with DiI. All membranes
were aligned into pearl chains by dielectrophoresis and
then released from dielectrophoresis and image recordings
started on video tape. Upper: When a faintly DiI-labeled
ghost and a FITC-dextran labeled ghost drifted apart, and
on the electric field axis, to a distance of about 2-3 um,
then a pulse was applied. Middle: Playback of the recorded
images and examination of the video images recorded about
100-120 msec after the pulse showed a blurry cloud of
fluorescence between both membranes. Lower: When this
cloud diffused into the background, the fluorescence level
was seen to increase in the ghost which originally did not
contain FITC-dextran (left member) and the fluorescence
level decreased in the ghost which did originally contain
FITC-dextran (from Sowers, 1988). This is consistent with
the occurrence of electroosmosis in electropores as
depicted in Fig. 11.

that this fact would not make large enough differences to matter in such
studies which are, for the most part, still characterizing the qualitative
aspects of the phenomenology. Future studies are likely to resolve this
experimental shortcoming.

According to Sukharev et al. (1987), there is an increase in fusion
yield with ionic strength, but Blangero and Teissie (1985) and Rols and
Teissie (1989), using mammalian cells, say that just the opposite occurs.
We have no explanation for this except that this may be due to other but
significant differences in the experiment. We also point out that in our
system (Fig. 7), there is relatively little change in fusion yield between
a buffer strength of 20 mM and 60 mM. However, in going from 20 mM to 10
mM, there is a large decrease in fusion yield.

Ionic conditions under which dielectrophoresis will occur has bene-
fited by more theoretical studies than experimental studies and it is felt
by the present author that much remains to be discovered about the inter-
play between dielectrophoresis, ionic strength, electroporation, and
electrofusion. An excellent and recent review of experimental studies of
dielectrophoresis may be found in Iglesias et al. (1989).

How electrofusion occurs is as unknown as for all other systems in which either a natural or an artificial fusogen are known. Early speculations that electropores are the obvious fusion intermediates (Pilwat et al., 1981) have not yet been supported by rigorous experimental data. Our finding (Sowers, 1986; Sowers, 1987) that the long-lived fusogenic state is not laterally mobile implies that if electropores are the intermediate, then they are not totally made of membrane lipids. There must be some membrane protein involvement and it must be linked with the cytoskeleton which restricts the lateral mobility of some membrane components.

On the other hand, a recent clever paper showed a similarity in the envelope of pulse characteristics for both electroporation and electrofusion (Zhelev et al., 1988). This envelope is replotted here as Fig. 14 using the same coordinate axes as for Figs. 8 and 9. It suggests, but

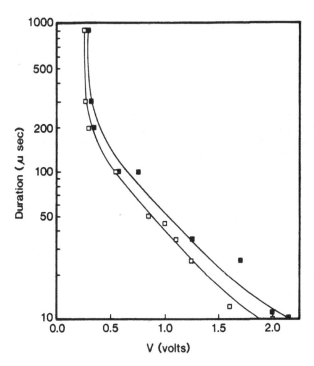

Fig. 14. Replot of data from Zhelev et al. (1988) to match form of presentation in Figs. 8 and 9 of this chapter. Note that overall non linear reciprocity is qualitatively similar between both figures. This data is from experiments which differ in several ways from the data from our experiments which are plotted in Fig. 9. This includes: i) the use of square wave pulses, ii) the use of plant protoplasts, and iii) different chamber, alignment, and scoring procedures. The abscissa is plotted in terms of induced transmembrane potential (Volts) calculated from formula of $V = 1.5rE\cos(a)$, where V is in volts, r is protoplast radius, E is bulk field strength, and $\cos(a)$ is set to 1.0. Contour lines are for initial thresholds for fusion (solid squares) and electroporation (open circles).

does not prove, that electroporation and electrofusion are closely related. Shortcomings in this paper include: i) the same membranes were exposed to multiple pulses, ii) there was no quantitation of electropores (the only measurement was of the pulse characteristic threshold for the initiation of the pores), and iii) fusion was measured in terms of the pulse characteristic threshold for the initiation of fusion. Electropore unknowns still include: number, size, geometry, location, and lifetime. Nevertheless, this study is an important advance because it showed isothreshold lines which generally and qualitatively agree with what we show in Figs. 8 and 9, above. Also, it shows that electrofusion does not have a linear reciprocal relationship with pulse strength and duration.

A second recent advance (Chang and Reese, 1989) was revealed when coordinated freeze-fracture electron microscopy was used to show that electropores are conical or volcano-shaped and have large (25 - 110 nm) diameters, and can be induced at least up to seven pores per square micron of membrane. This study used a "radio frequency" pulse in contrast to most studies which use a direct current pulse (either square or exponentially decaying) and, hence, the results may not be comparable. Nevertheless, this study represents an important advance which should stimulate further research. It is significant that Dimitrov and Jain anticipated a similar geometry for the electropore-type intermediate structure as part of the electrofusion mechanism (Fig. 10 in Dimitrov and Jain, 1984). Other relevant freeze-fracture electron microscopy studies have been done by Sowers (1983) and Stenger and Hui (1986).

A third area which is attracting interest is in the measurement of the membrane potential induced by bulk external electric field with potential sensitive fluorescent dyes (Ehrenberg et al., 1987; Farkas et al., 1984; Gross et al., 1986). In one paper by Kinosita et al. (1988) it was shown that the instantaneous induced transmembrane potential increased very rapidly during the pulse until breakdown and then decreased very rapidly after the pulse. Although this study concluded incorrectly (see Chernomordik et al., 1987) that the pores resealed equally quickly, more information about the induced membrane potential may help us understand more about what new forces are developed during and after the pulse.

In addition to our findings of biological variability and evidence (Fig. 8) that biological factors have significant effects on membrane electrofusibility, several experimental papers (Bliss et al., 1988; Weaver et al., 1988) have also suggested a large inherent biological variability in electroporation in response to a pulse and it may be possible that fractionation of whole populations of cells may lead to the discovery of membrane or cell components or properties which correlate with electroporation. These findings will also lead to advances in our understanding of both electrofusion and electroporation.

The long-lived fusogenic state which we first reported (Sowers, 1983a) has analogies in other membrane systems (Chernomordik et al., 1987; Stulen, 1981; Tomov and Tsoneva, 1988). Further study of these systems may reveal new knowledge which may unify these phenomena.

Although there is now good evidence that the permeabilization is similar in both hemispheres, there is also evidence that other forces are at play in the membrane system. We have seen tethers pulled out of osmotically-swollen, spherical shaped mitochondria inner membranes in response to a direct current pulse (Fig. 12 in Sowers, 1989).

There are major fundamental hindrances to the elucidation of the mechanisms of electrofusion and electroporation. Most of our fundamental knowledge of membrane properties deals with measurements with systems which are done at <u>equilibrium</u>. Since both electroporation and electro-fusion occur as <u>dynamic</u> changes with short lifetimes in very small areas of the membrane, there is little in the existing literature to be helpful. Also a very large number of physical effects are taking place in addition to biologically-relevant effects with which to relate to other experimental variables.

It is possible that electrofusion may proceed in a manner which in not unlike that induced by some biological entities. A short survey of the literature has revealed (Table 2) that the electric field strengths in the neighborhoods of certain proteins is much higher than the field strengths that induce fusion and permeabilization. It will remain for further studies to show whether electric fields are part of the cause of naturally-occurring fusion.

Table 2. Electric Fields and Biologically Relevant Effects

System	Field Strength	Reference
U-cyto c reductase	18,000 V/mm	Margoliash & Bosshard, 1983.
Membrane potential	14,000 V/mm	Woodbury, 1960
Helix-coil conf. change	22,000 V/mm	Kikuchi & Yoshioka, 1976
Lipid head group shift	180 V/mm	Lopez et al., 1988
Permeabilizing & fusing pulse	500-1,000 V/mm	Zimmermann et al., 1981

The use of vesicles as a model membrane has merit on fundamental grounds but we have experienced serious experimental difficulties in their use. First, vesicles do not align very rapidly. The approximate lower limit for particle alignment by dielectrophoresis is about 0.5 - 1.0 um. At those diameters, the particles experience such large Brownian motion forces that alignments are broken as fast as they are formed. When it was possible to make vesicles near the 1.0 um diameter, they were unstable and spontaneously broke down to smaller entities. Second, vesicles are usual-ly made with a wide range of diameters. This results in a heterogeneous pearl chain with different diameters and it confuses the distinctions between heterofusions and homofusions based on diameter. In fact, in agreement with theory, the smaller vesicles exhibited much more violent Brownian motion induced displacement from alignments than larger vesicles. Such diameter differences may play significant roles when one asks the question of what the diameter dependence of the induced membrane potential will be. Third, the vesicles must be unilamellar. The presence of multilamellar vesicles will also complicate matters. Hence, our efforts have been directed at the use of natural cell membranes whenever possible. It should be added that we have had better success with the electrofusion of osmotically-swollen, spherical-shaped mitochondrial inner membranes than vesicle membranes (Sowers, 1983; and unpublished).

ACKNOWLEDGEMENT

Supported by ONR Grant N00014-89-J-1715.

REFERENCES

Bates, G., Saunders, J., and Sowers, A. E., 1987, Electrofusion: princi-
    ples and applications, in: "Cell Fusion," A. E. Sowers, ed., Plenum
    Press, New York.
Berg, H., 1987, Electrotransfection and electrofusion of cells and elec-
    trostimulation of their metabolism, Studia Biophys., 119:17.
Blangero, C., and Teissie, J., 1985, Ionic modulation of electrically-
    induced fusion of mammalian cells, J. Memb. Biol., 86:247.
Bliss, J. G., Harrison, G. I., Mourant, J. R., Powell, K. T., and Weaver,
    J. C., 1988, Electroporation: the population distribution of macro-
    molecular uptake and shape changes in red blood cells following a
    single 50 us square wave pulse, Bioelectrochem. Bioenerget., 20:57.
Breckenridge, L. J., and Almers, W., 1987, Final steps in exocytosis
    observed in a cell with giant secretory granules, Proc. Natl. Acad.
    Sci. USA, 84:1945.
Chang, D. C., and Reese, T. S., 1989, Structure of electric field-induced
    membrane pores revealed by rapid-freezing electron microscopy,
    Biophys. J., 55:136a.
Chassy, B. M., Mercenier, A., and Flickinger, J., 1988, Transformation of
    bacteria by electroporation, TibTech., 6:303.
Chernomordik, L. V., Sukharev, S. I., Popov, V. F., Pastushenko, A. V.,
    Abidor, I. G., and Chizmadzhev, Y. A., 1987, The electrical breakdown
    of cell and lipid membranes: the similarity of phenomenologies.
    Biochim. Biophys. Acta, 902:360.
Dimitrov, D. S., and Jain, R. K., 1984, Membrane stability, Biochim.
    Biophys. Acta, 779:437.
Dodge, J. T., Mitchell, C., and Hanahan, D. J., 1963, The preparation and
    chemical characterization of hemoglobin-free ghosts of human
    erythrocytes, Arch. Biochem. Biophys., 100:119.
Ehrenberg, B., Farkas, D. L., Fluhler, E. N., Lojewska, Z., and Loew, L.
    M., 1987, Membrane potential induced by external electric field
    pulses can be followed with a potentiometric dye, Biophys. J.,
    51:833.
Farkas, D. L., Malkin, S., and Korestein, R., 1984, Electrophotolumi-
    nescence and the electrical properties of the photosynthetic
    membrane: II. Electric field-induced electrical breakdown of the
    photosynthetic membrane and its recovery, Biochim. Biophys. Acta,
    767:507.
Fernandez, J. M., Neher, E., and Gomperts, B. D., 1984, Capacitance
    measurements reveal stepwise fusion events in degranulating mast
    cells, Nature, 312:453.
Fricke, H., 1953, The electric permittivity of a dilute suspension of
    membrane covered ellipsoids, J. Appl. Phys., 24:644.
Gingell, D., and Ginsberg, L., 1978, Problems in the physical inter-
    pretation of membrane interaction and fusion, in: "Membrane Fusion,"
    G. Poste and G. L. Nicolson, eds., Elsevier/N. Holland, Amsterdam.
Glassy, M. C., and Pratt, M., 1989, Generation of Human Hybridomas by
    Electrofusion, in: "Electroporation and Electrofusion in Cell
    Biology," E. Neumann, A. E. Sowers, and C. A. Jordan, eds., Plenum
    Press, New York.
Gross, D., Loew, L. M., and Webb, W., 1986, Optical imaging of cell
    membrane potential changes by applied electric fields, Biophys. J.,
    50:339.

Hofmann, G. A., and Evans, G. A., 1986, Electronic genetic-physical and biological aspects of cellular electromanipulation, IEEE Eng. Med. Biol. Mag., 5:6.

Iglesias, F. J., Santamaria, C., Lopez, M.C., and Domingues, A., 1989, Dielectrophoresis: Behavior of microorganisms and effect of electric fields on orientation phenomena, in: "Electroporation and Electrofusion in Cell Biology," E. Neumann, A. E. Sowers, and C. A. Jordan, eds., Plenum Press, New York.

Inoue, S., 1986, "Video Microscopy," Plenum Press, New York.

Karsten, U., Stolley, P., Walther, I., Papsdorf, G., Weber, S., Conrad, K., Pasternak, L., and Kopp, J., 1988, Direct comparison of electric field-mediated and PEG-mediated cell fusion for the generation of antibody producing hybridomas, Hybridoma, 7:627.

Kikuchi, K., and Yoshioka, K., 1976, Electric field-induced conformational change of poly(L-lysine) studied by transient electric birefringence, Biopolymers, 15:1669.

Kinosita, Jr., K., Ashikawa, I., Saita, N., Yoshimura, H., Itoh, H., Nagayama, K., and Ikegami, A., 1988, Electroporation of Cell membrane visualized under a pulsed-laser fluorescence microscope, Biophys. J., 53:1015.

Knutton S., and Pasternak, C. A., 1979, The mechanism of cell-cell fusion, Trends Biochem. Sci., 4:220.

Knight, D. E., and Scrutton, M. C., 1986, Gaining access to the cytosol: The technique and some applications of electropermeabiliziation, Biochem. J., 234:497.

Liang, H., Prurucker, W. J., Stenger, D. A., Kubiniec, R. T., and Hui, S.-W., 1988, Uptake of fluorescence-labeled dextrans by 10T 1/2 fibroblasts following permeation by rectangular and exponential decay electric field pulses, Biotechniques, 6:550.

Lo, M. M. S., and Tsong, T. Y., 1989, Producing monoclonal antibodies by electrofusion, in: "Electroporation and Electrofusion in Cell Biology," E. Neumann, A. E. Sowers, and C. A. Jordan, eds., Plenum Press, New York.

Lopez, A., Rols, M. P., and Teissie, J., 1988, 31-P NMR Analysis of membrane phospholipid organization in viable reversibly electropermeabilized Chinese hamster ovary cells, Biochemistry, 27:1222.

Margoliash, E., and Bosshard., H. R., 1983, Guided by electrostatics, a textbook protein comes of age, TIBS, 8:316.

McLaughlin, S., 1989, The electrostatic properties of membranes, Ann. Rev. Biophys. & Biophys. Chem., 18:113.

McLaughlin, S., and Mathias, R. T., 1985, Electro-osmosis and the reabsorption of fluid in renal proximal tubules, J. Gen. Physiol., 85:699.

Mehrle, W., Zimmermann, U., and Hampp, R., 1985, Evidence for asymmetrical uptake of fluorescent dyes through electro-permeabilized membranes of Avena mesophyll protoplasts, FEBS Lett., 185:89.

Mehrle, W., Hampp, R., and Zimmermann, U., 1989, Electric pulse induced membrane permeabilisation. Spatial orientation and kinetics of solute efflux in freely suspended and dielectrophoretically aligned plant mesophyll protoplasts, BBA, 978:267.

Miller, J. F, Dower, W. J., and Tompkins, L. S., 1988, Proc. Natl. Acad. Sci. USA, 85:856.

Montecucco, C., Pozzan, T., and Rink, T., 1979, Dicarbocyanine fluorescent probes of membrane potential block lymphocyte capping, deplete cellular ATP and inhibit respiration of isolated mitochondria, Biochim. Biophys. Acta, 552:552.

Neumann, E., Gerisch, G., and Opatz, K., 1980, Cell fusion induced by high electric impulses applied to Dictyostelium, Naturwissenschaften, 67:414.

Neumann, E., Sowers, A .E., and Jordan, C. A., 1989, "Electroporation and Electrofusion in Cell Biology," Plenum Press, New York.

Neumann, E., Schaefer-Ridder, M., Wang, Y., and Hofschneider, P. H., 1982, Gene transfer into mouse lyoma cells by electroporation in high electric fields, EMBO J., 1:841.

Pilwat, G., Richter, H.-P., and Zimmermann, U., 1981, Giant culture cells by electric field-induced fusion, FEBS Lett., 133:169.

Pohl, H. A., 1978, "Dielectrophoresis," Cambridge University Press, London.

Pohl, H. A., Pollock, K., and Rivera, H., 1984, The electrofusion of cells, Int. J. Quant. Chem: Quant. Biol. Symp., 11:327.

Potter, H., 1988, Electroporation in biology: methods, application, and instrumentation, Analyt. Biochem., 174:361.

Rols, M.-P., and Teissie, J., 1989, Ionic-strength modulation of electrically induced permeabilization and associated fusion of mammalian cells, Eur. J. Biochem., 179:109.

Rossignol, D. P., Decker, G. L., Lennarz, W. J., Tsong, T. Y., and Teissie, J., 1983, Induction of calcium-dependent, localized cortical granule breakdown in sea-urchin eggs by voltage pulsation, Biochim. Biophys. Acta, 763:346.

Schwister, K., and Deuticke, B., 1985, Formation and properties of aqueous leaks induced in human erythrocytes by electrical breakdown, Biochim. Biophys. Acta, 816:332.

Senda, M., Takeda, J., Abe, S., and Nakamura, T., 1979, Induction of cell fusion of plant protoplasts by electrical stimulation, Plant & Cell Physiol., 20:1441.

Serpersu, E. H., Kinosita, Jr., K., and Tsong, T. Y., 1985, Reversible and irreversible modification of erythrocyte membrane permeability by electric field, Biochim. Biophys. Acta, 812:779.

Shikegawa, K., and Dower, W. J., 1988, Electroporation of eukaryotes and prokaryotes: A general approach to the introduction of macromolecules into cells, Biotechniques, 6:742.

Smith, T. C., Herlihy, J. T., and Robinson, S. C., 1981, The effect of the fluorescent probe, 3,3'-dipropylthiadicarbocyanine iodide, on the energy metabolism of ehrlich ascites tumor cells, J. Biol. Chem., 256:1108.

Sowers, A. E., 1989, The mechanism of electrically-induced fusion in erythrocyte membranes, in: "Electroporation and Electrofusion in Cell Biology," E. Neumann, A. E. Sowers, and C. A. Jordan, eds., Plenum Press, New York.

Sowers, A. E., 1988, Fusion events and nonfusion contents mixing events induced in erythrocyte ghosts by an electric pulse, Biophys. J., 54:619.

Sowers, A. E., 1987, The long-lived fusogenic state induced in erythrocyte ghosts by electric pulses is not laterally mobile, Biophys. J., 52:1015.

Sowers, A. E., 1986, A long-lived fusogenic state is induced in erythrocyte ghosts by electric pulses, J. Cell Biol., 102:1358.

Sowers, A. E., 1985, Movement of a fluorescent lipid label from a labeled erythrocyte membrane to an unlabeled erythrocyte membrane following electric field-induced fusion, Biophys. J., 47:519.

Sowers, A. E., 1984, Characterization of electric field-induced fusion in erythrocyte ghost membranes, J. Cell Biol., 99:1989.

Sowers, A. E., 1983a, Fusion of mitochondrial inner membranes by electric fields produces inside out vesicles: visualization by freeze-fracture electron microscopy, Biochim. Biophys. Acta, 735:426.

Sowers, A. E., 1983b, Red cell and red cell ghost membrane shape changes accompanying the application of electric fields for inducing fusion, J. Cell Biol., 97:179a.

Sowers, A. E., and Kapoor, V., 1988, The mechanism of erythrocyte ghost fusion by electric field pulses, in: "Proceedings of the International Symposium on Molecular Mechanisms of Membrane Fusion," S. Ohki, ed., Plenum Press, New York.

Sowers, A. E., and Kapoor, V., 1987a, The electrofusion mechanism in erythrocyte ghosts, in: "Cell Fusion," A. E. Sowers, ed., Plenum Press, New York.

Sowers, A. E., and Lieber, M. L., 1986, Electropores in individual erythrocyte ghosts: diameters, lifetimes, numbers, and locations, FEBS Lett., 205:179.

Stenger, D. A., and Hui, S. W., 1986, Kinetics of ultrastructural changes during electrically induced fusion of human erythrocytes, J. Memb. Biol., 93:43.

Stulen, G., 1981, Electric field effects on lipid membrane structure, Biochim. Biophys. Acta, 640:621.

Sukharev, S. I., Bandrina, I. N., Barbul, A. I., Abidor, I. G., and Zelenin, A. V., 1987, Electrofusion of fibroblast-like cells, Stud. Biophys., 119:45.

Teissie, J., and Rols, M. P., 1986, Fusion of mammalian cells in culture is obtained by creating the contact between cells after their electropermeabilization, Biochem. Biophys. Res. Comm., 140:258.

Teissie, J., Knutson, V. P., Tsong, T. Y., and Lane, M. D., 1982, Electric pulse-induced fusion in 3T3 cells in monolayer culture, Science, 216:537.

Tomov, T. Ch., and Tsoneva, I. Ch., 1988, Electroinduction of long-lived membrane potentials in yeasts, Bioelectrochem. Bioenerget., 19:397.

Tsong, T. Y., 1983, Voltage modulation of membrane permeability and energy utilization in cells, Biosci. Reports, 3:487.

Tsong, T. Y., and Kingsley, E., 1975, Hemolysis of human erythrocyte induced by a rapid temperature jump, J. Biol. Chem., 250:786.

Weaver, J. C., Harrison, G. I., Bliss, J. G., Mourant, J. R., and Powell, K. T., 1988, Electroporation: high frequency of occurrence of a transient high-permeability state in erythrocytes and intact yeast, FEBS Lett., 229:30.

Wojcieszyn, J. W., Schlegel, R. A., Lumley-Sapanski, K., and Jacobson, K. A., 1983, Studies on the mechanism of polyethylene glycol-mediated cell fusion using fluorescent membrane and cytoplasmic probes, J. Cell Biol., 96:151.

Woodbury, J. W., 1960, The cell membrane: Ionic and potential gradients and active transport, in: "Medical Physiology and Biophysics," T. C. Ruch, and J. F. Fulton, eds., Saunders, Philadelphia.

Zhelev, D. V., Dimitrov, D. S., and Doinov, P., 1988, Correlation between physical parameters in electrofusion and electroporation of protoplasts, Bioelectrochem. Bioenerg., 20:155.

Zimmerberg, J., Curran, M., Cohen, F. S., and Broderick, M., 1987, Simultaneous electrical and optical measurements show that membrane fusion precedes secretory granule swelling during exocytosis of beige mouse mast cells, Proc. Natl. Acad. Sci. USA, 84:1585.

Zimmermann, U., 1982, Electric field-mediated fusion and related electrical phenomena, Biochim. Biophys. Acta, 694:227.

Zimmermann, U., 1986, Electrical breakdown, electropermeabilization and electrofusion, Rev. Physiol. Biochem. Pharmacol., 105:175.

Zimmermann, U., Buchner, K.-H., and Arnold, W. M., 1984, Electrofusion of cells: recent developments and relevance for evolution, in: "Charge and Field Effects in Biosystems," M. J. Allen, and P. N. R. Usherwood, eds., Abacus Press, Normal, Illinois.

Zimmermann, U., Scheurich, P., Pilwat, G., and Benz, R., 1981, Cells with manipulated functions: New perspectives for cell biology, medicine, and technology, Angew. Chem. Int. Ed. Engl., 20:325.

MICROBIAL FUEL CELL STUDIES OF IRON-OXIDISING BACTERIA

H.Peter Bennetto, D.Keith Ewart,[*] Ali M.Nobar[*]
and Ian Sanderson

Bioelectrochemistry and Biosensors Group
Department of Chemistry
and Biosphere Science Division*
King's College (Kensington Campus)
Campden Hill Road, London, W8 7AH, U.K.

INTRODUCTION

In recent years the bioelectrochemistry of microorganisms has been variously applied to microbial electricity generation,[1,2] biomass assay[3,4] and biosensing.[5,6] Although these studies have been generally limited to the use of organisms in neutral or near neutral media, the possibilities for exploiting alkalophiles in high pH electrochemical systems was also investigated,[7] and the present study formed part of a further exploration of the potential uses of microbial fuel cell techniques under extreme conditions. It focussed on acidophilic species which promote bio-oxidation of arsenopryites (FeAsS) mineral concentrates. This reaction is of importance in a novel gold extraction process in which bacterial oxidation of the arsenical pyrites matrix of refractory gold ores enhances the yield of the metal.[8,9] These organisms derive their energy from the oxidation of reduced sulphur species to sulphate and of Fe(II) to Fe(III), and the latter process was exploited here using a concentration cell in which organisms were placed in the cathode compartment. The electrochemical effects are transmitted by the substrate Fe(II) ions and metabolite Fe(III) ions, which therefore act as mediators. Potentiometric, amperometric and coulometric investigations were conducted as part of a preliminary attempt towards solving one of the problems associated with this technology, namely the assay of bacterial biomass.

THE BIOLOGICAL SYSTEM

The chemolithotrophic ('rock-eating') bacteria studied in this work were from a mixed culture isolated from the acid leachate of a mine in Western Australia. As yet such organisms have been little studied and thus the contents of the culture are not characterised precisely. However, work by the Inorganic Microbiology Group at King's College, who supplied the culture, has identified both rod shaped and spherical bacteria, and the National Collection of Industrial Bacteria has confirmed the presence of a substantial proportion of *Thiobacillus ferrooxidans*.[8] The mixed culture (referred to as "M4") shows acidophilic and moderately thermophilic behavior, showing optimum growth in the pH range 1.3 - 1.6 at 40-45°C. Thus most of the electrochemical measurements were made at 40°C in media with pH 1.5.

The mode of action of *Thiobacillus ferrooxidans* is broadly representative of M4 as a whole, it is useful to consider briefly some aspects of its behavior as an individual organism.[10] It is found in several metal bearing ores including iron pyrites. The cells are straight rods approximately 1.0 μm long and 0.5 μm in diameter. It derives its energy for growth and cell maintenance from oxidation of reduced sulphur compounds to sulphate or Fe(II) to Fe(III). The SO_4^{2-} species is thus the predominant anion in the organism's natural environment, and it is the preferred anion for Fe(II) oxidation and cell growth. For this reason iron(II) sulphate is the ideal substrate for laboratory culture.

The organism is autotrophic and must therefore fix carbon dioxide and, when necessary, nitrogen. It is obligately aerobic, requiring oxygen in the Fe(II) oxidation process, and is quite unusual in that the oxidation of Fe(II) to Fe(III) represents one of the narrowest thermodynamic limits for which growth is known to occur. It has been calculated from bioenergetic considerations that the organism must oxidise 22.4 Fe(II) ions to fix one CO_2 molecule, but the measured values of Fe(II) oxidation are greater than this because CO_2 fixation is not the only anabolic process operating. Although the oxidation process is not well understood, study of it is facilitated by the short Fe(II) to O_2 section of the electron transport chain and the unusually high specific content of transport chain components in the cell membrane. The main components have been resolved by Ingledew[10] and comprise a cytochrome a_1-type oxidase, cytochromes c, the copper containing protein rusticyanin and polynuclear Fe(III).

The organism is an obligate acidophile, only able to grow in the pH range 1.0 - 3.5 (approx.) although it can survive at higher pH. Despite this it has an internal pH close to neutral, as evidenced by the 'normal' pH dependence of those *T. ferrooxidans* enzymes which have been tested and also by direct measurements. The maintenance of such a pH gradient across the external cell membranes requires a large amount of energy. Without substrate or air this energy cannot be generated, the neutral internal pH is not maintained, and the cell wall collapses. In consequence the organisms do not store well out of growth media, and a definite plan for culture growth and harvesting is necessary.

PRINCIPLES OF THE METHOD

The potential established at an electrode immersed in a solution containing ferrous and ferric ions depends on the activities (approximately concentrations) of these species according to the Nernst equation :

$$E_{Fe} = E^0_{Fe} + \frac{RT}{F} \ln [a_{Fe(III)} / a_{Fe(II)}]$$

It is convenient to study biologically induced changes in the redox ratio using a simple concentration cell formally represented by

⁻ Anode | Fe(III)/Fe(II) | | Fe(III)/Fe(II) | Cathode ⁺
+ bacteria

Where suffixes c, a, refer to anode and cathode respectively, and ignoring small liquid junction potentials, it can be shown that the e.m.f., E_{cell} of this cell is given by

$$E_{cell} = \frac{RT}{F} \ln \left(\frac{[Fe(III)]_c \cdot [Fe(II)]_a}{[Fe(II)]_c \; [Fe(III)]_a} \right)$$

Both anode and cathode initially contained equal concentrations of Fe(II) ions, but bacterial oxidation of Fe(II) to Fe(III) in the cathode chamber leads to higher Fe(III) ion concentration at the cathode, and under suitably chosen experimental conditions, the change of E_{cell} with time will give a measure of the rate of Fe(II) oxidation. Alternatively, if current is allowed to flow, Fe(II) ions from cathodic reduction of Fe(III) are re-oxidised by the bacteria (fig.1). Thus, by measuring the current flowing or the associated electrical yield, it is possible, in principle, to monitor the biological activity amperometrically or coulometrically. The maximum rates of Fe(II) oxidation observed in previous work[11] are about 0.09 mol hr^{-1} (g protein)$^{-1}$, equivalent to 2.5 A g^{-1}. This figure gives an estimate of the current-supporting capacity of he bacteria for fuel cell measurements, and may be compared with the value c.a. 1.0 A g^{-1} for a number of carbohydrate oxidising microorganisms.

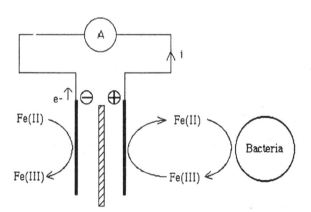

Fig.1.   Schematic representation of reactions occuring in the concentration cell operating amperometrically

Culture and Harvesting of organisms

Initially the bacteria were cultured on iron(II) sulphide pyritic ores, in heated and aerated shake flasks and harvested by centrifugation. However the tendency of M4 to bind to the ore surface led to low yields, whilst the large quantities of solid material in this medium led to difficulties with centrifugal separation. A standard method was therefore devised for culturing the bacteria which gave high yields of biomass with only small quantities of impurities in the form of Fe(III) precipitates.[12] Organisms from ore-grown cultures were sub-cultured in a growth medium (denoted 'GM') which containined $FeSO_4.7H_2O$ (180mM), $CuSO_4.5H_2O$ (8mM), KH_2PO_4 (0.2mM), $(NH_4)_2SO_4$ (1.0 mM) and H_2SO_4 (72mM, pH 1.5). To each litre was added 0.2ml of a solution containing 250ml/ litre of 36% w/w hydrochloric acid and trace metal salts ; MgO (1.2 M), $CaCO_3$ (10 mM), ZnO (5 mM), $CoCl_2.6H_2O$ (5mM), H_3BO_3 (5mM), $Na_2MoO_4.2H_2O$ (5mM). The presence of copper(II) ions in relatively high concentration ensured that heterotrophic organisms did not contaminate the culture. The cultures were grown at 42°C in shake flasks (200 r.p.m.), closed with sponge bungs but with no direct aeration.

The strategy for growth and harvesting of bacterial cultures was aided by an ancillary study in which the growth time of a culture(M4) was monitored over the whole growth cycle. A redox probe (consisting of a platinum electrode and a calomel

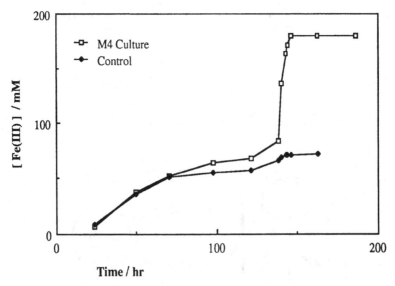

Fig. 2.   Monitoring of bacterial culture by redox potentiometry. The upper plot show characteristic lag,exponential and stationary phases.

reference electrode) was placed in the culture flask, and the redox potential generated by the Fe(II)/Fe(III) couple could then be related to the increasing Fe(III) concentration using a Nernst calibration curve. The Fe(III) concentration, which is a function of bacterial growth, was plotted against time as shown in Fig.2. The control flask contained only GM. The curve representing M4 clearly shows the expected growth pattern.

Harvesting was carried out when the culture was in stationary phase using a Beckmann J2-21 centrifuge. Using 500ml pots the culture solution was spun at 10,000 r.p.m. for 30mins giving a pellet of biomass and any Fe(III) precipitate which had formed in the flasks. This was re-suspended in small quantities (c.a.100ml) of 10mM sulphuric acid. This suspension was placed in 50ml pots and spun briefly at 5000 r.p.m. to remove the Fe(III) solids, then at 15,000 r.p.m. for 20mins to yield a pellet of M4 cells which was re-suspended in GM prior to use in the fuel cells.

Fuel cell measurements

The fuel cells and their manipulation were as previously described.[13,14] Pairs of cell compartments (20ml capacity) were placed 'well to well' separated by a sheet of cation permeable ion exchange membrane (BDH) sealed between 1.5mm thick silicone rubber gaskets. The design of the perspex sections incorporated inlet/outlet holes to allow separate aeration of the chambers and addititon of solutions during operation. The electrodes were made of reticulated 'glassy' carbon with a porosity of 100 pores per linear inch, and were bonded to platinum wire connectors with electrically conducting carbon cement (Ablebond 84-1LMI, Ablestik Laboratories).

Each electrode measured approximately 35 x50 x7mm. Stacks of 3-4 fuel cells were held between two perspex backing plates bolted together, and the whole fuel cell assembly was immersed in a constant temperature water bath (40°C).
The cells were connected to a purpose built variable passive load box which allowed selection of resistances in the range 100W to 1kW. The cell e.m.fs and currents were measured precisely, via the resistance box, using a Solartron 7066 Datastore Voltmeter with capacity for 16 channels. This was controlled by a Research Machines 380Z microcomputer, which also calculated the current and coulombic yield. The measured values were given as a print-out, and as an on-line graphic display which showed values made every 10s or as required during the course of the experiment.

Procedure

The fuel cells were assembled, filled with non-biological solutions, placed in the water bath and were connected to the monitoring equipment. The bacteria were then harvested and re-suspended in the appropriate medium before being injected into the fuel cells immediately before starting each experiment. Solutions were injected using various sizes of plastic syringe fitted with 0.8mm bore metal needles through inlets in the cell compartments. The time taken for final collection, re-suspension and injection of the M4 into the fuel cells was kept as short as possible (c.a. 20mins). The fuel cells compartments were aerated when required by a passage of air passed through 0.05M sulphuric

acid at 40°C and distributed to the inlets via a manifold fitted
with taps for control of the gas flow.

Dependence of Electrochemical Parameters on Biomass Content

A series of experiments was carried out in which GM (16ml)
was placed in each anode compartment, whilst freshly harvested M4
was re-suspended in GM in the cathode compartments at various
dilutions. The cells were loaded (1000W) and left to discharge
for several hours. Samples (0.5ml) of the cathode solution were
removed from each cell by syringe and were prepared for protein
assay as follows. An inorganic salt medium (300ml, pH 1.6) was
added to the sample solution (200ml) followed by 2M NaOH solution
(0.5ml) which solubilised the cells and precipitated the iron
present to prevent it interfering with the spectroscopic
measurements. This mixture was incubated at 90°C for 5 min. For
calibration, solutions of BSA in the range 0 - 100µg/ml were
treated identically. The protein assay was then performed by the
Lowry method[15] to provide a measure of the total protein content
of the system.

Previous work had shown that, under the conditions used, the
majority of bacterial cells were viable, and thus the protein
content of a solution could be related to the number of active
cells present, an important parameter commercially. Organisms
used were in the stationary phase; no growth took place in the
fuel cells, so that the point at which samples were taken from
the fuel cells for assay was not crucial, and it could be assumed
that the 'cell count' would not increase significantly during
measurement.

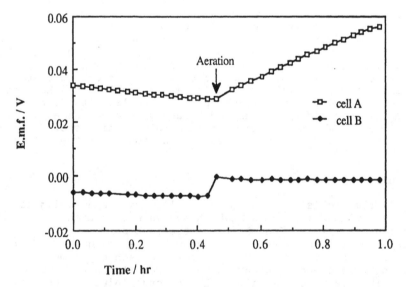

Fig. 3.   Operation of microbial fuel cell in the
           potentiometric mode. Cathode compartments of
           both cells aerated after 4.5 hr.
           (Cell A ; bacteria in the cathode compartment:
           cell B; control, no bacteria).

RESULTS AND DISCUSSION

Potentiometric and amperometric measurements

Preliminary experiments indicated that the behaviour of the organisms was more reproducible when the ratio Fe(II)/Fe(III) was large, and subsequent potentiometric studies were carried out using cells containing equimolar Fe(II) solutions in both cell compartments. Results shown in Fig.3 illustrate the electrochemical effects of bacterial oxidation typically observed. Cell A contained GM in the anode chamber and M4 suspended in an equal quantity of GM in the cathode chamber, whilst cell B contained GM in both chambers ([Fe(II)] = 180mM). Initially, the cell voltages (E_A, E_B) monitored on open circuit with no aeration were relatively constant Cell A showed a small 'bias' potential, probably attributable to minor differences in the cathode arising from the presence of organisms (e.g. adhesion of cells to the electrode, release of small quantities of Fe(III) or other electroactive materials from the cells). However, when both cells were aerated, E_A began to rise significantly as a result of bacteria-assisted oxidation of Fe(II) in the cathode compartment, whilst E_B remained almost unchanged.

To examine the feasibility of the amperometric approach, the cells A, B (above) were discharged after E_A had reached 0.06V. Current was allowed to flow in the external circuit for periods of

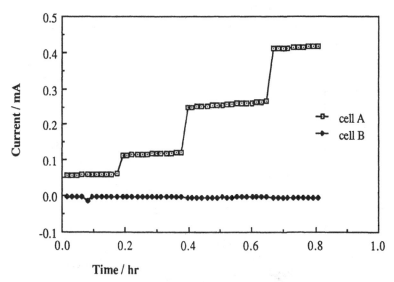

Fig. 4. Operation of microbial fuel cell in the
amperometric mode. Cathode compartments of
both cells aerated throughought.
Cell A (bacteria in the cathode
compartment); currents obtained using loads
of 1000, 500, 200 and 100 ohm: cell B
(no bacteria); current using 1000 ohm.

10-15 min successively through various resistances. Near steady-state (but slowly rising) currents were rapidly established in cell A, whereas the control cell B supported virtually no current (see fig. 4). Subsequently, when the cells were disharged through a 500W resistor for an extended period, a significant biologically-driven current from cell A was maintained for 18hr (after which period it was 0.044mA), peaking at 0.17 mA after 2.5hr. In a similar experiment the current was 0.04mA after 46hr, peaking at 0.128mA after 12 hr.

Coulometric measurements

The analysis of results suggested that the total electrical (coulombic) yield might give a better indicator of active biomass than either the e.m.f.s or the amperometric currents, mainly because of the difficulty in establishing reproducible conditions e.g. elimination of bias potentials, or variations in cell.
Figure 5 illustrates the results of an experiment in which a running total of coulombs from a series of four cells containing different quantities of organisms was observed over a period of 1hr (corresponding to between 2.5 and 3.5 hrs after the M4 pellet was resuspended in GM before injection into the fuel cells). All four fuel cells showed a substantially linear increase of coulombic yield with time.

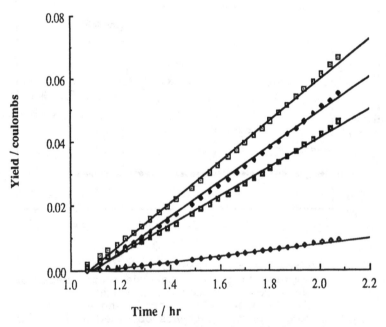

Fig.5. Cumulative coulombic yields from cells
containing different quantities of organisms

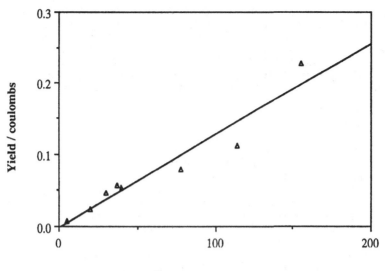

No. of Cells / 10^7

Fig. 6 Correlation of coulombic yields with cell
count. The gradient gives a measure of the
Fe(II)-oxidising power of the organisms over
the range studied.

Correlation Between Coulombic Yield and Biomass

The increases in coulombic yield in unit time (i.e. the
slopes of the lines in fig.5) give a measure of the different
quantities of bacteria contained in each catholyte solution.
Since the number of cells corresponding to a particular
amount of protein had been measured previously as 1mg protein
= 1.84×10^6 cells, the slopes could be directly related to
the 'cell count' of the solutions, as shown in Fig.6, which
includes the data from Fig.5, together with the results from
a separate series of experiments. The gradient, 1.3×10^{-10} C
hr^{-1} $cell^{-1}$, is equivalent to 2.4 C hr_{-1} (g protein)$_{-1}$ or
0.67 A g^{-1}. This figure compares with the peviously-mentioned
optimum values interpolated from the literature (2.5 A g^{-1}),
which relates, however, to stimulated oxidation rates of
organisms in the exponential phase. The value obtained from
the present work therefore appears to be in agreement with
predictions, and also indicates that the translation of
oxidising power to current in the fuel cells was probably
reasonably efficient.

Conclusions

It appears that the microbial fuel cell could be
usefully used for rapid assay in these biological systems,
although it is important that the effects of contaminants
should be studied fully in further work.(In particular, work
on the effect of arsenic on the system is in progress, since

arsenic in reduced form is commonly present in the ores used commercially). In addition, there is clearly potential for the wider use of these methods in the study of various aspects of the bacterial behavior. By suitable design of experiments and selection of conditions the voltage, current, and coulombic profiles of the fuel cells could yield information on the underlying biological processes.

REFERENCES

1. H.P.Bennetto, Microbial Fuel Cells, in: *Life Chemistry Reports*, Vol.2, no.4, eds. A.M.Michelson and J.V.Bannister, (Harwood Acad. Press, London, 1984), pp.363-453
2. G.M.Delaney, S.D.Roller, H.P.Bennetto, J.R.Mason,, J.L.Stirling and C.F.Thurston, Microbial Fuel Cells, in: *Charge and Field Effects in Biosystems,* eds. M.J.Allen and P.N.R.Usherwood (Abacus Press, Tunbridge Wells, 1984) pp. 507-514.
3. R.A.Patchett, A.F.Kelly and R.G.Kroll, Use of a microbial fuel cell for the rapid enumeration of bacteria. *Appl. Microbiol. Biotechnol.* 28, 26-31, (1988).
4. A.Swain, Rapid microbial assay technology, *Ind. Biotechol.* 8, 11-15, (1988)
5. H.P.Bennetto, J.Box, G.M.Delaney, J.R.Mason, S.D.Roller J.L.Stirling and C.F.Thurston, Redox-mediated electrochemistry of whole microorganisms: from fuel cells to biosensors, in: *Biosensors:Fundamentals and Applications*, eds. A.P.F.Turner, I.Karube and G.S.Wilson, (Oxford Univ.Press, Oxford, 1987) 291-314.
6. D.M.Rawson, Whole cell biosensors, *Ind.Biotechnol.,*8, 18-22, (1988).
7. T.Akiba, H.P.Bennetto, J.L.Stirling and K.Tanaka, Electricity production from alkalophilic organisms, *Biotechnol.Letts.*, 9, 611-6, (1987).
8. A.M.Nobar, D.K.Ewart, L.Alsaffar, J.Barrett, M.N.Hughes and R.K.Poole, Isolation and characterisation of a mixed microbial community from an Australian mine: application to the leaching of gold from refractory ores. In *Biohydrometallurgy*, Proc.Int.Symp., Warwick, 1987, eds.P.R.Norris and D.P.Kelly (STL, Kew, 1988) pp.530-1.
9. J.Barrett, D.K.Ewart, M.N.Hughes, A.M.Nobar, D.J.O'Reardon and R.K.Poole. The bio-oxidation of gold-bearing arsenical minerals. *In Research and Development for the Mineral Industry* (W.A.School of Mines, Kalgoorlie, in press)
10. W.J. Ingledew, Bioenergetics of an acidophilic chemolithotroph, *Thiobacilluss ferrooxidans. Biochimica and Biophysica Acta,* 683, 89-117, (1982).
11. S.B.Yunker and J.M.Radovich, Enhancement of growth and ferrous ion oxidation rates of *T. ferrooxidans* by electrochemical reduction of ferric ion. *Biotechnol.Bioeng.*, 28, 1867-75 (1986).
12. W.J.Ingledew and J.G.Cobley; A potentiometric and kinetic study on the respiratory chain of ferrous ion-grown *Thiobacillus ferrooxidans. Biochimica and Biophysica Acta,* 590, 141-58, (1980).
13. S.D.Roller, H.P.Bennetto, G.M.Delaney, J.R.Mason,, J.L.Stirling and C.F.Thurston, Electron-transfer coupling in microbial fuel cells,1; comparison of redox mediator reduction rates and respiratory rates of bacteria, *J.Chem.Tech.Biotechnol.*, 34B,3-12, (1984).

14. G.M.Delaney, H.P.Bennetto, J.R.Mason, S.D.Roller,
    J.L.Stirling and C.F.Thurston, Electron-transfer coupling
    in microbial fuel cells,2; performance of fuel cells
    containing selectred microorganism-mediator-substrate
    combinations, *J.Chem.Tech.Biotechnol.*, 34B,3-12, (1984).
15. O.H.Lowry, N.J.Rosebrough, A.L.Farr and R.J.Randall;
    Protein measurement with the folin phenol reagent.
    J.Biol.Chem. 193, 265-75, (1951).

IN VIVO VOLTAMMETRY WITH AN

ULTRAMICROELECTRODE

Kazuko Tanaka and Noriko Kashiwagi

Biophysical Chemistry Laboratory
The Institute of Physical and Chemical Research
Hirosawa, Wako, Saitama 351-01, Japan

INTRODUCTION

Studies of the transport properties of various substances through biomembranes give useful information in many aspects, particularly when they are
made in vivo.   Recent developments in electrochemistry have enabled us to
observe translocation of molecules and ions across a living cell membrane.
Ultramicroelectrodes fabricated from a single carbon fiber of micrometer
dimemsions have received much attention from both the perspective of
fundamental electrochemical work and their use as in vivo voltammetric
probes[1,2,3].   They possess unique and advantageous properties for in vivo
application, including extremely low capacitive current, high rates of mass
transport and much reduced ohmic effects.   Previouly we reported the suitability
of carbon fiber electrodes for the in vivo monitoring of the uptake of
electrochemically active substances such as an electron transfer mediator,
methoxy-phenaziniummethylsulphate[4].   In this paper we report uptake
behaviours of cadmium and mercury ions into living cells measured by the
ultramicroelectrode technique.

EXPERIMENTAL

Apparatus

Carbon fibre electrodes were prepared from M40-3K-99 (Toray Industries Inc.
Japan), having a diameter of 5-6 μm. The fibers were soaked and washed in ethanol,
and dried in air at room temperature. A glass tube was drawn using a pipet puller

(Narishige Scientific Instrumental Lab. Japan) to obtain a capillary with a tip diameter of less than 10 micrometers. A single fibre was then inserted into the capillary until the fiber passed through the capillary end and then sealed with epoxy resin. The electrode thus prepared can be used for several times when the disk surface has been polished prior to each run. A platinum wire electrode was used as the auxiliary electrode and the reference electrode was Ag/AgCl immersed in a 3M potassium chloride solutions. Differential pulse voltammograms (abbreviated as DPV) were measured with a Model 312 Polarograph (Fuso Seisakusho, Japan) connected to a Potentiostat Model 972 (Fuso Seisakusho)

Procedure

The single cell examined was a giant alga, Chara australis. An internodal cell isolated from the adjacent cells was placed in a polyacrylate vessel with a pool filled with artificial pond water (abbreviated as APW, an aqueous solution containing 0.1 mM of NaCl, KCl and $CaCl_2$). The microelectrode was inserted into a vacuole of the Chara cell in APW, and the auxiliary and reference electrodes were placed in the pool outside of the Chara cell. The APW in the pool was replaced by a sample solution after protoplasmic streaming of the cell was restored. All measurements were done in a faraday cage.

Preparation of an Hg Coated Electrode

Plating of a mercury film on the carbon fiber electrode proceeded by applying a potential of -1.0 V in a 10 μM $HgCl_2$ solution for about 30 minutes.

Determination of Cadmium Ion Concentration by Inductively Coupled Plasma Atomic Emission Spectrometry

Chara cells were immersed in 0.5 mM cadmium chloride or nitrate solutions for 10, 20, and 30 minutes, then, washed thoroughly to exclude the contamination from the surface of the cell and placed horizontally on a plastic plate. After the water on the cell surface was evaporated sufficiently, both ends of the cell were cut off with a pair of scissors and artificial vacuole solution which contained 90 mM KCl, 40 mM NaCl, 15 mM $CaCl_2$, and 10 mM $MgCl_2$ flowed into the vacuole from one end, pushing out vacuole solution to the other end. Vacuole solutions from four or five cells were collected for one analysis of cadmium ions by inductive coupled plasma atomic emission spectrometry (abbreviated as ICP-atomic emission spectrometry) to obtain the sufficient amount of vacuole solution for the measurement since the amount of vacuole solution drawn from one cell was about 25 μl.

RESULTS

Uptake of Cadmium by Chara Cells

Cadmium ions in aqueous solutions of cadmium nitrate and cadmium chloride show well defined DPV with a peak at around -0.8 V vs. Ag/AgCl with a carbon fiber microelectrode as shown in Fig. 1. An example of DPV measured in the Chara cell immersed in a 0.5mM cadmium chloride solution is shown in Fig. 2, where the time was counted since the pool solution was replaced from APW to a 0.5mM cadmium chloride solution. The Chara cell was left in APW for about 30 minutes after the microelectrode was inserted and a DPV taken during this period showed no peaks (curve a in Fig.2). A sharp current rise at about -0.8V was observed when DPV was measured at two minutes after the replacement of the solution, which was followed by a rounded peak with a peak potential at around -0.9V as shown in Fig. 2, curve b. Then, the peak current decreased and the shape of the DPV became more rounded as the time passed (curves c and d in Fig. 2). The distortion of the DPV shown in this figure may be an indication that cadmium ions undergo some modification reaction such as complex formation as soon as they have been translocated in the vacuole of the Chara cell. The protoplasmic streaming of the cell was still very active at 30 minutes after the immersion into a 0.5 mM cadmium chloride solution.

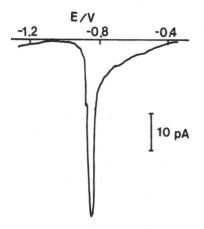

Fig. 1. Differential pulse voltammogram of cadmium ions (10 μM) in
aqueous solutions. Differential pulse
amplitude, 20 mV; scan rate, 10 mV/s.

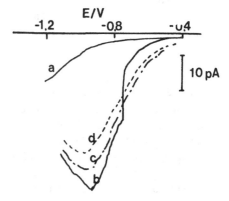

Fig. 2. Differential pulse voltammograms
in a Chara cell immersed in APW(a), and
those in a 0.5 mM CdCl$_2$ solution taken at
2 minutes (b), 4 minutes (c), and 6 minutes
(d) after the replacement.
Differential pulse amplitude, 20 mV;
scan rate, 10 mV/s.

No appreciable peak was observed when a Chara cell was immersed in a cadmium nitrate solution of the same concentration even though the Chara cell was left for over 10 minutes in the solution, indicating that the cadmium ion was not uptaken by Chara cells from cadmium nitrate solutions.

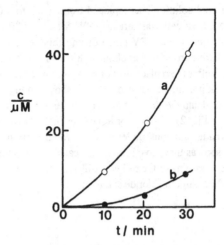

Fig. 3. Time course curves of the translocation of cadmium ions into
the vacuole of a Chara cell immersed in 0.5 mM cadmium solutions
of cadmium chloride (curve a) and nitrate (curve b) determined by ICP-atomic emission
spectrometry.

The voltammetric measurements show that cadmium ions in cadmium chloride
solutions are translocated in Chara cells.   In order to get more quantitative
information we applied ICP-atomic emission spectrometry to determine the amount of
cadmium ions translocated into the vacuole of the Chara cells, and the result was shown
in Fig. 3, where the amount of cadmium ions found in a vacuole solution of a Chara
cell was plotted against time since the Chara cells were immersed in 0.5 mM solutions
of cadmium chloride or cadmium nitrate. The cadmium ion concentration in vacuole
solutions increased significantly when the Chara cell were immersed in cadmium
chloride solutions, though from the voltammetric measurement decrease in peak
current was observed as the time increased(Fig. 2). A probable explanation for this
discrepancy is that the modified form of cadmium ions in vacuole solutions gradually
became electrochemically inactive or irreversible, hence the peak current was
decreased. Very slow uptake of cadmium ions from cadmium nitrate solutions as
shown in Fig. 3 curve b agreed to the result obtained by voltammetry.

Why cadmium ions in chloride solution were uptaken quickly and those in nitrate
media were not?   It is known that cadmium ions in aqueous solutions form
complexes and the pK values of the following equilibrium
$$Cd^{2+} + X^- = CdX^+$$
was reported to be 2.0 for chloride and 0.4 for nitrate[5]. Therefore, cadmium in
cadmium chloride stays in solution mainly as chloro-complexed forms such as $CdCl^+$,
$CdCl_2$, while cadmium in cadmium nitrate as Cd^{2+} ions. This difference in solvated
state may cause the difference in uptake behaviour of cadmium ions by the Chara cells,
who allow chloro-complexed cadmium ions to pass through the cell membrane.

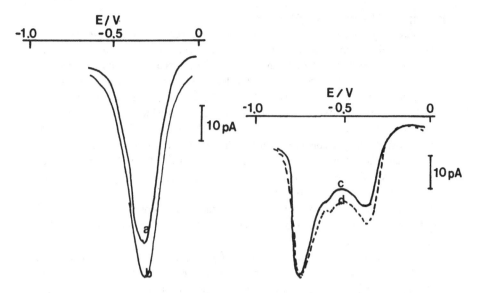

Fig. 4. Differential pulse voltammograms in a Chara cell immersed in a 0.5 mM HgCl$_2$ solution, taken at 3 minutes(a) and 5 minutes(b), 7 minutes(c), and 9 minutes(d) after the replacement of the pool solution. Differential pulse amplitude, 20 mV; scan rate 10 mV/s.

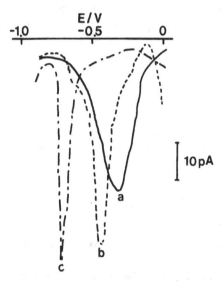

Fig. 5. Differential pulse voltammograms obtained at an Hg coated electrode in APW(a), in a Chara cell immersed in APW taken at 2 minutes(b), and 4 minutes(c) after the insertion. Differential pulse amplitude, 20 mV; scan rate, 10 mV/s.

Behaviour of Mercuric Ions in Chara Cells

An evidence of the translocation of mercury ions into a Chara cell was observed when the Chara cell was immersed in a mercuric chloride solution as shown in Fig. 4. A peak appeared several minutes after the pool solution was changed from APW to an o.5 mM $HgCl_2$ solution, then, the peak was devided in two with the new peak at around -0.8V about seven minutes later. Thus, Chara cells seem to uptake mercuric ions very quickly and then modify them to form some complexes which have more negative redox potentials than that of mercuric ions. The amount of mercuric chloride translocated into the vacuole of the Chara cell, however, could not be estimated from the peak current of the DPV, since the surface of a carbon fiber disk electrode was easily coated with mercury and the peak current was not proportional to the concentration of mercuric chloride in solutions.

It is known that the most of mercuric chloride in solution is present as undissociated molecules, therefore, this is another evidence that a chloro-complex can penetrate through the cell membrane.

A similar behaviour was observed when an Hg coated electrode was inserted into a Chara cell immersed in APW. A peak with a peak potential at about -0.3V obtained in APW (curve a in Fig. 5) was shifted to negative potentials as shown in Fig. 5, where curve b was obtained two minutes after and curve c four minutes after the Hg coated electrode was placed in the vacuole of a Chara cell. The characteristic voltammogram of an Hg coated electrode was restored when the electrode was withdrawn from the Chara cell and left it in APW solution for several minutes, although the peak current was decreased slightly.

Our experiments clearly show the usefuleness of the voltammetric measurements using ultramicroelectrode to measure in vivo uptake behaviour of electrochemically active substances such as heavy metal ions. It is surprising how fast metal ions are translocated into the cell across a living cell membrane, and undergo some modification reaction in the cell. The characterization of a modified form of these metal ions in the cell is now in progress

References

1. F. Gonon, M. Buda, J. F. Pujol, Treated Carbon Fiber Electrodes for Measuring Catechols and Ascorbic Acid, in "Measurement of Neurotransmitter Release In Vivo", C. A. Marsden, ed., John Wiley & Sons Ltd (l984).
2. R. M. Wightman, Microvoltammetric Electrode, Anal. Chem., 53: 1125A(1981).
3. A. C. Michael, J. B. Justice Jr., Oxidation of Dopamime and 4-methylcatechol at Carbon Fiber Disk Electrodes, Anal. Chem., 59: 405(1987).
4. K. Tanaka, N. Kashiwagi, In Vivo Voltammetry with an Utramicro-electrode, Bioelectrochem. Bioenerg., 21:95(1989)
5. E. C. Righellato, C. W. Davies, The Extent of Dissociation of Salts in Water, Part II Uni-Bivalent Salts, Trans Faraday Soc., 26: 592(1930).

MEASUREMENTS OF THE ELECTRICAL IMPEDANCE OF LIVING CELLS

IN THE FREQUENCY DOMAIN

Robert Schmukler

Center for Devices and Radiological Health
Food and Drug Administration
12721 Twinbrook Parkway
Rockville, MD 20857

Introduction

Electrical impedance measurements of biologic materials have been performed for at least 75 years. The first systematic studies were done by Hober (Schanne and Ruiz-Ceretti, 1978) in 1910. Later investigators include Fricke (Fricke, 1924, and Fricke, 1925), Oncley (Oncley, 1938 and Oncley, 1942), K.C. Cole (Cole, 1928, and Cole, 1968), and Schwan (Schwan, 1957 and Schwan, 1985). All of these earlier studies utilized cell suspensions and the suspension equation. This equation was first derived by Maxwell (1873) for dielectric spheres in a conducting medium and modified by Fricke (Schanne and Ruiz-Ceretti, 1978) to account for the non-spheroidal geometry of human erythrocytes (HRBC), viz,

$$\frac{\dfrac{\sigma}{\sigma_1}-1}{\dfrac{\sigma}{\sigma_1}+1} = \rho' \; \frac{\dfrac{\sigma_2}{\sigma_1}-1}{\dfrac{\sigma_2}{\sigma_1}+\gamma} \qquad [1]$$

Where, σ_1 = medium conductivity, σ_2 = suspended particle conductivity, σ = conductivity of resulting suspension, ρ' = volume concentration of particles, and γ = form factor dependent on suspended particle geometry. These studies were primarily concerned with the existence and physical basis of the electrical characteristics of biologic material. In the last few years, it has become apparent that the processes and the kinetics of the processes occurring at the cell membranes are important to the understanding of the electrical properties (impedance) of living cells. Schwan (Schwan, 1985) speaks of the counter ion relaxation at cell membranes, which from an electrochemical viewpoint, is similar to ion adsorption kinetics (specific and non-specific) at the cell membranes. The difference between the two processes is that counter ion relaxation has ion movement tangential to the cell membrane and adsorption generally involves ion movement perpendicular to the cell membrane. In addition, electrochemists consider specific adsorption due to affinity sites on the membrane (i.e. receptor sites).

A higher sensitivity technique, than the earlier work on suspensions has been previously described (Schmukler, 1981, Schmukler, et al., 1982 and Schmukler, 1986), by which the impedance of isolated living cells can be evaluated under physiologic conditions over a wide frequency range (1Hz-5MHz) will be presented in detail here. The increased sensitivity derives from the reduction of current shunt pathways around the cells, during the measurement, by imbedding the cells in a polycarbonate (insulating) filter with well defined cylindrical pores. A pseudo-epithelium is achieved by the entrapment of cells from a suspension in these pores by using a small (20-40mmHg) hydrostatic pressure.

Impedance Chamber

Impedance measurements are performed in a modified electrochemical conductance chamber with six electrodes. The chamber provides: a means of maintaining the cells under physiologic conditions during the entire measurement, the ability to vary the temperature, a high resistance seal to the porous polycarbonate film, electrolyte resistivity compensation, and access for introducing drugs or chemical agents. The chamber, (see figure 1) consists of a vertical cylinder of polycarbonate with a 1 cm inside diameter, and 2 cm length. The upper and lower half of the chamber each contains : 1 axially placed porous carbon electrode for current input, 2 radially placed voltage measuring electrodes of porous tantalum, 2 ports (1 radial, 1 axial), 1 water jacket, and 1 teflon flat ring for sealing between the two halves. All six electrodes, two carbon and four tantalum, have large surface areas which reduce electrode polarization effects at all the electrodes (see section on Electrodes). The carbon electrodes, for the current input, form the upper and lower ends of the cylinder and are 2 cm apart. The vertical separation, between the upper and lower voltage measuring electrodes is 2 mm, while that between one voltage electrode and one resistivity compensation electrode is 1.5 mm. All tantalum electrodes are recessed into the polycarbonate so that they lie completely outside the main chamber. This is done in order that perturbations to the current distribution in the chamber at the point of voltage measurement are minimized (Schwan, 1963). When the current electrodes are separated by a distance 2x the diameter of the chamber, the current flux distribution at the point of voltage measurement is uniform and linear (Eisenberg, 1970, Rall, 1969, Robillard, 1979 and Schwan, 1963).

Between the upper and lower halves of the chamber, a polycarbonate Nuclepore (TM) filter is clamped between the two teflon rings. When the cells are embedded into the filter a pseudo-epithelium is created. The filter is a sheet of approximately 13 μm thick polycarbonate, transversed by true cylindrical pores. The resistance of the polycarbonate film is very high (10^{12} Ω), so that the only pathway for current conduction between the upper and lower halves of the chamber is the electrolyte filled pores of the filter. Edge sealing of the filter to the teflon is checked by using a polycarbonate sheet with no holes. When this is measured the resistance at the edge seal was found to be >10^{9} Ω. Both halves are water jacketed so that temperature control of the cells and the electrolyte inside the chamber is possible. The access ports provide a means of changing or adding to the electrolyte around the pseudo-epithelium.

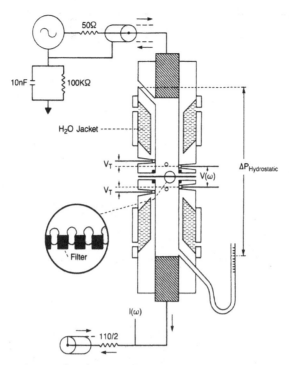

Figure 1 Schematic of Impedance Chamber

Care is taken to insure that the filter is fully wetted and that no air bubbles exist in the system. First a measurement is taken without the filter. This determines R_o, the ohmic (IR) drop between the measurement electrodes due to the electrolyte (buffer) in the chamber. The chamber is carefully disassembled and then reassembled with a filter between two halves and another measurement taken. The last measurement determines R_L^0, which is the electrolyte filled filter resistance. Both Measurements are used to determine the system stray capacitances and C_f, the filter capacitance (Chiabrera, 1985). The measurement of R_o and R_L^0 provides a check on the linearity of the system. (see figure 2) There are no frequency dependent phenomena within the frequencies of measurement in an electrolyte solution impedance, just a simple resistance, so that the voltage and current should be exactly in phase and the amplitude constant.

The final step in preparation of the pseudo-epithelium and subsequent measurement of the cell impedance is to occupy the pores of the filter with the cells. A suspension of washed cells containing approximately 1×10^6 cells, (3 x total pores present) is placed in the upper half of the chamber and a hydrostatic pressure head applied from below. The data presented here are measurements on Human Red Blood Cells (RBC), in which no anti-coagulants are used. No anti-coagulents were necessary because a few drops of blood were directly diluted into 15 ml of buffer. Other cell types can also be employed. The cells are allowed to move under the influence of gravity (sedimentation) and the pressure differential (connective volume flow) until they occupy the pores of the filter. The pressure head forces the cells into the pores and since the current pathways across the filter are restricted to the pores, this forces a majority of the current through the cells. Thus there is an increase of several fold, in impedance measurement sensitivity compared to suspension techniques, due to the large increase in leak resistance around the cells.

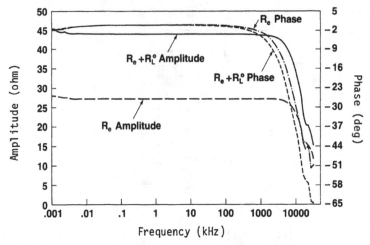

Figure 2 Plot of Amplitude and Phase of R_e, and $R_e + R_L^0$ vs log frequency

The determination of the relative contribution of multiple cell layers to the measured electrical impedance of the pseudo-epithelium is extremely important because of current pathway considerations (Richardson, 1972). The multiple cell layers are formed as a result of possible sedimentation of cells on top of the filter, and are in series with respect to current pathways with the pore-embedded cells. This can result in the inclusion of a multiple cell layer component to the measured impedance. If, however, the conductivity of the leak pathway around the cells in the multiple layers above the pseudo-epithelium is sufficiently high, the relative contribution to the measured impedance will be small. This was determined by allowing a very large quantity of cells (100 x's normal amount) to sediment onto the pore embedded cells without the presence of the hydrostatic pressure head. The presence of these multiple layers of cells above the filter resulted in an identical impedance to that seen without any cells present. During an experiment the multiple cell layers above the filter are subjected to an isotropic pressure due to the depth of fluid above the filter. Since the pressure head is attached from below, the isotropic pressure is small and this condition is analogous to the absence of a pressure head. Thus, the cell layers above the filter during an impedance measurement do not contribute to the measured impedance and can be neglected (Schmukler, 1981). The instrumentation used to measure the impedance is a Model 1260 Solartron Impedance / Gain-Phase Analyzer. The current and voltage amplifier are AC coupled (-3dB, 1Hz).

Electrodes

One of the major problems associated with electrical measurements that involve an electrode/electrolyte interface is the problem of electrode polarization. For any irreversible sensing electrode, polarization produces a non-linear voltage drop which is a function of current density (not total current) and frequency. Additionally, measurement problems are associated with faradaic and non-faradaic currents. Faradaic currents result from oxidation/reduction reactions on the electrode surface. Specific adsorption/desorption processes (usually involving hydrogen and oxygen) produce non-faradaic currents depending on the composition of the electrode.

Two complementary approaches are used here to reduce current densities at the electrodes and provide a wider linear frequency range for the measurements. The first is the 4-electrode technique, where the input current is carried by the 2 current electrodes (carbon), and the measurements of voltage are made at 2 separate electrodes (tantalum). The current densities at the voltage electrodes are greatly reduced by the high input impedance (10^6 Ω) of the measuring amplifiers. By this approach the current at the points of voltage measurements is reduced by 4 orders of magnitude when compared to the input current. This produces a concommitant reduction in current density at the voltage electrodes. In order to prevent the bias currents from the measuring amplifiers from polarizing the voltage electrodes, a decoupling (series) capacitor is used. The low frequency AC cutoff (-3dB point) is 1Hz. By AC coupling the current measuring amplifier with the same cutoff point, most of the effects of AC coupling at low frequencies are negated. (see figure 2)

The second method used to further reduce current density, is to increase the active surface area (SA) of all the electrodes for the given current load. This is why platinizing a platinum (Pt) electrode is done, the active surface area is increased with a tiny increase in geometric size. By exploiting the technology of powder metallurgy to produce totally porous voltage electrodes, extremely high surface areas can be achieved in small electrodes (Beard, et al. 1972, DeRosa, 1969, DeRosa, 1974 and Schmukler, 1976). The technique uses a high pressure (50 klb/in^2) uniaxial compression of high surface area Tantalum (Ta) powder onto a roughened end of an annealed Ta wire to produce the porous electrode. Both the Ta powder (5000 $\mu F/g$) and wire are metallurical grade (Cabot Corporation, Boyertown, Pa.). The high compaction pressure cold welds the Ta particles together. A high mechanical material (green) strength is not necessary and, since sintering to increase green strength substantially reduces the porous electrode SA, the electrodes were used without further processing. From previous work on porous Pt electrodes, axial compression reduces the SA of the powder by 1/2-1/3 (DeRosa, 1969, Derosa, 1974). The surface areas for these porous Pt electrodes, measured by BET gas adsorption, was between 7 and 10 m^2/g (Beard et al., 1972, DeRosa, 1969, DeRosa, 1974, and Schmukler, 1976). The resultant porous Ta electrodes are small in size (.9mm diameter, 1mm long) with a capacitance (related to surface area) of 15-25 μF.

Cyclic voltammetry (Pilla, 1974) on electrodes of differing composition, i.e., Pt, Ta, titanium (Ti), and stainless steel (SS), in various buffers, i.e., Ringers and McCoy's tissue culture medium with fetal calf serum, demonstrated that Ti and Ta had the widest inert potential range. This inertness (the absence of faradaic and non-faradaic currents) for Ta, provides the largest stable, non-interfering potential range (2V) for reproducibility. The absence of non-faradaic processes such as specific adsorption prevents compositional changes in the electrolyte.

Resistivity Compensation

Most of the techniques, including this one, for the measurement of the electrical impedance of biological tissues and cells employ a conductance type chamber. Classically in electrochemistry, conductance chambers are used to measure the conductivities ($\sigma_e = 1/\rho_e$) of electrolyte solutions. These measurements are normally performed at low concentrations of solute(<.01M) and extrapolated to infinite dilution. The extrapolation to infinite dilution allowed comparison of measured values of conductivity to theoretrically derived values. The theories developed to explain solution conductivity utilized certain assumptions that permitted the theoretical expressions to be derived but limited their application to unique conditions (such as infinite dilution). From a more practical standpoint, conductance chambers with a fixed geometry are calibrated by standardizing with a known solution at a specific temperature. In an analogous manner, all measured values in this study are referenced to the resistivity ($= 1/\sigma_e$) of the electrolyte filling the chamber of a fixed geometry. For this reason, it was determined in an unpublished study (Schmukler, 1982), that a method of compensating for changes in electrolyte resistivity due to changes in solution, composition, temperature, etc., was critical to the accuracy of electrical impedance measurements.

From theoretical derivations using the Stokes-Einstein and the Nernst-Einstein equations it can be shown to a first order approximation that resistivity is related to temperature by:

$$\rho_e = 1/\sigma_e \sim T/D = C\eta \qquad (2)$$

where ρ_e = resistivity, σ_e = conductivity, D = diffusion constant, η = viscosity of the solvent, C = proportionality constant containing the ion radius, and T = temperature. Therefore:

$$\rho_e = F(T, \text{Ionic Strength, Solution} \\ \text{Composition, Concentration,etc}) \qquad (3)$$

which has been found to exhibit a non-linear dependence on the variables, so that to correctly find the relationship, for example between ρ_e and T, it is necessary to employ higher order fits, viz:

$$\rho_e = C_1 + C_2 T + C_3 T^2 \qquad (4)$$

where constants C_1, C_2 and C_3 must be determined empirically for each solute at a fixed concentration.

As stated earlier, one of the parameters to be determined from measurements is $R_e = (\rho_e L)/A$, the voltage (IR) drop between the voltage measuring electrodes. Where ρ_e = electrolyte resistivity, L = vertical separation distance between voltage electrodes, and A = cross sectional area of the chamber (1 cm diameter, A = .785 cm^2). The position of the electrodes and the chamber diameter are permanently fixed. This implies the R_e will only change as ρ_e changes. As can be seen in the following section other parameters which are cellular in origin (i.e. R_i) are also measured relative to ρ_e. Any change in ρ_e during an experiment, that is not taken into consideration, will result in the introduction of errors in the impedance measurement.

In order to avoid the extensive calibration necessary to compensate for resistivity changes that result from changes in T, electrolyte composition and concentration including evaporation effects in the course of an experiment, an additional conductance chamber was set up on both sides of the pseudo-epithelium (see figure 1). Since the biological sample is constrained to the zone between the voltage measuring electrodes, resistance measurements $V(\omega)/I(\omega)$ at low frequencies (<5KHz) taken outside this zone would only measure electrolyte resistivity. The changes in the resistance between one of the resistivity compensation electrodes (RCE) and the voltage measuring electrode (VE) in the same half of the chamber, $V_T(\omega)/I(\omega)$, allows normalization of any changes due to electrolyte resistivity shifts back to the original (starting) or reference resistivity level.

This is illustrated for temperature changes by the following equation (see figure 3):

$$R_c = R \times (R_T^0/R_T) \qquad\qquad (5)$$

where:

R_c = resistivity compensated resistance between both VE,
R = resistance between both VE at an arbitrary temperature point = $V(\omega)/I(\omega)$,
R_T^0 = Resistance measured between one RCE and its VE on the same side at the initial temperature = $V_T^0(\omega)/I(\omega)$, (F $\leq$ 5KHz, t = $^1/F \geq$ .002 sec), and
R_T = resistance measured, at the time R is taken between the same RCE and VE the initial reading was taken = $V_T(\omega)/I(\omega)$, (F $\leq$ 5KHz, t = 1/F $\geq$ .002 sec.).

The measurements shown in figure 3 were performed in the time domain using pulses, where the pulse width was $\geq$ .01 sec.).

Then, for a simple electrolyte from equation 5:

$$R_c = R \times (R_T^0/R_T) \times (R^0/R^0) \qquad\qquad (6)$$

where R^0 = Resistance at initial temperature between both VE.

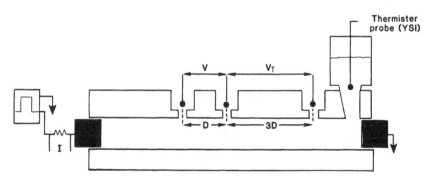

Figure 3. Schematic of experimental setup for resistivity compensation experiments.

Rearranging:

$$R_c = (R/R^0) \times (R^0_T/R_T) \times R^0 \qquad (7)$$

Since the ratio $(R/R^0) = (R_T/R^0_T)$, then $(R/R^0) \times (R^0_T/R_T) = 1$ and therefore:

$$R_c = R^0 \text{ at all temperatures.} \qquad (8)$$

This was checked by varying T from 10° to 30°C and is illustrated in the following graphs (figures 4 and 5) for initial temperatures of 21.1 and 25.1°C (Schmukler, 1982). The temperature was measured with YSI thermister probe (Y.S.I.,Y. Springs, OH.) continuously throughout the experiment (see figure 3). The maximum deviation for the 21.1°C (see figure 4) experiment was 1.14% at 10°C, and for most of the T range it was less than .2%. the maximum deviation for the 25.1°C experiment was 1.64% at 10°C, and for most of the range it was less than 1.2%. Three additional experiments were performed and exhibited the same behavior as the 21.1°C experiment. The slightly larger deviation in the 25.1°C experiment was attributed to a non-uniform temperature distribution in the conductance chamber. To eliminate this problem in the chamber described, the separation distance between the RCE and the VE was reduced (see figure 3 and section on Impedance Chamber).

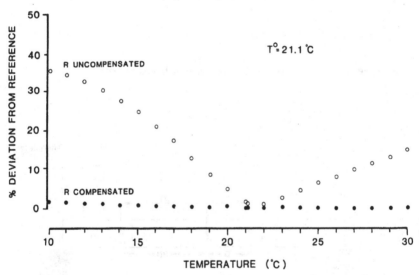

Figure 4. Plot comparing R and R_c vs T for R^0 at 21.1°C.

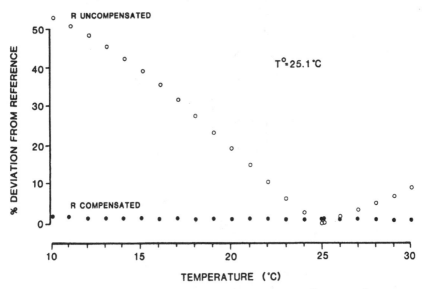

Figure 5. Plot comparing R and R_c vs T for R^o at 25.1°C.

Modeling of Pseudo-epithelium

From the physical construction of the filter (see figure 6) with the embedded RBC there exists a resistance R_s which is always present in series with pseudo-epithlium model due to the resistivity of the solution between the measuring electrodes and the filter surface. In the model there is, in parallel to the cells in the pores, a leak pathway R_L due to the unfilled pores, if any, and the leaks between the RBC and the pore wall. In series with both the leak pathway around the cells and the cells themselves is the resistance R_f due to the unfilled pore length, if the cells do not penetrate the filter pore completely (normal behavior). In addition there is a small capacitance C_f, (Chiabrera, 1985) from the polycarbonate dielectric in parallel to $((R_L \mathbin{/\!/} R_i) + R_f)$. The theoretical value of this capacitance = 180 pF (Kirtland, 1985).

An important point to remember is that R_f can never be larger than R_L^0, the filter resistance without cells and must be some fraction of it. The cell circuit model consists of a resistance due to the cytoplasmic resistance R_i in series with the membrane impedance Z_c (see figure 7). The surface area of the upper portion of the imbedded cell, is at least 13x the surface area of the lower portion. Thus the lower smaller area has 13x higher impedance compared to the upper area and is the limiting series impedance. The smaller surface area is always defined as a hemisphere with radius = pore radius. (Schmukler, R and Pilla, A.A. 1982).

365

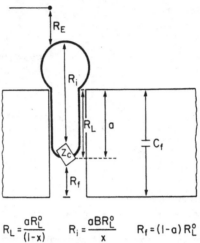

$$R_L = \frac{aR_L^o}{(I-x)} \qquad R_i = \frac{aBR_L^o}{x} \qquad R_f = (I-a)R_L^o$$

a = Penetration of RBC into pore

x = Fraction of cross sectional area obstructed by RBC

B = Ratio of relative resistivities of cytoplasm to extracellular electrolyte = ρ_c/ρ_e

R_L^o = Resistance of electrolyte-filled filter without cells

C_f = Filter capacitance

Z_c = Cell membrane impedance

Figure 6. Schematic representation of a living cell embedded in the cylindrical pore of the insulating filter. See text for details.

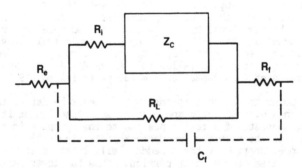

Figure 7 Equivalent electrical circuit representing the impedance of a living cell embedded in the pore of an insulating filter.

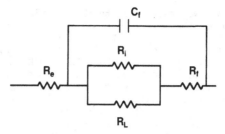

Figure 8. Equivalent electrical circuit representing the high frequency limiting behavior of the living cell/pore complex.

As stated in the section on the Impedance Chamber, R_e as well as R_L^0, which is the filter resistance without cells, can be determined from a measurement so the Z without cells $= (R_e + R_L^0)//C_f$. The resistance of a single electrolyte filled pore, R_p, is given by: $R_p =(\rho_e L)/A$, where ρ_e is the resistivity of the electrolyte in ohm-cm, L is the path length in cm ($= 13 \mu m$), and A is the cross sectional area in cm^2. R_L^0 is simply R_p/N, where N is the total number of pores in the filter filled with electrolyte.

Inspection of figure 7 and figure 10 shows that the limiting behavior at both high and low frequency limits can be used to evaluate some of the components of the model. The equivalent circuit high frequency limit is shown in figure 8. C_f is determined during system calibration, and only effects measurements at the highest frequencies since it is in the range of 10^{-10} F. The equivalent circuit at the low frequency limit is shown in figure 9.

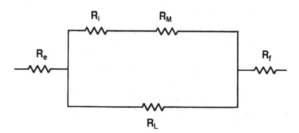

Figure 9. Equivalent electrical circuit representing the low frequency limiting behavior living cell/pore complex.

To relate the leak pathway resistance to the cytoplasmic resistance and the unfilled portion of the filter pore, it can be shown (Schmukler, et al., 1982 and Schmukler, 1981) that $R_L = (aR_L^0)/(1-x)$, $Ri = (aBR_L^0)/x$, and $R_t = (1-a)R_L^0$ where a = fraction of total pore length, L, into which the cell penetrates, or the ratio of the path length the current traverses inside the cell to the path length of a pore; x = fraction of cross sectional area for current conduction of the filter obstructed by the cells; $B = \rho_c/\rho_e$ the ratio of relative resistivities of the cytoplasm to the extracellular fluid electrolyte (see figure 6). The value for C_f can be most accurately determined by calibration using a 4-port technique and measurements on R_e, R_L^0 from a normal filter, and one additional R_L^0 from a high resistance filter (Kirtland, 1985).

By using a sensitive (to .01 ml) buret as the end of the hydrostatic pressure head, the entrapment of the cells by the filter can be measured from the volume flow at steady state pressure before and after the cells are added to the impedance measurement chamber. If the filter has pores left open, i.e., unfilled, this will change the value of x by a small increment, but will not add any error to the data analysis using the model. This is because the maximum available conduction pathways through the filter are predetermined by the measurement of R_L^0.

Membrane Model

Zc (see figure 7) contains at least one capacitance due to the membrane dielectric structure and one or more due to processes such as specific adsorption of various ions. In addition, there is in parallel a membrane resistance R_m (or D.C. pathway) for ionic phase transfer across the membrane ion channels so that, assuming one specific adsorption time constant, the membrane impedance model can be represented by Figure 10.

Both specific adsorption and counter ion relaxation involve the counter ion associated with the cell membrane double layer. Counter ion relaxation is generally described as movement along the membrane surface (Schwan, 1985) by the counter ions in cell suspensions. Specific adsorption describes ion binding to ion specific sites and it usually thought of as movement occurring more perpendicular to the membrane. Specific adsorption does not involve chemical association, like metal: ligand bond, between the adsorption site and a particular ion. Non-specific adsorption is membrane counter ion association, but strictly from an electrostatic viewpoint without any chemical specificity, in the same sense as water dipole orientation. Both specific and non-specific adsorption kinetics do not disallow tangential counter ion movement to satisfy membrane and field charging requirements. In either case the resulting circuit analog would be a parallel RC pathway.

If the value of Rm is much greater than R_L, the low frequency circuit limit (see figure 9) becomes $R_e + R_t + R_t$. The value for Rm for the HRBC is approximately 10^{12} Ω based on the smallest available surface area of 5.4 μm^2 and a ρ_m of from 10^5 Ω-cm^2 (Schanne and Ruiz-Ceretti, 1978), the

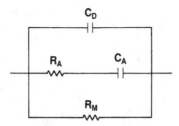

Figure 10. Electrical equivalent circuit representing the specific adsorption impedance model of a cell membrane.

lower limit for the measurement resistance of one cell is still 1.9×10^{10} Ω. The resistance of a single fluid-filled 1.85 μm pore is $Rp = (\rho_e \, L \, /A) = 3.4 \times 10^6$ Ω where ρ_e = 70 Ω-cm, L = 13μm, and A = 2.7μm^2. The R_L value is normally 10x the R_L^0 value, so that the resistance for a single pore is approximately 3.4×10^7 Ω (= R_L^0 measured x N). Thus, it can be seen from figure 9 the pathway containing R_m will have a resistance of at least 3, but probably closer to 5 orders of magnitude greater than the R_L pathway. This means that less than .001% of the current at low frequencies will go through R_i and R_m (see figure 9), which is the limit of the experimental accuracy of the measurement. The model for the RBC filled filter then becomes that shown in figure 11.

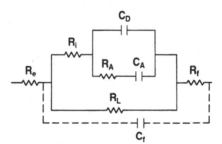

Figure 11. Equivalent electrical circuit representation of the cell/pore complex assuming a simple specific adsorption impedance model.

From the low frequency data, the factor x can be determined if R_f is known. R_f can be determined by measuring the cells in the filter with either the scanning electron microscope (SEM) (Schmukler, et al. 1982, Brailsford et al., 1977) or the transmission electron microscope (TEM) after the impedance measurement, to determine the relative penetration of the cells into pores. Once x has been determined, it is possible to calculate the factor B due to the relative resistivity of cytoplasm with respect to electrolyte solution. This provides a check on the assumptions made, since R_i has been determined in several other studies (Cole, 1968, Pauly, 1959, and Pauly and Schwan, 1966) for HRBC.

It can be seen from the above that using the high and low frequency limiting impedance behavior with a physical measurement of the depth of penetration of the cell into the pore, and the calibration to determine C_f, it is possible to evaluate R_e, R_L, and R_f (see figure 11).

Results and Discussion

The chamber is connected to a Solartron 1260 Gain/Phase Impedance Analyzer using cable lengths as short as possible. Initial experiments were run without controlling the electrolyte temperature. The input voltage level was 10 mV, and therefore the voltage change across an entire cell is less than 5 mV. The frequency was scanned in logarithmic steps from 1 Hz – 32 MHz. Both the voltage and current amplifiers were AC coupled (-3db, 1Hz). Using AC coupling for both the current and voltage amplifiers resulted in almost complete cancellation for impedance measurements of the low frequency rolloff, which normally starts at 10 Hz. This can be seen in figure 2 from the measurement of R_e and $R_e + R_L^0$ for both amplitude and phase. The amplitude response of the measurement system can be seen (see figure 12) to exhibit essentially flat behavior over the range from 1Hz – 7MHz for R_e and from 1Hz – 4MHz for the measurement of $R_e + R_L^0$. The phase shift for R_e is flat ($<5°$) from 1Hz – 2MHz and from 1Hz to slightly greater the 1 MHz for $R_e + RL$ The additional phase shift for $R_e + R_L$ appears to be due to an additional capacitance the filter introduces. The measurement system appears to be extremely well behaved over more than 6 decades of frequency. The system time constant due to parasitic elements for both R_e and $R_e + R_i$ is 13.4 MHz.

Inspection of Figure 12, shows a definite relaxation at 42 KHz (α') and a second relaxation in the MHz range (β?). The center frequency for this relaxation is at 9.4 MHz. Unfortunately, because it appears that the β relaxation, normally at 2–4 MHz (Schwan, 1957,1963), overlaps somewhat with the system time constant at 13.4 MHz, it is not possible in these early experiments (Bao,1989) to separate the two. It is evident upon inspection of the phase plot in figure 12 that there is another time constant in the low MHz range. Work is currently ongoing to determine the mechanism for the α' relaxation by looking at the impedance as a function of electrolyte ionic strength and to improve the low MHz system response. The author feels that because of the filter geometry, counter ion relaxation might be severely restricted since the tangential flux along the membrane is severely restricted. Thus, it may be possible in this system to distinguish between counter ion relaxation and specific adsorption mechanisms for the observed α' relaxation.

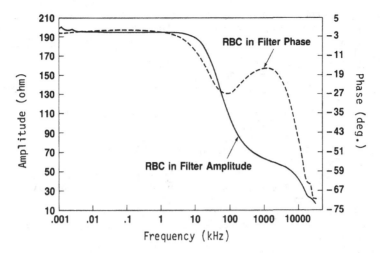

Figure 12. Plot of Phase and Amplitude of RBC filled filter vs. Log Frequency.

In comparison with standard suspension measurements, this technique provides certain advantages. The area of membrane measured is defined much more precisely and the cellular geometry and orientation with respect to the imposed field is known. The current leak pathways around the cells under measurement are reduced by several orders of magnitude, resulting in a concomittant increase in sensitivity. Electrical properties of living cell membranes are measured under physiologically appropriate conditions due to the technique improvements, and the existence of the α' relaxation at 42 KHz for the human RBC is clearly evident.

References

Bao, J., Schmukler, R., Davis, CC., " Impedance Measurements of Living Cells," J. Electrochem Society., 136, 3, 1409, 1989.
Beard, R.B., DeRosa, J.F., Koerner, R.M., Dubin, S.E., and Lee, K. (1972): "Porous Cathodes for Implantable Hybrid Cells". IEEE Transactions on Bio-Medical Engineering, vol. BME-19.
Brailsford, J.D. et al. (1977) Blood Cells, 3: pp. 25-38.
Brailsford, J.D., Personal Communications.
Bull, B.S. et al. (1977). Blood Cells, 3: pp. 39-54
Chapman, T.W. and Newman, J.,(1968). University of California, Berkley, AEC contract W-7405-eng 48
Chiabrera, A., Schmukler, R., Kaufman, J., Pilla, A., 1985 Personal Communications.
Cole, K.S. (1968). Membranes, Ions and Impulses, pt. 1, University of California Press Berkely, L.A.
Cole, K.S. (1928). J. Gen. Physiol., 12: pp. 29-36.
Cole, K.S. (1928). J. Gen. Physiol., 12: pp. 37-54.
DeRosa, J.F. (1969): "Fabrication and Evaluation of Cathode and Anode Materials for Implantable Hybrid Cells". M.S. Thesis, Drexel University.

DeRosa, J.F. (1974): "Linear A.C. Electrode Polarization Impedance Studies on Solid and Porous Noble Metals". Ph.D. Thesis, Drexel University.

Eisenberg, R.S. and Johnson, E.A., (1970). Prog. Biophys. Mol. Biol. 20: pp. 1-5.

Fricke, H. (1924). Phys. Rev. 24: pp. 575-587.

Fricke, H. (1925). Phys. Rev. 26: pp. 678-681.

Kirtland, G.M. (1985). Living Cell Impedance Identification, M.S. Thesis, Columbia University.

Lassen, U.V. (1977). In Membrane Transport in Red Cells, eds., Ellory, J. and Lew, V., Academic Press, N.Y. pp. 137-174.

Oncley, J.L. (1942). Chem. Revs. 30: pp. 433-450.

Oncley, J.L. (1938). J. Am. Chem. Soc. 60: 1115-1123.

Pauly, H. (1959). Nature 183: pp. 333-334.

Pauly, H. and Schwan, H.P. (1966). Biophysical Journal, 6: pp. 621-639.

Pilla, A.A. (1974). Electrochemical Information "Transfer at Living Cell Membranes", Ann. NYAS, 238, pp. 149-169.

Rall, W. (1970). Biophysical Journal 9: pp. 1509-1541.

Richardson, I.W. (1972). J.Membrane Biol. 8: pp. 219-236, Springer Verlag, N.Y.

Robillard, P. and Poussart D. (1979). IEEE Transactions in Biomedical Eng. 26:8 pp. 465-470.

Schanne, O.F. and Ruiz-Ceretti, e.a. (1978). eds.: Impedance Measurements in Biological Cells, J. Wiley and Sons, N.Y.

Schmukler, R. (1976). "Characterization Studies for an Optimal Hybrid Fuel Cell Cathode," M.S. Thesis, Drexel University.

Schmukler, R., Kent, M. and Chien, S., (1982)an unpublished study.

Schmukler, R. (1981). A New Technique for Measurement of Isolated Cell Impedance, E.Sc.D. Thesis, Columbia University.

Schmukler, R., Pilla, A.A., Gerf, G. and Lee, M. (1982). IEEE Proceeding 10th Northeast Bioengineering Conference, ed. E.W. Hansen, pp. 213-216.

Schmukler, R. and Pilla, A.A. (1982) Journal of Electrochemical Society 129:3 pp. 526-528.

Schmukler, R., Kaufman, J.J., Maccaro, P., Ryaby, J.T. and Pilla, A.A. (1986). " Transient Impedance Measurement on Biological Membranes: Application to Red Blood Cells and Melanoma Cells", Electrical Double Layer in Biology, ed. Blank, M., Plenum.

Schwan, H.P. (1957). Electrical Properties of Tissue and Cells Suspensions, ed. Lawrence J.H. and Tobias, C.A., Advances in Biological and Medical Physics, vol. 5, Academic, N.Y.

Schwan, H.P. (1963). Physical Techniques in Biological Research, vol. 6, ed. W.L. Nastuk, Academic., N.Y.

Schwan, H.P. (1985). "Dielectric Properties of Cells and Tissues", Interactions Between Electromagnetic Fields and Cells, ed., Chiabrera, Nicolini, and Schwan, Plenum Press, N.Y.

Acknowledgements

The author would like to thank: Mr. Gary Johnson of Columbia University for his help with the design and for the construction of the impedance chamber, Dr. Shu Chien of Columbia University for his advice, support and inspiration, Jian Boa, of the University of Maryland for the experimental data, Dr. A. Pilla of Mt. Sinai Medical Center for all his help and Professor A. Chiabrera of the University of Genoa, for his help, inspiration, and discussion above and beyond the call of duty.

This work was partially supprted by : Office of Naval Research, Contract #N 00014-85-K-0682.

MEMBRANE ELECTROPORATION:

BIOPHYSICAL AND BIOTECHNICAL ASPECTS

Eberhard Neumann and Elvira Boldt

Physical and Biophysical Chemistry, Faculty of Chemistry
University of Bielefeld, P.O.Box 8640
D-4800 Bielefeld 1, F.R.Germany

(I.) INTRODUCTION

Transient membrane permeabilization by short high voltage pulses, nowadays called electroporation [1], as opposed to irreversible electric breakdown leading to cell lysis [2] or vesiculation, was at first recognized in 1972.[3] Transient membrane electroporation is the fundamental basis of a series of modern electric techniques for the direct gene transfer [1] into all types of cells and microorganisms, for cell electrofusion or for appreciable stimulation of cell growth and cell regeneration. For a review see reference 4. Recently, it was reported that electric pulses mediate the insertion of proteins into electroporated membranes.[5]

Although the electroporation techniques become enormously popular not only in cell biology and biotechnology but also in medicine, the mechanisms of the various electric field effects are not known at the required molecular membrane level. Therefore, data analysis and optimization strategies for cell electrotransfections, electrofusions, electrostimulation and electroinsertion are by and large empirical.

The electrotransformation of intact bacteria and other microorganisms requires low-conductive media to reduce Joule heating.[6] The interpretation of the low-conductivity data necessitates an extention of the analytical framework in terms of the external solution conductivity (λ_0).

Using the green algae cells of the species Chlamydomonas reinhardtii (Ch.reinh.) as an example, a simple procedure is outlined to determine the critical electroporation voltage $(\Delta\varphi_{M,c})$ and the membrane/envelope conductivity (λ_m).

373

Finally, a strategy is presented that permits the determination of optimum values of electric field strength, pulse length, pulse number, etc. for the transient permeabilization of viable cells, thus providing the basis of optimization strategies for cell biological and medical applications.

(II.) ELECTROPERMEABILIZATION AND CELL INHOMOGENEITIES

It was recently recognized that the transient membrane electroporation may be viewed as a cycle of electropore formation and resealing resembling a relaxation hysteresis.[4,7] In brief, in the presence of the field pulse of limited duration ($\Delta t = 5 \, \mu s - 10 \, ms$), the cell membrane (initial state C) becomes porous (state A) and, at the same time, fusogenic. The state transition

$$C \xrightarrow{(E \geq E_c)} A \tag{1}$$

in the electric field $E \geq E_c$, above a critical value E_c, is unidirectional.[1,7]

If the electric pulse is switched off before rupture or lysis of the cells occur, the electropores or electrocracks slowly anneal at $E = 0$:

$$C \underset{(E = 0)}{\xleftarrow{\hspace{1cm}}} A \tag{2}$$

At $E = 0$, the membrane resealing is also unidirectional.

The electroporation cycle may be represented by the scheme

$$C \underset{(E = 0)}{\overset{(E \geq E_c)}{\rightleftharpoons}} A \tag{3}$$

The electroporated (and fusogenic) membrane state A usually is long-lived, in particular at low temperatures (4°C).

In Figure 1, the relaxation hysteresis of a single cell is shown in terms of the degree ξ of structural changes of the membrane phase. Electroinsertion[5] of proteins at field strengths $E < E_c$ appears to be a subcritical phenomenon.

If the electroporation hysteresis is coupled to other processes (state X), e.g. material release or uptake, or lysis, the scheme (3) has to be extended to:

$$C \rightleftharpoons A \longrightarrow X \tag{4}$$

Usually for short pulse times Δt, the transitions A → C and A → X occur at $E = 0$; they are after-field effects.

Inhomogeneity of the Electrosensitivity

It is known that a biological cell population usually is in-homogeneous in cell size, in the state of growth, or metabolic conditions. Non-spherical cells such as bacterium rods have different positions relative to the external electric field. All these factors cause a distribution of the critical electropora-tion field strength E_c.

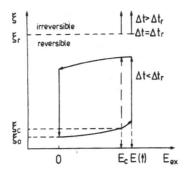

Fig. 1. The relaxation hysteresis of the membrane electroporation. The cyclic change in the extent of structural rearrangements ξ (at $E > E_c$ and $\Delta t(E) < \Delta t_r$) are displayed as a function of the external field E_{ex}. The cycle comprises reversible as well as irreversible elements. The subcritical changes of ξ between ξ_0 and ξ_c ($E < E_c$) are reversible. At the supercritical field strength $E(\uparrow) > E_c$ the structural rearrangements associated with the electroporation process are unidirectional and irreversible. If the field duration is larger than the rupture threshold Δt_r, the membrane ruptures. If however the field pulse is switched off at $\Delta t < \Delta t_r$, we remain in the reversible electroporation domain. The (slow) return at $E = 0$ from the upper branch to $\xi = \xi_0$ is unidirectional; the resealing or annealing processes are irreversibly occurring relaxations to the initial state ξ_0.

Figure 2 shows examples of such a distribution. Here, the state X in scheme (4) is the coloured cell G, obtained by the uptake of a dye by the electroporated cell. The number of coloured cells or the percentage

$$G(\%) = [G]/[C_T] \tag{5}$$

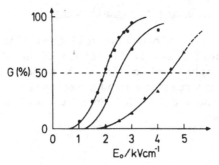

Fig. 2. Electrosensitivity G(%) of a suspension of green algae
cells Chlamydomonas rheinhardtii (wild type 11-32c, Göt-
tingen) to quasi-rectangular electric pulses of the
initial field intensity E_0 and of the pulse length Δt =
0.2 ms at different medium conductivities λ_0: $\bullet$: λ_0 =
$3.5 \pm 0.1 \cdot 10^{-4}$ S cm^{-1}; $\blacksquare$: $\lambda_0 = 1.5 \pm 0.1 \cdot 10^{-4}$ S cm^{-1}; $\blacktriangle$: λ_0 =
$5.6 \pm 0.5 \cdot 10^{-5}$ S cm^{-1}. G(%) is the percentage of cells which
were critically permeabilized such that the (lethal) dye
Serva Blue G (M_r = 354, largest dimension 2.5 ± 1 nm) was
taken up, thus visibly colouring these cells.

of the total cells is a measure for the critically permeabilized cells; see scheme (13).

For practical purposes we may take the field strength E_0(50%)
where G(%) = 50 is representative for the cell population and define a mean value by

$$\bar{E}_c = E_0(50\%) \qquad (6)$$

The width of the electrosensitivity of the cell population may
be given in terms of a variance ($\pm \overline{\Delta E}_c$). Since E_c depends on the
cell size, the mean value $\bar{E}_c$ corresponds to a mean value $\bar{a}$ of the
effective radius of spherical cells. Here, too, we have a variance: $\bar{a} \pm \overline{\Delta a}$.

(III.) INTERFACIAL POLARIZATION AND SOLUTION CONDUCTIVITY

The electroporation data suggest that the membrane electropermeabilization results from an indirect electric field effect.
The structural changes ξ of the membrane phase are preceded by
the ionic interfacial polarization.[7] See also Table I.[4]

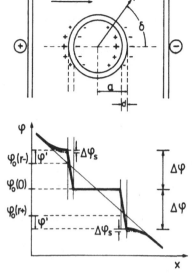

Fig. 3. Interfacial polarization of a spherical nonconducting shell of thickness d and outer radius a in a constant external field $\vec{E}$. The stationary electric potentials are given in polar coordinates of the radius vector $\vec{r}$ and the angle δ, such that the conducting interior of the cell has the constant reference potential $\varphi_0 = 0$ for $0 \leq r \leq (a - d)$. For $r > a$, $\varphi_0 = - \vec{E} \cdot \vec{r} = - E \cdot r \cos \delta$. The total potential $\varphi(r)$, relative to $\varphi_0(0) = 0$, is given by $\varphi(r,\delta) = \varphi_0(r,\delta) + \varphi'(r,\delta)$, where $\varphi'(r,\delta)$ is the contribution of the interfacial polarization. The $\Delta\varphi$ terms are the interfacially induced crossmembrane potential differences in the absence of fixed ionic groups and adsorbed ions (surface potential $\Delta\varphi_s = 0$). The dash/point line models schematically the potential profile in the presence of fixed surface charges (here negative).

If, as usual, the polarization time constant τ_p is small against the pulse duration Δt, the buildup of the stationary value $\Delta\varphi(E)$ of the interfacial potential difference across the membrane is practically instantaneous.

Figure 3 shows the profile of $\Delta\varphi(E)$ in the direction of the external electric field vector. For low-conductive membranes of thickness d of cells of radius a, the stationary value is given by

$$\Delta\varphi(E) = -1.5 \cdot f(\lambda)a\,E\cdot|\cos\,\delta| \qquad (7)$$

where δ is the angle between the membrane site considered and the direction of E. The conductivity factor $f(\lambda)$ is a function of the specific conductances or conductivities of the external solution ($\lambda_0 \geq 10^{-4}$ S cm^{-1}), of the cell interior ($\lambda_i \approx 10^{-2}$ S cm^{-1}) and of the membrane ($\lambda_m \approx 10^{-7}$ S cm^{-1}), respectively, and of the ratio d/a.

Usually $\lambda_m \ll \lambda_i,\lambda_0$ and d $\ll$ a such that [4]

$$f(\lambda) = \frac{1}{1 + \lambda_m(2 + \lambda_i/\lambda_0)/(2\lambda_i d/a)} \qquad (8)$$

If in low-conductive media $\lambda_0 \ll \lambda_i$, but still $\lambda_0 \gg \lambda_m$, eq. (8) reduces to

$$f(\lambda) = \frac{\lambda_0}{\lambda_0 + \lambda_m\,a/2d} \qquad (9)$$

describing an increase of $f(\lambda)$ with increasing λ_0 up to $f(\lambda) = 1$ at saturation of the ionic interfacial polarization in high-conductive media.

Obviously, E_c corresponds to a critical cross-membrane electroporation voltage $\Delta\varphi_{M,c}$. For cell lysis, $\Delta\varphi_{M,c} \leq 1$ V.[2] When the contributions of fixed surface charges and the associated ionic atmospheres can be neglected, the total cross-membrane voltage $\Delta\varphi_M(E)$ in the presence of the externally applied field E is a function of the intrinsic membrane potential $\Delta\varphi_m$ (e.g. $\Delta\varphi_m = -70$ mV) and of the interfacial polarization term $\Delta\varphi(E)$:

$$\Delta\varphi_M(E) = fct\ [\Delta\varphi_m,\ \Delta\varphi(E)] \qquad (10)$$

The total potential profile (Fig. 3) in the direction of E is explicitly given by

$$\Delta\varphi_M(E) = -\left(\frac{3}{2}\,f(\lambda)aE + \frac{\Delta\varphi_m}{\cos\,\delta}\right)|\cos\,\delta| \qquad (11)$$

Oftenly $|\Delta\varphi(E)| > |\Delta\varphi_m|$, hence $\Delta\varphi_M(E) \approx \Delta\varphi(E)$. The pole caps of spherical membranes are the sites of maximum interfacial polarization. With $|\cos\,\delta| = 1$, eq. (11) can be used to relate the mean values $\bar{E}_c$ and $\bar{a}$ by $\bar{E}_c = -\Delta\varphi_{M,c}/[1.5\,\bar{a}\,f(\lambda)]$. Applying eq. (9) for low-conductive media we obtain:

$$\bar{E}_c = -\frac{\Delta\varphi_{M,c}}{1.5\,\bar{a}}\ \frac{\lambda_0 + \lambda_m\,\bar{a}/2d}{\lambda_0} \qquad (12)$$

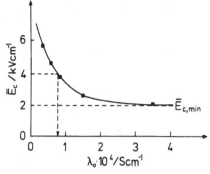

Fig. 4. The population mean value $\bar{E}_c = E_c(50\%)$ of Ch.reinh. cells (Fig. 2) decreases with increasing conductivities λ_0 of the pulsing medium. The λ_0-value at which $E_c = 2\,\bar{E}_{c,min}$ (i.e. $f(\lambda) = 0.5$) is given by $(\lambda_0)_{0.5} = \lambda_m \cdot \bar{a}/2d$ and permits a simple graphical estimate of the λ_m-value. From $\bar{E}_{c,min}$ we obtain $\Delta\varphi_{M,c} = -1.5\ \bar{a} \cdot \bar{E}_{c,min}$.

When $\bar{a}$ is given by the geometrical radius of the most abundant cell size, the data in Fig. 4 can be used to determine $\Delta\varphi_{M,c}$ and λ_m for the Ch.reinh. cells ($\bar{a} = 3.5\ \mu m$): $\Delta\varphi_{M,c} = -0.8$ V, $\lambda_m \approx 5 \cdot 10^{-7}$ S cm^{-1}.

(IV.) PRACTICAL ASPECTS

The longevity of the electroporated cells may be measured by the after-pulse addition of a dye (which colours the cell) at various times t_{add} after the electroporation pulse. Due to annealing processes more cells loose the permeability property at larger after-pulse addition times (Fig. 5).

For the data analysis the scheme (4) is specified to

$$C \underset{k_R}{\overset{(E)}{\rightleftharpoons}} A \overset{+D}{\underset{k_D}{\longrightarrow}} G \qquad (13)$$

where k_R and k_D are the rate coefficients for resealing (R) and dye uptake (D) and X = G is the (lethally) coloured cell state.

The data suggest that $k_D \gg k_R$, such that G is a quantitative measure of the cells in the critically electroporated state A where [A] = [G].

The rate equation for scheme (13) under these conditions is given by

$$\frac{d[G]}{dt} = -\frac{d[A]}{dt} = k_R [A] \qquad (14)$$

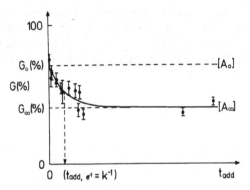

Fig. 5. Time course of the recovery of electroporated cells (resealing), measured by the uptake of the dye Serva Blue G which was added at various times t_{add} after the electroporation pulse. For this example $k_R = 2.6 \pm 0.9 \cdot 10^{-2}$ s^{-1}.

Integration for the boundary conditions $t \to 0$: [A] = [A]$_0$, and $t \to \infty$: [A] = [A]$_\infty$ yields:

$$[A(t)] - [A]_\infty = ([A]_0 - [A]_\infty)e^{-k_R \cdot t} \qquad (15)$$

Eq. (15) permits the evaluation of the coefficient k_R of cell recovery from the observed exponential decay of G(%) in Fig. 5.

The values G_0(%) and G_∞(%) can be used to determine, at a given pulse length, pulse number, etc., the optimum range of E_0 for the electroporation experiments (Fig. 6). This range shifts when the medium conditions (λ_0, temperature, etc.) are changed.

TABLE I

FUNDAMENTAL PROCESSES OF THE ELECTROPORATION HYSTERESIS OF MEMBRANES

Physical-chemical processes	Electric terms

I. REVERSIBLE PRIMARY PROCESSES

1. Primary electric events

1.1 Electric dipole induction and dipole orientation	Dielectric polarization
1.2 Redistribution of mobile ions at phase boundaries membrane/ solution,including (a) ionic atmosphere shifts (b) local activity changes of effectors, e.g., H^+- (pH-changes) or Ca^{2+}-ions.	Ionic-dielectric interfacial polarization (Maxwell-Wagner, ß-dispersion)

2. Structural rearrangements   Electro-restructuring

2.1 Conformational changes in protein and lipid molecules	
2.2 Phase transitions in lipid domains, resulting in pores, cracks (via pore coalescence) and percolation	Electroporation Electropores, electrocracks Electropercolation
2.3 Annealing and resealing processes	

II. IRREVERSIBLE SECONDARY PROCESSES

1. Transient material exchange   Electropermeabilization

1.1 Release of internal compounds, e.g. hemolysis	Electrorelease
1.2 Uptake of external material, e.g. drugs, antibodies	Electroincorporation, Electrosequestering
1.3 Transfer of genetic material, e.g. DNA, mRNA, viroids, ... with stable cell transformation transfer	Electrotransfection Electrotransformation Electroporative gene

2. Membrane reorganizations

2.1 Cell fusion (if membrane contact)	Electrofusion
2.2 Vesicle formation (budding)	Electrovesiculation Electrobudding
2.3 Electromechanical rupture	Dielectric breakdown

3. Tertiary effects

3.1 Temperature increase due to dissipative processes	Joule heating, dielectric losses
3.2 Metal ion release from metal electrodes	Electroinjection
3.3 Electrode surface, H and O in statu nascendi	Electrolysis

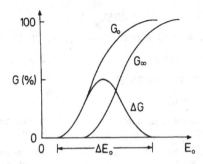

Fig. 6. Scheme for the optimization of electroporation conditions; $\Delta G = G_0 - G_\infty$, representing the transiently permeabilized surviving cells. The field strength range ΔE_0 is
shifted when the electroporation conditions are changed.

REFERENCES

1. E. Neumann, M. Schaefer-Ridder, Y. Wang, and P.H. Hofschnei-
    der, Gene transfer into mouse lyoma cells by electroporation
    in high electric fields, EMBO J. 1:841-845 (1982).
2. A. J. H. Sale and W. A. Hamilton, Effects of high electric
    fields on micro-organisms, III. Lysis of erythrocytes and
    protoplasts, Biochim. Biophys. Acta, 163:37-43 (1968).
3. E. Neumann and K. Rosenheck, Permeability changes induced by
    electric impulses in vesicular membranes, J. Membrane Biol.
    10:279-290 (1972), Potential difference across vesicular
    membranes, J. Membrane Biol. 14:194-196 (1973).
4. E. Neumann, The relaxation hysteresis of membrane electro-
    poration, in: "Electroporation and Electrofusion in Cell
    Biology", E. Neumann, A. Sowers, C. Jordan, eds. Plenum Press,
    New York (1989).
5. Y. Mouneimne, P.F. Tosi, Y. Gazitt, and C. Nicolau, Electro-
    insertion of xeno-glycophorin into the red blood cell
    membrane, Biochem. Biophys. Res. Comn. 159:34-40 (1989).
6. H. Wolf, A. Pühler, and E. Neumann, Electrotransformation of
    intact and osmotically sensitive cells of corynebacterium
    glutamicum, Appl. Microbiol. Biotechnol. 30:283-289 (1989).
7. E. Neumann, The electroporation hysteresis, Ferroelectrics
    86:325-333 (1988).

Acknowledgements. We thank Mrs. M. Pohlmann for careful typing
of the manuscript and the Deutsche Forschungsgemeinschaft for
Grant Ne 227/4 to E.N.

382

CONTRIBUTORS

Cytochromes (continued)
  cytochrome c (continued)
    break temperatures of, 84-88
    cyclic voltammetry of, 70-75,
      82-83, 87, 91, 93-94, 98
    effects of pH, 84-85
    effects of temperature on,
      83-88
    electron transfer kinetics, 73,
      77, 130
    electron transfer reactions of
      irreversibly adsorbed
      cytochrome c, 69-79
    interfacial electron transfer
      reactions of, 81-90
    oligometric forms of and
      electron transfer, 81, 94
    orientation of tin oxide
      electrode, 77
    quasi-reversible electron
      transfer kinetics of, 130
    surface-enhanced resonance
      Raman scattering, 43, 45-47,
      49-54, 57
    temperature dependence with
      indium oxide electrodes,
      81-90
    thiobacillus ferrooxidans, 340
  cytochrome c$_3$
    electron transfer kinetics of,
      130
    NMR spectra of, 59-61
    Raman study of, 59, 65
    redox potentials of, 59, 61-66
  cytochrome c$_7$, electron transfer
    kinetics of, 130
  cytochrome c$_{553}$, quasi-reversible
    electron transfer kinetics
    of, 130
  cytochrome c peroxidase, 40, 70
  cytochrome oxidase
    charge translocating reactions
      of, 13-17
    electron transfer reaction with
      magnetic field effects on,
      162-164
    thiobacillus ferrooxidans, 340
  cytochrome P-450
    and silver substrates, 43-46,
      54-57
    surface-enhanced resonance Raman
      scattering, 43, 45-45, 49,
      54-57

Derivative cyclic voltabsorptometry,
    (DCVA), 132-134
Desulfovibrio desulfuricans, 59-64
Desulfovibrio Vulgaris Miyazaki,
    59-64
Diazoluminomelanin
  chemiluminescence of, 293-299

Diazoluminomelanin (continued)
  phosphorescence of, 297
Dipalmitoylphospotidycholine, 109-114
Dystrophia muscularis, 115

Effective polarizability equation, 47
Electric fields (see also AC electric
    fields)
  and antibody regulation of ion
    channels, 21-29
  and biological tissue coupling,
    251-255, 258
  effect on protein adsorption and
    desorption processes, 69-79
  frequency effects on calcium-45
    uptake, 103-108
Electrical impedance measurements,
    see Impedance measurements
Electrical organizations, of cell
    membranes, 266-267
Electroconformational coupling,
    167-177, 179-190, 193
Electrode
  carbon fiber, 351
  4-electrode technique, 358, 361
  indium oxide, 81-83, 91, 94, 129
  methylene blue modified platinum,
    129
  promoter modified, 91-98
  optically tranparent thin layer, 60
    97-98, 130
  polarization, 360-361
  tin oxide, 70-77
  ultramicroelectrode, 351-352, 356
Electrofusion
  in erythrocytes, 320-321, 323
  in erythrocytes ghosts, 319-323,
    325-330
  membrane electrofusion mechanisms,
    303, 315-334
  protocols of, 317
Electrogenic reactions
  in bc complexes, 7-12
  in cytochrome oxidase, 13-17
Electrokinetic field interaction,
    diagram of, 257
Electromagnetic fields,
  biological responses to, 267-268
  effects on
    calcium binding, 269-270
    cellular level, 211-221, 245-246
    gap junction communication,
      277-278
    living systems, 241-244
    membranes, 245-246
  effects used in cell biology, 303
    304, 312
  and pulsating electromagnetic
    induced current, 303-307,
    309-312
  mechanisms of action with living

Printed in the United States
By Bookmasters